COLLATERAL DAMAGE

TARA MEIXSELL

Copyright © 2010 Tara Meixsell
All rights reserved.

ISBN: 1451557582

ISBN-13: 9781451557589

This book is a personal journal written by the author, a resident of Garfield County Colorado—the site of a huge natural gas boom in the years during which the book was written. The ideas put forward in this book reflect the personal opinions and recollections of the author, and no other entity or individual.

Dedicated To

Chris and Steve Mobaldi
Dee Hoffmeister
Rick Roles
Susan Haire Babbs
Karen Trulove
Laura Amos

SPECIAL THANKS TO

Al Laurette
Anne and Mike Meixsell
Morgan Ulloa
Vicky Elliot
Lance Astrella
Deb Anderson
Nick Isenberg
Josh Fox
Molly Gandour

Amy Mall, author

Drilling Down: Protecting Western Communities form the Health and Environmental Effects of Oil and Gas Production, Natural Resources Defense Council

Michelle Nijhuis, author

"How Halliburton is Wrecking the Rockies"
OnEarth magazine, Natural Resources Defense Council Quarterly

Contents

Trouble In Paradise	1
First Trip To The Gas Fields	9
The Chris Mobaldi Story	31
Making Movies	63
Take One: Film Crew Arrives	91
Take Two	121
Whistleblowing: Weston Wilson Of The Epa	145
The Beginning Of Split Estate	163
Pit Violations: The Runaround	197
Forced Pooling	217
Reality Check	247
Lawsuit Filed Over Underpaid Federal Oil Royalties	285
Legislations, Politics And Explosions	xxx
Front Page News	347
Making More Movies	363
Bad Times	415
The Cogcc Tour: The Fight	421
Congressional Oversight And Governmental Reforms Hearings: Washington D.C.	431

MobalDi Legal Case Ends	457
Mit And The Bbc	461
Rulemaking: The Legal Action Committee	475
Hollywood Here We Come	485
The Big Time	517

Trouble in Paradise

On May 1, 2001 a Ballard company subcontractor, BJ Services, was doing work to three natural gas wells. The wells were located in a mountainous rural area of western Colorado where natural gas is abundant, in a landscape of breath-taking beauty. Just ten minutes from the Interstate highway, a simple country road climbed the hills between the rocky knolls and juniper-studded acreage. Hay fields, forty-acre ranchettes and large tracks of ranch lands composed the area known as Dry Hollow. Strangely out of place in this picturesque western area were the drilling dereks that towered 150 feet into the air right beside homes and barns. What happened that day at one ranch was to forever change the life of one Colorado woman and her family, and would be a catalyst in the lives of many others.

As the employees from BJ Services were working on a gas well in Dry Hollow, suddenly water gushed into the air, erupting like a geyser from a neighboring property. There was a problem; something had gone wrong and it was likely related to the work being done deep underground at the gas wells. That day the wells were being hydraulically fractured, a stimulation process invented by Halliburton that creates mini-seismic events deep underground in gas and oil bearing formations to allow for the release of the fuels. Fracturing uses a high-pressure blast of fluids to crack open the formations. The ingredients of each gas company's fracturing mixes are proprietary—and it is widely known that toxic chemicals are regularly included in the secret mixes.

The Amos family owned the property where the explosion occurred, and they were out of state at the time. Laura Amos was visiting her parents in Kansas with her husband Larry and their infant daughter Lauren. The Amoses, outfitters and hunting and fishing guides, received a phone call from one of their employees. He told them that their water well had exploded, the cap of the well had blown off and the surrounding pasture had flooded with a fountain of murky, fizzing water. The Amoses immediately suspected a connection between the seismic fracturing work done on the neighboring gas well that was some 700 feet from their kitchen and the eruption of their water well. Something had caused the well to explode, as water wells aren't known to erupt on their own.

After returning home from Kansas, the Amoses found the water that ran from their home faucets to be filled with dark gray sediment, and it fizzed like soda water. Tests were done on the water by the state regulatory entity, the Colorado Oil and Gas Conservation Commission, and high levels of methane were discovered in the water. The Amoses were told that the water was safe to drink, but the high concentration of the methane gas could accumulate within their house to an explosive level. The family took heed of the state's warning to keep their windows open, and ventilate the home. Although the family was convinced that the gas well work had destroyed their water well, neither the gas company nor the Colorado Gas and Oil Conservation Commission agreed. The methane in the Amos water well, they said, could have been naturally occurring in its origin, as Garfield County is rich in methane-containing rock formations. A geologist from the Colorado School of Mines did not agree. He stated that water wells don't explode like the Amoses' well did, and he speculated that pressurized fracturing fluids pushed gas or other liquids through a much shallower leak in the side of one of the gas wells. While the fracturing was targeted for more than a depth of one mile underground, the Amoses and the geologist believed that the fracturing had caused the gas to leak through a naturally occurring fissure in the area's fragile geology that contains many natural fissures—thus communicating with and contaminating the water well.

Ballard Company began hauling drinking water to the Amos family two weeks after the explosion. Then, when in late August of 2001, further tests concluded that the levels of methane in the water had greatly dropped, it was concluded that there was not enough gas present any longer in order to find the source of the methane contamination. Although the water still stank and was visibly discolored, within two months Ballard Company ended its delivery of potable water to the household. The Amoses were on their own, left to deal with their ruined water well. Laura Amos refused to accept the COGCC's decision, and she began doing her own research. Soon she learned that many other families living in gas fields that had been hydraulically fractured also had run into problems with their domestic water wells—not only in Colorado, but also in Alabama, Virginia, New Mexico, Wyoming and Canada. Soon Laura began to realize that what others asserted was true: the reason why there was no proof—according to the COGCC- of drinking water contamination from hydraulic fracturing was because no regulatory entity cared to work very hard to seek any evidence.

The gas well that was closest to the Amoses' home, the one that had been hydraulically fractured on the day when the Amoses' water well exploded, was cemented in and considered a loss. No governmental entity ever cared to proceed further with conducting any investigation of what may have gone wrong at that gas well. The Amoses were stuck with no proof and a fouled well. They hauled water in for household use, and continued to worry about what they and their young daughter may have been contaminated with.

Two years after the water well explosion, Laura fell suddenly and severely ill while vacationing in Florida. She felt swollen, was extremely thirsty, and one day while snorkeling she could barely get her breath. She was so faint and short of breath that even with a life preserver she barely made it back to land. Medical tests revealed that Laura was suffering from extremely high blood pressure and very low potassium levels. A specialist eventually diagnosed her with Conn Syndrome, a benign tumor in one of her adrenal glands. Laura underwent surgery and had her entire adrenal gland with the tumor removed. After the procedure, both her blood pressure and her potassium levels became normal again. Conn Syndrome is an extremely rare condition, affecting fewer than 200,000 people a year in the United States. Laura, busy raising her young daughter and her first child, put the event behind her as a fluke of fate.

Some months later Laura read a scientific memo describing the possible health effects of a solvent 2-butoxyethanol (2-BE) sometimes used in hydraulic fracturing mixes. Theo Colburn, a world-renowned scientist who researches hormone disrupting chemicals, linked the chemical 2-BE with unusually high rates of adrenal tumors in laboratory rats and mice. Colburn discussed the possibility of 2-BE used in hydraulic fracturing of gas wells leading to the contamination of domestic water wells. The thought that her own water well may have been contaminated with 2-BE, and perhaps had caused her rare adrenal tumor, stopped Laura Amos dead in her tracks. After all, their well had exploded and the water had been fouled....Could this chemical 2-BE possibly be the cause of her illness? Night after night she stayed up until 2 a.m. or later doing research on her computer. She was determined to seek justice for the damages that she was certain that Encana Gas Company had done to her water well and her family. It seemed so blatantly obvious, and it was preposterous that no one was being held accountable. Laura Amos was furious and determined to prove the truth, no matter how long or what it took.

For several years Laura voiced her concerns about her health and the possible link to the mishaps at her family's water well. She contacted every possible local, state, and federal agency to voice her concerns over the dangers of hydraulic fracturing chemicals. Numerous newspaper articles were written, and a documentary film focused on the issue and filmed the Amoses the day they moved off their ranch in Silt, Colorado. They had long ago decided to relocate away from their contaminated ranch and start over somewhere else; their fears over continued health risks from the fumes that permeate the air in Dry Hollow plagued them. Wells continued to be punched in the valley at record rates, and Laura fretted over the burning eyes and frequent nosebleeds that her family suffered with. They had finally had enough of it.

Laura Amos knew how important it was to others in the future to know what contaminations and illnesses might occur because of gas drilling in close proximity to homes and drinking water sources. Because she was determined to get the truth out and seek justice, many other impacted landowners read about her and pursued their own issues that were similar. It was crucial that Laura's issues were well covered in the local media. Senator Jeffords of Vermont told her story on the floor of the United States Senate, when he introduced the Hydraulic Fracturing Safety Act of 2005 that would have required hydraulic fracturing mixes to contain non-toxic ingredients. In spite of Senator Jeffords's efforts, the Federal Energy Bill that passed in 2005 allowed for the exemption of hydraulic fracturing fluids from the Safe Drinking Water Act. The landowners were no match for the powerful industry lobbyists from Halliburton and other fracturing giants.

Finally, in 2005 - four years after the explosion of the Amoses' water well—the Colorado Oil and Gas Conservation Commission continued their investigation of the Amoses' contaminated water well. Conclusive tests were done which proved that the same methane being extracted from the shallow Williams Fork gas formation well contaminated the Amoses' water. This put to rest the notion that the methane in the Amoses' water was naturally occurring—on the contrary, now it was evident that the natural gas being extracted from the gas well seven hundred feet from their home had communicated underground with their water well. If it were not for scientist Theo Colborn's questions to the COGCC and the EPA regarding Laura's situation, it is almost a certainty that nothing would have transpired. Behind closed doors in a private meeting that even the Amoses themselves could not attend, the scientist presented her case before the regulatory entities to push them toward further investigating what really happened to the Amoses' water well, and subsequently to Laura's health. Prior to that time the COGCC had asserted that the methane in the Amoses' well water was "naturally occurring". It was also suggested that Laura's rare adrenal tumor was linked to using Windex! I was told that personally by a female member of the COGCC. I countered with the question as to why all the Hispanic housekeepers in the ski area resorts of Colorado did not also have Conn Syndrome and rare adrenal tumors linked to 2-BE? Housekeepers certainly used far more cleaning products than a hunting guide's wife on a rural ranch property did. It was one of many frustrating and fruitless calls…

Encana Gas and Oil Company, which bought Ballard Petroleum in 2001, finally reached an undisclosed settlement with the Amoses in July of 2007. The family moved and resettled on a large ranch property far from Silt, Colorado. The State of Colorado fined Encana $176,800 for methane contaminations occurring at the Amoses' and a neighboring well. No further findings were reported linking the fracturing work on the gas well to the fouled water wells, and the EPA failed to proceed with investigations into finding the cause of the mishaps at the Amos water well—in spite of several senior EPA officials' formal requests to conduct further studies. Those in power at the federal level ignored the requests; there was no interest in finding out what caused these mishaps and contaminations in Silt.

In the rare instances when large industrial mishaps occur and pay-offs or settlements are made, the story does not go public. For all practical purposes the companies pay the damaged parties to keep quiet. No one needs to know more. Often the company buys the property, either directly or through another party. Encana now owns the Amoses' old house in Silt. It continues to sit vacant, and the contaminated water well is sealed.

Many people in the gas fields suffering from health impacts they suspect to be linked to the gas wells and the noxious odors had given up calling governmental entities for help. They had called the gas companies, the local county offices, the Department of Health, senators and representatives, and the EPA. Time and time again they were told that the regulations allowed these things to happen, and there was no recourse for their issues. Many eventually give up and move away, realizing there is no solution to their problems and that no regulatory entity is going to help them. It is better just to take the losses and just get out—away from gas wells.

Laura Amos set the example that the media was one outlet available to use to begin working toward a solution and toward changing regulations. And that is what is happening right now in Colorado in the fall of 2008 as a result of state legislation passed in 2007—the state is changing the rules that dictate regulations for natural gas exploration and extraction. It has been a long fight, a hard fight, and the fight is nowhere near over.

For many of those who are living in the bulls-eye of gas development, there is a sense of abandonment. Unless it is happening in your back yard, there is a collective ignorance and blindness to the realities of what can and does go wrong with gas and oil exploration. It is easier to take the ostrich-with-the-head-in-the-sand approach, and not pay attention. After all, we need the gas...

Thankfully, there are others who realize that we are on a dangerous path currently in terms of human and environmental health problems from gas and oil development poorly done. There are the landowners such as Laura Amos who have never abandoned the cause to speak out about the dangers of hydraulic fracturing chemicals, and the environmentalists, politicians, lawyers, and scientists who work ceaselessly to right the wrongs of industrial practices and put into place better regulations and safer technologies.

First Trip to the Gas Fields

April 20, 2008

I'm often asked about my involvement in the oil and gas controversy.

It all began with a drive in the country about five years ago to buy some affordable horse hay during a Colorado drought period. I saw the flier at the local feed store for cheap hay and I jumped on it. The previous summer, my first living in Garfield County after moving down from the high mountains near the Continental Divide, had been a frightening and stressful time. My husband had gotten a good job with the city and I came along and settled into our new life in what seemed like the tropics of Colorado at 6,000 feet, where you could see green grass for more than three months a year. We used to live at 9,000 feet in Fairplay where it snowed as early as September and could continue into June and sometimes even July.

This year, the fire danger was extreme and a number of large wildfires burned fairly close to our ranch. In addition to worrying about my horses and llamas and our home potentially being in the path of a wild fire, I was trying to find a new job and I also ran into the problem of acquiring hay during a drought season. To my horror, the price of hay was now as high as $10 a bale—if you were lucky enough to find it.

So, when I saw the flier for hay at under $4 a bale I got right on it. Soon I was venturing up mountain roads south of Silt, just off I-70 close to where we lived. As I meandered further from developed areas, driving past ranch homes, rolling fields and dry rocky outcroppings sprinkled with gnarled junipers, I realized that I was seeing a part of the West I had never experienced before. It looked like true mountain lion country. As the truck rose over each new vista, each sight was more spectacular than the one before and I was excited that this adventure had taken me to such a beautiful new part of the land I now called home. I later learned that Teddy Roosevelt had ridden over this very same land on some of his Colorado bear hunting expeditions.

As I negotiated the s-curves at the crest of a hill, I saw something off the shoulder of the road that I could not comprehend. Rising like a monolith from the dense trees, a huge metal derek towered 150 feet into the air like an Apollo launching site. A large area was bulldozed around it and there were temporary modular structures and vast pits filled with liquids surrounded by countless trucks and men with hard hats. I really had never seen any thing like that out in the country before and I was puzzled.

Continuing toward my destination, driving up through the beautiful desert mountain country, I began noticing more and more large, heavily graveled roads with metal green gates and black and white signs that looked industrial. The gates were locked and the signs were definitely not family ranch entrances; it was something else all together. There were also pipes and other industrial type machines; large metal boxes with placards and more strange little fenced-off areas. What, I wondered, was all of this doing here up in the middle of a semi-rural wilderness ranchland?

By the time I took a hard left after descending down the mountain nearing my destination, I had had quite an eye full. I pulled into a ranch drive at the mailbox with the number written on my notepaper.

There was a small white farmhouse under some big trees and a man and a woman saw me and came over.

I got out of the truck.

"What is going on up here?" I asked.

"Our ranch sale is next weekend. We're getting out," the woman said.

She and her husband introduced themselves and then proceeded to tell me that they were leaving the ranch they had restored and had hoped to live at during their retirement years, raising a closed herd of Black Angus cattle.

They were leaving because their land was being incorporated into a federal unit for natural gas development. Not only would they be increasingly surrounded by gas development but they would also have to endure drilling on their own surface property even though they owned 50 percent of their mineral rights. It was similar to having your property taken over by eminent domain for the good of the national interest. In this case, the interest was for the extraction of natural gas. They said that one day they came home from work and found survey stakes in their pasture plotting out where the first gas well would be located. Furious over this invasion of their ranch, they pulled out all the survey stakes.

"We hired a lawyer and after a year he told us to give up, sell and leave," the woman explained. "There was nothing we could do, so we sold the place and our ranch auction is next Saturday."

We all continued to talk for a long time by the small stucco house under the big trees before we drove down to the backfield through a bog where the haystack was. The only reason this hay was for sale was because they were giving up their ranch, selling their cattle and moving away. As we loaded the hay onto the truck, I learned that the woman was a scientist and her husband was an accountant. They were going back to Aspen, where they had lived before.

For the time being they were abandoning their dream of having a closed Black Angus cattle herd on the ranch they had worked so hard to bring back to life. They had redone the irrigation systems and cleared the ditches and had brought back the recently abandoned pastures and the hay fields. The buyer for the ranch could likely profit from the mineral takings from the gas and subdivide the large acreage into 35-acre ranchettes. Regardless of the lucrative future fate of the land they had planned to retire on, they were done with it.

"I came home one day and I told my husband we have to leave, we have to get out. They're going to wreck this place," the woman said. After the truck was loaded down with hay, I said goodbye to the couple. Driving home to our little ranch north of the river, I was stunned and saddened by what I had seen and learned. It was a day seared into my memory.

COLLATERAL DAMAGE

One morning about a year later, I was going through a pile of mail that my husband, Al, had left on the kitchen table. I opened a small flier that said, "Gas development is coming to Silt Mesa and Peach Valley."

As I held the flier in my hands and re-read it, I froze. We lived very close to both Silt Mesa and Peach Valley.

The flier went on to explain that there was going to be an informational meeting about the proposed development at the middle school in Rifle the following week.

On the appointed night, Al and I drove over to the middle school auditorium by the Rifle fairgrounds and joined over 200 other locals, mostly blue collar middle income and ranching types like us. There were not a lot of suits in the audience. It was mostly Carhartt coats and blue jeans.

Everyone packed into the school gymnasium was anxious to learn what type of industrial event was about to hit the community and, most importantly, their own homes and land. Tensions ran high and the nervousness was palpable from the people sitting in the rows of folding metal chairs that filled the basketball court as they waited to hear what the people at the podium had to say.

A number of people spoke about the future possibilities for natural gas development in the area. They talked about community mobilization, education about surface use agreements, legal rights of landowners and how to best go about becoming informed. Detailed, complex descriptions were given of mineral rights and the split estate between the surface and the mineral owners. This mining law dates back to the days of the gold and silver boom in Colorado when a law was passed that severed the below-ground mineral rights from the surface rights of a property. A simple way to understand it is to think of the surface owner living on the top of the land, while a mineral owner is able to extract minerals from below—such as an underground coal or silver mine. But, when it comes to gas and oil extraction, the simplicity fades and becomes much more complicated because the gas and oil companies actually operate on privately owned land, pre-empting the surface owner's right to use the property because the law gives the companies the right to extract the minerals below the land. To many citizens this reality seems outrageous and unbelievable; it flies in the face of the concept of private land ownership. The people in the audience were full of questions.

Two attorneys with practice in dealing with landowner and energy industry negotiations spoke, and they spoke very well. As I drifted off to sleep that night, I continued to think about what they had said.

After that, I became better educated about landowner issues. I went to all the meetings, met staffers from the local grassroots organizations and became interested in helping with training sessions and meetings to give the landowners the resources and tools to best protect their homes and interests in the event that drilling came near or onto their properties. We held meetings in the Silt fire station and in people's living rooms, discussing the issues and passing out Xeroxed sheets with contact information for attorneys with experience in drafting surface use agreements as well as water testing companies who could perform baseline studies on water wells at cheaper group rates for neighborhoods.

I remember one meeting in a living room on Silt Mesa where about eight people showed up.

One lady became angry. "We need to get a group together and stop them," she said as she listened to the plans for her area.

I don't believe she had gone to the big meeting in Rifle because what was said there should have erased any idea that a simple neighborhood group—no matter how fired up—could keep the gas industry from moving forward with their drilling plans.

It was common knowledge that energy company landmen were soliciting many landowners. Landmen are the first to approach the landowners and inform them that the company has an interest in drilling for gas on their properties.

Some landowners take for granted what they are initially told by the company representative and sign the first agreement they're handed.

COLLATERAL DAMAGE

Others scrutinize the details, educate themselves or hire an experienced lawyer for assistance in drafting surface use agreements that can greatly reduce the impacts they will endure on their own property as the gas is being extracted. By drafting a good surface use agreement, a landowner can retain use of the surface property and allow for directional underground drilling, minimize or eliminate pits, pads and infrastructure, which can include man camps.

Shortly thereafter, I started regularly attending the Grand Valley Citizens Alliance monthly meetings, plus numerous other events relating to natural gas drilling. My normally homebound way of life was suddenly changing and I think Al was quite surprised. Aside from our annual vacations to Mexico or to visit family back in Massachusetts, plus a whitewater rafting and camping trip or two each summer, I was typically very intent on being at the ranch. Town held little attraction for me after work or on the weekends; it was at home on the small mountain ranch with my animals where I wanted to be. Cleaning the barns, caring for the animals, writing, cooking and gardening consumed me.

In March of 2005 when legislative bills regarding the surface owner rights versus mineral owner rights were introduced to the state legislature, I joined in a few of the trips to the Capitol with others from the Western Slope. Representative Kathleen Curry of Gunnison introduced a bill in an effort to level the playing field between the landowners and the all-powerful mineral owners. The first time I went to Denver, I made my mind up fairly spontaneously to join some other people who were driving down in a blizzard. After making a few quick phone calls I was told that there was an extra hotel room booked in Denver at the Ramada that I was welcome to use and that the group would leave about midday, which was just a few hours away. I got permission from my boss to take the next two days off and I told a very surprised Al that I was going to Denver that afternoon to attend the hearing for the Surface Owner's Rights Bill and would return the following evening.

I hastily packed some respectable clothes for some of the events we would be attending, plus the items I would need for the overnight stay. Within a few hours I was on my way to Denver with Orlyn and

Carol Bell from Silt. Carol was a retired chemist and Oryln had been a state water engineer. We were driving on icy, treacherous roads in their ranch pickup truck over Vail Pass and the Continental Divide. There were some white-knuckle moments as we made our way through the worst of it.

That first trip to the Capitol was memorable in many ways. After we arrived in Denver we attended a fancy reception put on by the Republican representatives environmental group and we enjoyed hobnobbing with staffers, representatives and environmentalists downtown at a high-end location with ornate wood-paneled rooms tastefully decorated with hunting prints and gleaming leather furniture. It's what I imagined the Harvard Club must look like.

Next we met the entire group from Garfield County at our hotel right near the Capitol. We assembled for dinner at a Mexican restaurant off the lobby and the service was abysmally slow. The restaurant was not only short-staffed, but they were also out of salsa. Finally our orders arrived and we began to eat. Out of the corner of my eye, I saw a man in a black sweatshirt with the hood up wearing a ski mask walk into the restaurant. As it was freezing cold outside, I thought little of it. From the seat where I sat at the end of a long rectangular table I had a prime view of what was to come next. The man went over to the cash register, leaned over and began pulling at the receipt tape that flew out in white ribbons. He was rifling about the register and I froze as I realized that he was robbing the place not more than 20 feet away from where I sat. He ran from the restaurant and I instantly got up and walked hurriedly toward the kitchen and, hopefully, an exit.

"This place just got robbed," I said under my breath to the people dining with me as I hurried out.

It seemed that I was the only one who saw what happened due to the position of my seat at the table. I ran into the kitchen and pushed open the swinging doors. In the brightly lit, white industrial interior, the mostly Hispanic kitchen help in white smocks and hairnets looked surprised as I repeated,

"This place just got robbed!"

COLLATERAL DAMAGE

I lingered in a hallway behind the kitchen for a while; my greatest fear was that a hail of bullet fire would follow the robbery as the thief exited the restaurant.

Needless to say, the already poor service became even slower as we finished our meal. Also, we had to wait for the police to arrive so they could write down our eyewitness accounts.

After the female officer took my statement, I asked her,

"Are there many murders here?"

She stared at me.

"Don't you know that this is the drug and prostitution center of Denver?" she asked.

Actually, I had no idea. But that night, in our hotel, you can be sure that all the doors were double-bolted and chained and windows were locked. I sat on the bed, stuffed a few pillows behind my back, leaned against the headboard and wrote a draft of a speech on hotel paper to read tomorrow at the state Capitol. Little did I know that this was the first of many such future trips for me, preparing for gas hearings and writing or revising drafts of speeches the night before in some hotel.

As I looked out the metal-grated window in the dark night to the lights shining from the inner city streets far below, I felt very far from home and from the little ranch. Very far...

The next morning when we all assembled to get over to the Capitol for the beginning of the hearings on the split estate bill, the Western Colorado Congress organizer, Deanna, looked somewhat distressed as she explained her choice in booking us into that particular hotel.

"It was cheap and near the Capitol. I was trying to save the non-profit some money. But I'll never do that again!"

Then we were off for a full day of hearings with testimony from the gas companies, industry groups, landowners and environmentalists.

Hours of sitting in rows of chairs mostly filled with people in suits, reviewing speeches and taking it all in.

That was the first of many trips to Denver for me and although environmentalists and landowners lobbied the state representatives to pass House Bill 051219, the Oil and Gas Surface Damage Compensation Reform, the bill failed. When Kathleen Curry sponsored a similar bill the following year, I went again with a group to lobby for our cause. Another night in a Denver hotel room writing drafts of speeches in a spiral notebook, another long day at the Capitol listening to testimony in front of the Colorado House of Representatives. And another two days spent away from home logging long hours driving across the state.

Soon I had a whole network of acquaintances as a result of my involvement with natural gas issues. I began to refer to them as my gas friends and I spent many hours in their company. I met landowners, environmentalists, lawyers and assorted politicians. One woman I met, Laura Amos from Silt, was in the midst of seeking monetary damages for contamination of her family's water well. After the water well at their home exploded like a geyser one day when hydraulic fracturing work was being done to a gas well on a neighboring property, the Amoses' water became foul and discolored and fizzed like a glass of Sprite. The well water was highly contaminated with methane. Later, Laura was diagnosed with a rare adrenal tumor and she and other experts believed that a chemical used by the gas industry on a neighboring property might have caused her tumor. Scientific studies have linked exposure to the chemical 2-butoxyethanol, known as 2-BE, to rare adrenal tumors.

Laura and I struck up a friendship and we often attended meetings together. Prior to the passage of the 2005 Federal Energy bill, both of us lobbied hard against the exemption of hydraulic fracturing chemicals from the Safe Drinking Water Act. Laura believed that the chemical 2-BE (often used in fracturing mixes) contaminated her family's drinking water and caused her illness. But despite meeting with senators and congressmen and conveying concerns over hydraulic fracturing chemicals, the 2005 Federal Energy bill exempted these undisclosed and often carcinogenic chemicals from the Safe Drinking

Water Act. Perhaps I had been naïve, but I thought that putting in countless calls to Washington DC politicians prior to the vote on the energy bill might actually make a difference. I had lists of key representatives and congressmen and each morning at 7 a.m. MST, I would begin my calls to Washington DC when the offices were just opening up. As I held the phone and dialed with one hand, I brushed my hair and awkwardly got dressed with the other. Those 30 minutes spent each morning before work was the best time for me to make these calls. If I did the ranch chores before 7, I could get more calls in before leaving for work. The staffers and legislators were often sympathetic to the issue, but the majority of them stated that the votes were already stacked in favor of the exemption.

And they knew what they were talking about. On April 21, 2005, the U.S. House voted 249-183 to pass the Federal Energy Bill, which included exemptions for hydraulic fracturing chemicals from the Safe Drinking Water Act. That was a very depressing day.

TWO WORLDS COLLIDE

During the summer of 2005, Al and I took a trip to Boston to visit my family. Before I left Colorado, Laura and I discussed the upcoming Colorado Oil and Gas Conservation Commission hearing on the Divide Creek seep moratorium. In 2004, the Colorado Oil and Gas Conservation Commission imposed a moratorium on drilling within a two-mile radius of the sight of the seep after 115 million cubic feet of natural gas escaped from an Encana gas well and contaminated West Divide Creek south of Silt. Nearby landowners suspect that hydraulic fracturing, an underground seismic technique that shoots highly pressurized chemical mixes out from the borehole, led to the disaster. The seep contained benzene, a cancer-causing substance that was linked to the gas well in question. The COGCC fined Encana $371,000 for the violation. It was the largest fine ever levied in Colorado history. A two-day hearing starting on July 11, 2005, was planned to discuss the possible lifting of the drilling moratorium in the two-mile radius area of the seep. A number of landowners planned to speak to the COGCC about their concerns over various impacts from the gas wells near their homes. I would be flying back from Boston on the first day of the hearing and would miss it, but Laura and I agreed to keep in touch about meeting up to attend the second day of the hearing together.

DIVIDE CREEK SEEP HEARINGS

On July 11, 2005, as we waited at the gate at Boston's Logan Airport for our plane to depart, I found a phone booth and called Laura to get an update on the hearing. She wasn't home so I left a message and said I would call her tomorrow morning. Then I found a quiet spot at an airport restaurant counter and wrote a draft of a speech for the next day's hearing.

Al and I got home late and the next morning as soon as I woke up I called Laura. The hearing had been cancelled for today; it was going to be continued at a later date in Denver. Then Laura described the goings-on of the previous day's hearing.

A number of people had spoken about concerns over their health and several relayed some pretty stunning symptoms that they had suffered after exposure to unknown fumes from the gas wells. Laura said that she had questioned the Colorado Oil and Gas Conservation Commission about their obligation to look after people's health and well-being, and she stated that her own family's welfare had been at stake as a result of the gas wells close to their home.

Laura continued to describe the previous day's testimony and then she said that a woman who used to live in Rulison had talked about her health problems that she believed were also caused by the gas wells. Laura asked me if I knew of her—Elizabeth Mobaldi. The name did not ring a bell.

Elizabeth had suffered from two pituitary brain tumors and a plethora of other strange symptoms including a rare speech disorder known as acquired foreign accent syndrome. As Laura continued to talk, it suddenly hit me—the woman she was describing had been a close co-worker of mine three years ago. However, I knew her as Chris, not Elizabeth, so I didn't recognize the name immediately.

COLLATERAL DAMAGE

Chris and I had been hired right at the same time in the fall of 2002 for the large nonprofit agency where I still work. I was hired as her supervisor and we had trained and worked closely together for about five months. During that time Chris had suffered increasingly debilitating pains, especially in her hands. Sometimes she could not even hold a pen. She had headaches so severe that she couldn't work and her vision was bothering her, too.

One awful afternoon we had been together in a conference room upstairs at the agency after Chris had just gotten a phone call from a doctor's office. It was a brain surgeon who had previously removed a pituitary tumor from Chris's head.

Chris put her face into her hands and began to cry.

"It's back. The tumor is back. He just told me," she said. "I knew it because the pain is so bad and I can't see well. I have to have surgery again."

Later I went downstairs to the front office and my co-workers asked if I was okay. It was obvious that something was wrong. I don't remember if I was crying then or if I had been crying before or both.

"Chris has a brain tumor and she has to have surgery," I told them. I was terrified for what my friend and co-worker would have to endure. They would have to saw her skull open for this surgery.

Two days after the procedure, my boss and I went to visit Chris at St. Mary's hospital in Grand Junction. We had been at a training session all day at a nearby conference center, so we stopped in on our way home.

Before we went into Chris's room, my boss asked me, "Are you going to be okay?"

I said yes. I knew she was worried that I was going to fall apart when I saw Chris.

We walked into the room after the nurses at the station showed us the way. There was Chris, propped up by some pillows and obviously sedated from the painkillers. The large incision across her forehead where her skull had been sawed open was held together by a series of large metal staples. It was a frightening sight.

Although she was woozy, Chris spoke in a faint voice. We didn't stay for too because we didn't want to wear her out anymore. Before we left I placed a tiny brown teddy bear on her bed stand. I had bought it some years ago when I was a teacher in Breckenridge.

"Here, I am going to leave Teddy with you. He's going to stay with you while you're getting better," I told Chris. Then my boss and I left and drove back to Garfield County.

In the weeks that followed I had a number of phone calls with Chris at the hospital as well as with her husband, Steve. She was not recuperating well and had developed a serious case of pneumonia. Sometimes it took a long time to reach her room before we could speak because they moved her around the hospital a few times. At one point she was so weak they put her on a respirator, her voice barely louder than a whisper. Once she told me that she had come close to dying, that she had gone to another place and she was afraid.

Finally, long after the initial surgery, she was released from the hospital and returned home to Rulison. One sunny day I drove out to their country home for a visit and to pick up some documents I needed for work. We sat outside and Chris said how happy she was to be home. Before I left, she placed something in my hand. It was the little brown bear I had given her.

"Here, I'm giving him back to you. I'm all better now," she said.

Later Chris came back to work at the agency but she couldn't work in the same job as before due to permanent partial loss of vision following the surgery. Instead, she worked in a different area from my office and we lost touch. Sometime later I heard from her neighbor that they had left Rulison and moved to Grand Junction, about 80 miles west of here. That had been over two years ago.

COLLATERAL DAMAGE

When Laura told me about Chris coming forward with her health complaints at the COGCC meeting yesterday, I couldn't believe it.

"I do know Chris Mobaldi. I worked with her three years ago. I was her supervisor and then she had to leave to have brain surgery. I went to see her after the surgery at the hospital when she had metal staples in her head," I said, still stunned.

Memories of my visits to the Mobaldi's home in Rulison began to race through my mind. I remembered seeing the strange brown metal boxes and wheels in an industrial looking area on a cement pad at the bottom of their driveway. When I asked them what it was, they said it was a gas well and that the company flared it off sometimes and huge flames burned from the well pad. At the time, I thought nothing about it.

Laura told me more about Chris's strange symptoms and said that she spoke with a foreign accent. This made no sense to me whatsoever; Chris had had no trace of a foreign accent just a couple of years earlier.

After ending my call with Laura, I called information and got the Mobaldi's new number in Grand Junction. After dialing the number, the phone rang and a woman with a strange voice answered.

"A-low?"

It was Chris, but she was speaking with a bizarre accent that was very difficult to understand. We talked for some time after I described to her how Laura had told me about the previous day's hearing in Rifle. Long after I hung up the phone, I still couldn't believe that was the same Chris Mobaldi I had worked with and knew and she was now testifying at COGCC hearings regarding very serious health impacts that she and her husband believed were caused by nearby natural gas development. Two of my previously unconnected worlds were colliding.

Not too long after that, the Mobaldis made contact with the law firm of Astrella and Rice in Denver and began pursuing a legal case for their damages. It was the Mobaldi's belief that toxins from the gas wells surrounding their home were the cause of Chris's illnesses. The toxins were either airborne or waterborne or both. After reviewing the information from Chris's medical records and discussing her symptoms with the Mobaldis, apparently the lawyers felt that they had a case. I was pleasantly surprised, as I had been told that it would take a large number of people to bring a case forward.

I kept in touch with the Mobaldis, but then as fall turned to winter, some months went by before I called them again. It was early in March of 2006 and I was curious to know how the legal case was going for them.

"Hi, how are you? I called to see what's going on." I said to Steve, who answered the phone.

"Well, we're going down to Denver tomorrow for a press conference and to lobby for a gas bill. We're going with the Western Colorado Congress," he told me.

"Really! I've done that a few times," I said.

"Why don't you come with us? Call Matt and see if there's room," Steve suggested. So I called Matt Sura, the energy organizer of the Western Colorado Congress and inquired about going along on the lobby and press trip.

In less than an hour the plans were made for me to meet the group as they went to Denver. I would leave my truck at City Market in New Castle and travel the rest of the way with them. My hotel room was already booked. How coincidental that I happened to call them that afternoon. Once again, I was off to Denver for a gas trip.

Later that afternoon they picked me up. Matt was driving with the Mobaldis and a woman named Susan Haire, who lived west of Rifle. Chris Mobaldi was reclined in the back of the SUV. She was quite ill and not talkative. Susan sat in the passenger seat and Steve Mobaldi and I sat in the seats behind. As we drove east on I-70, we talked about gas issues and the plan for tomorrow.

COLLATERAL DAMAGE

Susan began describing the symptoms that she had experienced one evening while irrigating on a neighboring ranch. I was on the edge of my seat. After smelling a strange odor in the air, she experienced an incredible pain in her head as if she were being hit with two 2 by 4's. She barely managed to crawl into her truck and drive home, where she was violently ill for the entire evening. She said that someone came by during the night and tested the air inside her home, but she was so ill she didn't even know who they were. She then talked about the permanent damage that had been done to her health after that exposure to the unknown airborne toxins, and it was stunning.

From that time on Susan could not breathe the air on her ranch without falling ill. She experienced severe and debilitating headaches, nausea, dizziness, skin rashes, burning eyes and more. Even inside the house the fumes caused her problems so she installed three air scrubbers. Her horse fell ill with strange symptoms and had to be put down. In order to spend time outside on her ranch without feeling sick, she took to wearing a respirator. She felt like a prisoner in her own home and was extremely frustrated that no governmental entity seemed to care about what had happened to her.

This trip to Denver was her last effort to bring attention to the problems that were being ignored in western Colorado from gas and oil development. But either way, Susan had had it—after speaking at the press conference and with our state representatives at the Capitol, she was getting in her car, turning her back on her ranch and leaving for Arizona. She had the ranch up for sale and had no intention of ever living in or near Garfield County again. She was getting out—she felt it was a lost cause.

As Susan relayed her story, to my horror I realized that what happened to her was almost identical to what happened to the Canadians living near the sour gas fields in Alberta. I had just finished reading a book about the issue and it was not a happy story. Up there, the sour gas exploration released hydrogen sulfide into the air, among other various toxins such as toluene and benzene. Animals and people living near the wells experienced high rates of miscarriages and stillbirths, cancer, nausea and skin rashes. Animals and humans were born with physical deformities. The trip from western Colorado to the

Continental Divide, usually a tediously long drive, passed in what seemed like minutes.

The trip had a huge effect on me. It was all because of the people I met, particularly the other landowners who also had stories of their sad health impacts. Things I learned chilled me, both on the way down to Denver and the next day at the Capitol press conference.

We stopped to eat dinner midway through the trip and Steve assisted Chris into the restaurant. She was unsteady, convulsing mildly and almost completely incoherent. We ate our food in the noisy chain restaurant filled with happy, red-faced après-skiers drinking beer in the bar area. It made me slightly nostalgic because that area had been my stomping ground not too long ago. In fact, we were just a few miles from the school where I used to teach on Lake Dillon. And I, too, had spent many days barreling down the frozen mountain slopes of Arapahoe Basin and Breckenridge.

Chris barely ate a few spoonfuls of the cottage cheese Steve ordered for her. When we left the restaurant and went out to the snowy parking lot, Chris leaned heavily on her husband who all but held her up. She was still convulsing and barely verbal. I saw the look of suppressed panic on Matt's face. I honestly thought he was afraid she would not last the trip. Or maybe he was wondering how she would come across at tomorrow's press conference at the Capitol building.

Finally we arrived at our hotel in Denver and Steve helped Chris inside. She had been incoherent and in pain the entire trip. Hopefully she would be better tomorrow.

The next morning we assembled in the hotel lobby and thankfully, Chris was looking quite a bit better and was coherent in spite of her bizarre accent. Off we went to the Capitol where we met the others in the cafeteria of an adjacent building. Matt Sura told us how the day was going to go: we were going to start with a press conference where specific people were going to give statements regarding their impacts from gas and oil development and then we were going to split into two groups and go on pre-arranged lobbying tours and visit with representatives and senators in their offices. Matt wanted us to each take a minute and introduce ourselves and give a short summary of what we wanted to speak about at the press conference.

People got up one by one. When an attractive but tired looking woman in her late 40s with brown hair dressed in a business suit spoke, I was riveted. Her name was Karen Trulove and she wanted to talk about the health problems she was experiencing at her home in Silt, which was very close to numerous gas wells. She described inhaling strong fumes wafting over from the gas wells then being overcome by intensely painful and debilitating headaches and nausea. Her ability to function seemed to be severely affected after the "odor events" and she was drained of all energy. She just felt plain sick. She said it was like having a terrible hangover - but all the time. Unable to breathe the air outside her ranch home, she took to wearing a respirator when she did her ranch chores.

In the few minutes she spoke in the cafeteria, her eyes filled with tears.

"I don't even plan my days anymore. I just try to get through them," Karen said.

THE CRYING ROOM

Soon we all made our way to a pressroom lined with dark, velvet drapes. There was a podium up front and on a film screen nearby, a slide show depicted oilfield mishaps, spills, black air emissions and aerial photos of prairies covered with snaking lines of roads connecting gas pads for miles and miles. The seats were filled with landowners and as well as members of the press poised with pens and reporter's pads.

For over an hour, landowners gave their testimonies. People came to the podium with tear-streaked faces after hearing the stories of those who spoke before them.

Susan Haire talked about what happened that night when she was irrigating a neighbor's field and how living the ranch life was no longer what it had been. There was no quality left in her once pleasant country life, where she had planned to retire after her career as a schoolteacher.

Karen spoke tearfully and with great difficulty. The circles under her eyes blazed under the ceiling lights as she described how she had been having massive and debilitating headaches as well as other symptoms.

She had realized that as soon as she got a whiff of fumes from the wells surrounding her home, the headaches came immediately. The once energetic and savvy businesswoman spoke about how she no longer had the stamina to work—she barely made it through each day just trying to function minimally. She was deeply concerned about the long-term effects of whatever was causing her to be so ill, and she knew the life she had once enjoyed was now gone.

Chris Mobaldi's barely intelligible, yet powerful testimony moved the room. Duke Cox, the president of the Grand Valley Citizens Alliance, slipped out the door when Chris began to speak in her strange accent, her voice barely audible. Her story brings him to tears.

As I sat there, I thought, "The Crying Room." That's what it was.

The press jumped on the health concern issues, including the Denver Post, Rocky Mountain News and a reporter from The Associated Press.

For the next five hours we were expertly directed around three floors of the beautiful Capitol building, with its gold railings and trim, flags and marble. Attractive pages and smartly dressed staffers marched the halls from office to office and dignitaries in suits with their blue lapel nametags came and went from large ornate rooms as committee sessions unfolded.

We went from senators' and representatives' offices one after another and discussed concerns about the impacts of natural gas development on the property owners. The legislators seemed happy to talk with us but they didn't all have a very deep understanding of the issues surrounding gas development. A few seemed shocked by the pictures we showed them from the newspaper that showed a huge derek towering beside a rural home in Silt.

One jovial legislator said "Come see us more often," as we prepared to leave his inner office. I think we would have liked to be able to do just that, but the reality of the situation was that many of us had to skip work and take two days to drive four hours each way to go to Denver. It wasn't that easy to get down to Denver to visit the Capitol. It was a big deal.

By the end of the afternoon, Chris Mobaldi was becoming visibly tired. Sometimes we had to wait for up to 30 minutes to be shown inside the offices for our pre-arranged appointments. We found places where Chris could sit down as we waited so she could rest. When introduced to the last several legislators, Chris's attempt to say hello and say her name resulted in gibberish. The legislators merely shook her hand and smiled, unsure of exactly what her difficulty was and what language she was speaking.

Toward the end of the afternoon, the group I was with passed the other Western Colorado Congress group. We were having a break for ten minutes in the huge hallway and Karen Trulove had found a phone booth and was having an impassioned phone conversation with someone.

As I passed by her, I distinctly heard her say, "I want to move. I want to get out of there!"

After a pause, where she listened to the person on the other end (presumably her husband), she cried,

"I'm the one who's getting sick!"

Although Karen seemed to have started the day rather timidly, it was evident that she was hitting her stride as we made the rounds that afternoon. I watched her enter one representative's office with her face held high, confident and determined. Additionally, she was motivated to act on her frustrations because of the living conditions and noxious fumes from the wells around her ranch. She wanted to get out.

The last part of our day at the Capitol was a private meeting with State Senator Jim Isgar. We sat around a long rectangular table in a large conference room and discussed the future of House Bill 1185. We were there for over an hour discussing the nuts and bolts of the bill as well as its current stage and final direction.

It was politics.

The evening ended with a cocktail hour and buffet at the swanky Warwick Hotel where environmentalists, elected officials and staffers hobnobbed and mingled with drinks and hors d'oeuvres in hand. We talked, ate, drank and left in the already dark night for Garfield County.

Although exhausted, we continued to discuss the day's events on the long drive home up and over the Continental Divide and again over the steep and long Vail Pass. It was very late by the time I finally arrived back at the ranch and at last fell into bed for a short night's sleep.

The Chris Mobaldi Story

In 2001 my husband and I moved from the high mountains of Central Colorado to Garfield County, some two thousand feet lower in elevation. We bought a twenty-eight acre ranch with a little old farmhouse surrounded by beautiful huge elm trees, and my life began anew. After getting through the first rather grueling summer of job searching and re-doing pasture fences while massive wildfires burned uncomfortably close to our place, I finally found a job that paid the bills. When not working I absorbed myself with my favorite pastimes, which included spending time in the barns and pastures with the llamas and horses, writing historical fiction, gardening, and cooking.

We were unaware that Garfield County was soon going to be at the epicenter of a huge natural gas boom, and in fact I knew absolutely nothing about natural gas development. Nothing. Natural gas was the blue flame that ignited on your stove, or heated your home. How it got there was a mystery to me, I had never taken a moment to ponder that question. Nor did I ever think it would be something I needed to know about. Well, things change—and sometimes they change in a very big way.

Let me explain how it has come to pass that I have spent almost four years of weekends writing this book—if I can. The quiet country life I was planning on enjoying took a sharp turn—all because of learning about natural gas development and the problems that accompanied it. The reality of what was happening to the landowners here was shocking, and it was equally shocking that nothing was being done to stop it. When a friend asked me to write a book about what happened to her, I said I would. I had written one before, so there was a good possibility that I could do it again.

I thought long and hard before making the decision to move from the high country, for a number of reasons. It was an area I had gotten to know well in the ten years I lived in the shadow of the continental divide. Many things pulled me to stay there, and I struggled with the decision to move. One day on a neighbor's ranch, I was talking to a friend who is a Native American about my quandary. He offered some advice, saying

"The Indians believe that you follow the red road. It is the road you are supposed to travel in life."

Confused, I asked

"Is it possible to get on the wrong road?"

"Yes," he replied.

"How do you know if you are on the red road, the right road?" I asked.

"You are on the red road if it feels right, if it is where you are supposed to be," he said.

The series of events that have transpired over the past six years are proof enough to me that I am on the red road. It has taken me to some places I never would have imagined myself to be in, some good, and some bad—but never have I doubted that it was the right thing to do. If you pursue something to help another, then it must be the right thing to do. The pursuit has landed me in the Colorado State Supreme Court chambers, in Boston at MIT, in Hollywood on Sunset Boulevard, and in the company of documentary film producers, prominent lawyers, politicians, scientists, and many, many members of the press. This book tells the story of what the journey has been like, and what the struggle entails. It all begins with one woman, and her name is Chris Mobaldi.

It was August of 2002 when I first met Chris Mobaldi, the friendly red-haired woman in her fifties who had been hired at the very same time I had. We attended the numerous required training classes together, and then proceeded to work side by side for the next four months. We both worked in the same department and I was her supervisor. I enjoyed Chris, she was friendly and had a good sense of humor, and as the demands of the job were at times overwhelming, we developed a good partnership and we helped one another out a lot.

But, as autumn turned to winter, Chris began having health problems. Sometimes she complained of horrible headaches or dizziness, other times her fingers were too stiff to hold a pen and she could not write. Then the eyesight problems came, and the pain between her eyes. Eventually Chris had to take a leave from work for brain surgery: she had been diagnosed with a recurrent pituitary brain tumor.

Nothing was ever the same after that, and the tragedy of what happened to Chris was astounding. Chris never regained her health, and eventually she and her husband moved several hours west to Grand Junction, Colorado. We drifted out of touch for several years, until a strange twist of fate brought us back into contact.

A newspaper article had been written about Laura Amos's health problems and the Amoses' belief that their damaged water well had been contaminated not only by huge amounts of methane gas, but also by hydraulic fracturing chemicals—specifically 2-BE. Long before the Amoses received monetary compensation from Encana gas company and signed a non-disclosure agreement with the company that forbid them to discuss their damages, Laura had said that an Encana employee finally verbally admitted that 2-BE had been part of the chemical mix used to fracture the gas well some 750 feet from the Amoses' kitchen.

The Mobaldis, now living in Grand Junction Colorado, read the article about Laura Amos and put two and two together. There were so many similarities: blown out water wells, tainted, fizzy, smelly water coming from the taps, and strange and rare symptoms and tumors. Laura Amos had also lived surrounded by natural gas wells in Garfield County. Both families had experienced the "regular" symptoms endured by many living near gas development: burning eyes, respiratory difficulties, dizziness, nausea, massive headaches, joint pain, pain in the hands and feet, recurrent and massive nosebleeds, and more. But

COLLATERAL DAMAGE

Laura, similarly to Chris, also was stricken with very unusual symptoms including carcinogen-related tumors.

The Mobaldis made contact with Laura Amos, and they began communicating about their suspicions. Laura told the Mobaldis that the Colorado Oil and Gas Conservation Commission was holding a public hearing soon in Rifle, Colorado. Laura and a number of other landowners with health problems planned to speak before the commission. The Mobaldis decided to come too, and ask the commission for help.

The meeting that day led to my reconnection with Chris Mobaldi. I was not in attendance at the meeting; I was flying back to Colorado from Boston where Al and I had been visiting my family. But when I called Laura Amos the next morning to see how the meeting went, that is when she told me about what had happened to Chris. I was dumb-founded—and within hours I was on the phone with Chris. We had reconnected after more than three years; it was a shocking phone call and I was horrified to hear what had happened to my once healthy co-worker. It was beyond strange…

After we re-kindled our friendship, soon Chris asked me to write her story—she knew that I had published a book before. So I put my current project aside, and diligently began to write "The Chris Mobaldi Story." After three and a half years now I find myself working on the chapter outlines and submission material for the completed manuscript. The first publisher I queried with asked for a second submission. There is nothing simple about editing and submitting for publication—and I have to do it while holding down my full-time job.

This book was written in real time about real people and events. It was written on behalf of Chris Mobaldi and all the others who have ever suffered damages to their health and lives as a result of gas and oil development. Most importantly, it was written in an effort to keep such tragedies from happening to others in the future.

In 1989 Barrett Corporation gas company was drilling wells above Chris and Steve Mobaldi's property, and apparently something went very wrong at a nearby neighbor's place. Water spewed high into the air, and the Chris and Steve Mobaldi recalled a large number of gas company employees swarming over the scene. The Goad's water well at a neighboring property was blown out due to the drilling of a gas well nearby.

For the next three months Barrett Corporation trucked water into the small valley neighborhood. The Mobaldis and others were told not to drink the water. After the water wasn't trucked in any more, they were told it was all right to drink from their wells, which the Mobaldis did. Much to their regret today.

Colorado Oil and Gas Conservation Commission public hearing, Rifle, Colorado:

On July 11th, 2005 the state regulatory body that oversees gas and oil extraction in Colorado held a hearing in Rifle, Colorado. Items on the agenda included discussion on the largest known natural gas mishap in Colorado's history known as the Divide Creek seep—when a faulty cement job on a well allowed 115.5 million cubic feet of natural gas to escape through underground fissures. The subsequent gas bubbled up into the waters of Divide Creek and was eventually discovered by a rancher who watered his cattle from the creek. The landowners were even able to light the gas on fire as bubbles boiled from the creek. The mishap led to benzene contamination of the creek and the surrounding soil. Although the Encana Gas Company initially denied responsibility for the event and claimed that the gas bubbling out of the creek was naturally occurring, the Colorado Oil and Gas Conservation Commission eventually laid the blame for the contamination on the gas company. Members of the public were given the opportunity to speak before the commission, and a number of landowners with complaints lined up to speak. Among the group were two women: one was Laura Amos and the other was Chris Mobaldi. Laura spoke about the health problems that plagued her after her water well exploded when gas well work was operating some seven hundred feet from her home. She also pressed the commission members to explain how they were upholding their mission to protect the public's health and well-being. Commission members skirted the questions and stated that that was not the nature of the hearing today. Laura continued to pepper the commission with questions as to the damages to her water and to her health and that of her family.

COLLATERAL DAMAGE

When Chris Mobaldi's turn came to speak, she referred to typed pages that she held in her hands. The room fell silent when she uttered her first words in a barely intelligible strange accent.

The frail and ill-looking redheaded woman began to describe the horrific health symptoms she had endured while living surrounded by natural gas wells, pits, and flaring. The skin rashes, headaches, massive nose-bleeding, debilitating headaches, joint pain and more were dwarfed by two recurrent pituitary brain tumors that required her to undergo invasive brain surgery. Her health has drastically deteriorated, and now she suffers a plethora of strange symptoms including acquired foreign accent syndrome, and constant illnesses from vomiting and diarrhea to exhaustion and overall body pain.

The room was overcome with stunned silence as the stooped and ill-looking woman spoke that day.

Others testimonies followed, but Chris Mobaldi's horrific health impacts eclipsed all others.

The hearings were to be continued at a later date, and they would be moved to Denver.

April 28, 2006

Today, April 28, 2006 Jonathan Vigliatti from Channel 8 and his cameraman came to the Mobaldi's home in Grand Junction to do another interview. This morning Chris was not doing well. It was one of her many "sick" days. I came down from Garfield County to meet them for lunch before the interview. Chris and Steve met me in the Applebee's on Horizon Drive. Chris looked unsteady on her feet and she was barely intelligible. I had great difficulty understanding anything that she was saying for some time. She had to make three attempts to successfully get into her high seat at the small round table. Steve shrugged his shoulders and held up his hands in a gesture indicating helplessness.

"She's not feeling good today; I haven't understood anything she's been saying all morning…" He said. He turned to his frail and stooped wife who was chattering in an incomprehensible tongue as she fussed with the high stool she was seated at.

"What do you want, Chrissy? Do you want to be closer to the table?" he asked.

She continued to chatter away to herself but nodded yes, and then finally she climbed with help onto the stool.

"How was she yesterday?" I asked. When I had called last night I had been able to have a good conversation with Chris, and I understood almost everything she said, which is rare.

"Not good, she got sick last night. And she's been sick this morning," said Steve.

I turned to Chris, who was making a valiant attempt to speak clearly. We could make out that she was concerned about her lack of ability to speak, and she was worried about the upcoming interview in an hour and a half.

"Dey wont bey ah-bool ta umba stahnd mey," she said slowly.

"It's OK, it's going to be fine," Steve said.

"They still have the tape from last time, they can use that..." I said.

Two weeks ago the first filming had taken place, but for some reason our faces had turned green during fade in and fade-outs. The bright lights that the cameraman set up had bothered Chris, and part way through the interview they had reduced the lighting at her request. Perhaps that's why the film didn't turn out well.

A waitress arrived to take our orders, but we were not ready so we asked for more time to peruse the menu. It was a thick, glossy menu with numerous pages and photos of some of the choices. Chris was unable to say what she wanted, so she stabbed a finger at one of the pictures. It was a piece of chocolate cake with ice cream and fudge sauce. Steve laughed.

"You want chocolate cake?" he said.

She nodded firmly, her mind made up.

"She's not eating anything these days, she losing so much weight." His face had a concerned look.

"Well, it doesn't help that she's throwing up three or four times a week," I added.

Steve and I placed our orders. I was a bit worried too that Chris would not be understandable for the re-taping with Channel 8. What a shame it would be, because she did a great and valiant job for the first taping. When you hear her telling her own story in her own now baffling tongue it is impossible to re-create the impact coming from another person.

While we were at the restaurant we talked about a number of things, and Chris could not verbally communicate to any extent. She began trying to draw letters and numbers with her hands on the tabletop instead.

"What size did you used to be, Chris?" I asked.

Her garbled response was gibberish, so she slowly drew a shaky 1 and 4, and then a plus. She garbled some more unintelligible speech, and I knew as she drew the numbers 1 and 6 with her stiffed fingers that she was trying to say she had at one time worn 16s. We then discussed her current and past weight, and I was astonished to hear that she had gone from 165 to just over 100 pounds. I glanced over at my friend's pant leg; it was not much bigger around than my bicep if that.

We finished eating. Chris had not eaten half of her cake, merely a few shaky bites.

When we got back to the Mobaldi's home the reporters pulled up right behind us in the white Channel 8 vehicle with the logo on the side. We all shook hands, they unloaded the camera and recording gear, and we went in. Jonathan briefed us on how to go about the re-film as the camera and lights were set up. He wanted us to answer the same questions as last time, and he had everything all laid out.

Once things got rolling, it went very well. Steve and I jumped in and interpreted for Chris, as we were familiar with her speech patterns. Soon it became evident from the reporter that he was catching what she was saying also—at least the majority of the time.

Once again, Chris was putting everything she had into being able to describe her condition and what had happened with her health when she lived at their country home in Rulison, Colorado surrounded by gas wells.

"I know I am not what I used to be like…" she said.

When asked to describe a typical day she answered:

"I get up in the morning, and I am in pain. I eat a little for breakfast, then I soon throw it up. I try to make it till noon, then I eat some more and it usually stays down. Then I am exhausted and I have to sleep."

"What keeps you going?" the reporter asks.

Chris points at her husband as she says with a wan smile, "He does."

The interview ends, the reporter and camera man leave, and Chris lies down on the couch after we come back inside from shooting the B-roll film and saying goodbye to the reporters. Steve covers her with a blanket and she closes her eyes, obviously exhausted. Toughy, the Mobaldi's miniature poodle, jumps onto the sofa and snuggles besides Chris's knee.

"Toughy protects Chris, see?" said Steve.

We talked about the re-filming, and congratulated Chris again on such an incredible job of summoning her strength to force herself to be intelligible. Chris was all but asleep when I left shortly afterwards.

COLLATERAL DAMAGE

I drove back to Garfield County along Interstate 70 that hugged the banks of the Colorado River. As I got closer to Garfield County I noted that the density of wells along the highway had been nowhere near that number of rigs just last summer. Now they were everywhere, and closely packed. How in God's name could the people in Garfield County continue to live so close to these fumes and pollutants with no one caring about what was happening to their health? Money, money, money...

Chris and Steve had arrived at my house unexpectedly last weekend. They were going to Rifle to check on their old house that now stood vacant, and they called at noon and asked if they could stop by. In the early afternoon they pulled in the gravel drive of our old ranch. They both had big smiles on their faces as they came up to the house, and Steve called out "Don't pay attention to the pointy things under my shirt!" It was obvious that he had a box stuffed under his black tee shirt; it seemed to be about the size of a twelve pack.

I gave Chris a hug and we exchanged greetings.

"I can understand you really well today, and yesterday on the phone you sounded clearer than usual! I could understand everything you said, or just about everything," I said.

Chris was beaming at me with a merry twinkle in her eyes. In spite of her frail body and stooped posture she looked so much better than she had the last times I had seen her over the past year. Since the Mobaldis moved away over a year ago to Grand Junction due to Chris's increasingly failing health, she had not been back to Garfield County too many times. I had only seen her four times since then, all at events where she was speaking publicly or to the press about her gas-related health issues. Each time she had appeared a bit worse for the wear, and had managed to summon her limited amount of strength and mental focus to get through a speech or an interview, after which she would all but collapse from exhaustion. It was so good to see her looking energetic and happy for once, a pleasant reminder of the healthy and vibrant woman I had worked with three years ago.

Chris and Steve came inside our small ranch house, through the cluttered mud room filled with bags of horse feed, bridles, halters, lead ropes, tools, and coats and boots and into the living room. They continued to beam at me with excitement.

"We have a present for you, Tara. A Christmas present!" said Steve as he pulled a brown corrugated box out from under his shirt.

"Ees for evedy ting you hab doon for us....You ah sush a good fend to mey," said Chris.

It took me a moment to realize that the Mobaldis had brought me a brand new laptop computer. I stared with disbelief at the brown box with the TOSHIBA logo printed on it. I was stunned and speechless as Steve handed me the box.

"It won't take you so long to start up the computer now." he said. "This will be a lot faster than your old computer."

"Trow dat ode ting away!" Chris said, indicating my five-year old Dell clunky monitor that sat on my cluttered desk. "We trew ahs ah-wey, we gave it to mey dah-tur. Den she trew it ah-wey too, and got a new bettah one."

"You guys, I can't believe you did this! Oh my God!" I finally stuttered, overcome with disbelief.

I kneeled on the floor and began to pick carefully at the complicated box, trying to neatly open it with out wrecking the box. Steve finally helped me, and we carefully drew out the slim, neat black laptop.

"Now I really have to write your story! I was just working on it when you came in, here, sit and read what I started yesterday!" I said to Chris.

She sat at the old computer on my cluttered desk in the living room and tried to read from the monitor, but she couldn't. It was too hard for her to make out. She now has difficulty reading, too.

COLLATERAL DAMAGE

We all went into the kitchen and Steve helped me get the computer set up. Chris didn't say much as Steve and I talked about the computer. But she was happy and smiling as she sipped at her bottled Desani water. She only drinks bottled water now, after so many years of drinking the almost certainly contaminated well water from their Rifle home. Sometimes the water had an oily film on the top and it fizzed, or it was clouded with particulate matter. At one time they had actually been able to light a glass of their tap water on fire. The Mobaldi's water filter had constantly become plugged with sand and they had had to replace it numerous times on short order, far before a typical time span for replacement. They had gone to the extent of installing two water filters, but even so the filters had clogged so fast with sand they had had to replace them every few weeks. Thus, the bottled water now.

Steve installed my personal set-ups on the Toshiba, and gave me a quick description of the basics that I very much needed. We were sitting at the kitchen table for a good half an hour before I noticed that Chris didn't look so good anymore. She had a look of discomfort on her face, and seemed to be getting unsteady even in her chair. Steve and I asked her how she felt.

"Not good" she said as she rose to her feet and began to unsteadily tack about the small kitchen like a sailboat with no one at the tiller.

I jumped up to assist her as she grabbed for the kitchen counter. I feared that she was going to fall and hurt herself.

"Are you getting sick?" Steve asked.

Chris nodded and I helped her to the bathroom, holding her arm as she took the step down from the kitchen. She was sick for some time.

"It happens all the time, she'll be fine," Steve said.

"How often?" I asked.

"Oh, four or five times a week," he said.

When Chris came out from the bathroom she sat at the table and held herself up with effort. She sipped a little water and the Mobaldis decided it was time to go soon. They still had to stop at the Rifle house and check it before heading back home to Grand Junction. Before they left we made a quick trip out back to see the horses and the llamas. Chris fed the horses carrots and petted Scout, the rambunctious young black llama who sniffed at her curiously.

"Ah mees deh llamaaas," she said, with regret in her voice.

The Mobaldis had been raising llamas at their old house, but the young ones had sadly died with pneumonia-type symptoms. A typical illness for animals raised near gas wells…not a typical cause of death for young llamas.

In the driveway I hugged Chris goodbye and waved as they drove down the road. She smiled weakly, but looked beat and ill.

A year ago Chris had asked me to write her story. I can make no promises as to what I will produce, but I will give it my best effort possible. So here I go…

Today Judith Kohler from the Associated Press called and we talked for close to two hours. I told her what I knew of the others in the area who had health issues that they believed were related to the gas industry and the wells near their homes. I met Judith down in Denver at the Capitol when there was a press release regarding the Surface Owner's Rights Bill. She gave the Mobaldis her card and said she would be in touch. She called me several weeks ago and let me know that she was coming up to Garfield County to cover Senator Salazar's visit. We set up an early morning interview at a local café with several others who had come forward with health-related issues.

A woman named Dee Hoffmeister came, and Tim and Karen Trulove. Judith set a small tape recorder on the table and asked if she could record the interview. Then Dee began telling of her problems that began when she was knocked unconscious after being enveloped in a cloud of fumes surrounding their home—presumably from the gas well and open pits containing chemical mixes and undisclosed liquids that were some 700 feet from her home. Karen Trulove broke down into tears as she described her health problems, which she soon linked to strong odors that came and went from the wells surrounding her home. As soon as she got a big whiff of fumes, she was hit with an instantaneous headache. She remained continually incapacitated and ill to the point that she could no long work, or even function around the ranch. "I don't even plan my days anymore, I just try to get through them."…After the interview I went to work, and Judith said she would be in touch. She accompanied Karen and Tim Trulove up to their home for a look at ground zero.

During our phone call today Judith relayed what had occurred during her visit to the Trulove's property south of Silt.

"There was a lot of activity around the gas well, and it smelled. Karen went into the house and Tim and I went to talk to the people he knew that were working on the pad. There was some problem with a valve that had been shut off, and there was a leakage of paraffin; that was what caused the smell." Judith paused.

"Tim seemed to know the guy he was talking to, who worked for the gas company. We asked him how the valve got shut off — was it vandalism? He didn't seem to have an answer. I found it strange that this happened just at the time when I was here. How often does it really happen?" she asked.

"It was probably just someone screwing up who didn't know what they were doing, there's no one watching them, so anything can happen," I said.

As we talked I went in and out of the house with the phone a few times to complete some ranch chores and to enjoy the beautiful, balmy spring Colorado day. The elm trees were budding out in brilliant chartreuse green and the sheep pastures were deep green and as lush as a carpet as far as the eye could see. It seemed implausible to be discussing the industrial activities occurring just across the Colorado River on ranches once so similar to where we lived. I sat on the small back porch and watched the lambs across the creek jumping like popcorn as their mothers relaxed in the pasture, dozing and flipping their wooly ears at the occasional fly.

Out of the corner of my eye I noticed one of my cats hunched to the ground with a struggling bird in her mouth.

"Hold on! My cat caught a bird! I have to get it away from her!" I said.

Quite a few times I have been able to get to a bird in time to save it from the jaws of death, but as I held the tiny sparrow in my hand I knew this was not going to be one of those times. The dark red blood began to spread across the bird's chest feathers; the wound was deep and fatal. Still talking to Judith, I carried the tiny dying bird up to a sagebrush bush beside the vegetable garden on the hill above the house.

The blood of the bird was warm as bathwater on my hand as I gently laid the little body into the grass. I could not shake this thought from my mind long after my phone call with Judith ended.

There seemed to be a correlation my mind was reaching for: trying to save the dying bird seemed about as impossible as trying to save the people who were falling ill, seriously ill, from the gas wells beside their homes. Nonetheless, talking to the suddenly interested press was the best thing that could be done. The stories of what was really happening to the people needed to be told, and told in a big way.

Talked to the Mobaldis last night and tonight.

Chris went down on the fourth to see Dr. Gerdes in Denver, a specialist who did a number of toxicity tests. He was the same doctor that the Mobaldi's neighbor had seen. The neighbor was diagnosed with severe levels of natural-gas-related toxins in his system. The doctor informed him that he needed to move away from the area if he wanted to recuperate from his symptoms. Tests were run on his toxicity levels and it is reported that they topped the charts at over 200 times the allowable level for lead, mercury and other heavy metals. The tests only go up to 200 times the allowable level of the different toxins, so his actual levels are most likely unknown. Ted went to see his nearby neighbors and showed them the results of his blood test. All of them had endured intense impacts from natural gas drilling over the years, and they were suspicious about the quality and safety of their drinking water after the first water well erupted during a drilling procedure. The gas company had tested the water wells, and trucked in drinking water for a period of time—but that soon ended. Ted wanted his neighbors to be aware of the reality of what he believed he had been exposed to. He said that the neighbors really didn't know how to respond to the information, Over the years many quietly moved out of the area, abandoning the Red Apple Subdivision.

The neighbor followed the doctor's advice and moved away, out-of-state, in fact. Neighbors and friends speculate that the gas company bought him out, either directly or through a middleman. When recently asked for specifics of the chemical levels found in his medical tests, after a consultation with his lawyer he said that without a subpoena that information could not be disclosed.

Apparently the entire family experienced health problems, not unlike some of Chris's. Their dogs also died. Their decline was described as similar to chronic wasting syndrome.

I was very curious over what had transpired in the subdivision, and I spoke with a number of the residents. One of my co-workers used to live there during the time of the blowout at the Goad's water well. She described the neighbor's valiant and unsuccessful attempt to stop the drilling, and also said they were afraid of what they were potentially being contaminated with—especially after the water well exploded.

"There were big trucks all the time, like tanker trucks. They used to drive up the dirt roads above the Mobaldi's, and finally one day I went up there to see what was going on. It was awful, there was a huge pit filled with stinking liquid. You know, when all this started, I said to my husband 'We're all going to die of brain cancer...'"

As we stood in the office at work, with the sun streaming through the window on that beautiful day—after she uttered those words—a deep chill struck us both. We both immediately thought of Chris Mobaldi, her old neighbor, who fatefully did end up with exactly that—a brain tumor. I know because as we stared at each other, no more words needed to be said…it was too poignant, and so incredibly tragic and real.

I had drawn a crude map of the subdivision, and as I learned of different neighbors who had some strange ailments occurring I would add them to the map. I kept it with me in my folder with the phone numbers and names of people with health complaints who lived in the gas fields. And there were some strange ailments: One woman's voice became high and squeaky, and when she complained to the gas company operating around her ranch, they thoughtfully moved her into a motel in Rifle that summer. People referred to her as "the lady who talked like Mickey Mouse," no joke. Even the lawyers were stunned when I relayed some of the symptoms to them.

I took to thinking of the Red Apple subdivision as "The Old Hell Hole." For that truly seemed to be what it was—for some, anyhow…

The lawyers called Steve today and wanted to know information about the wells that the Mobaldis needed to get from the previous owner of the property. Steve couldn't find it on the Internet, so he called me and we found it in the local phone book.

We talked about reporter David Frey's extensive article in the Mountain Journal section of the Aspen Daily News that came out last Wednesday. Also about the *Salon* Internet spread that included information from Lance Astrella, one of the Mobaldi's lawyers, and his take on the whole scenario here in western Colorado in regards to the gas industry's unregulated practices. Exposés of other people like Chris with serious health issues who live near wells were in that layout. The invaluable views and commentary of Weston Wilson, the whistleblower from the EPA who has been ignored was also included.

COLLATERAL DAMAGE

May 13, 2006

I went to a friend's house on Silt Mesa for a barbeque to celebrate her new business's opening. Sue had recently purchased and installed a water therapy unit for dogs, and more excitingly (and costly) a hydrotherapy equine unit complete with an enormous customized horse trailer she had designed herself. This was the first ever mobile equine water therapy unit in the US, and probably the world.

The company she purchased the units from had flown out their veterinarian and technicians from Virginia to train Sue Schmidt and her team on the operations of the equine hydrotherapy unit. Several horses had been brought over that day, and successfully run through the water tank that resembled a huge aquarium for horses inside the massive trailer. Once the horse entered the unit, the water began to fill and soon the horse was swimming! Tonight's open house was a large party by invitation for local veterinarians, kennel club members, and animal owners to see the dog and horse units in action.

As I drove over the Hog Back Mountain Ridge and down the winding country road from Harvey Gap reservoir, I noticed a bright flame leaping skyward some seven miles away. It was obviously flaring from a gas well, or some type of mishap yet again. I kept my eye on the leaping column of yellow orange flame that rose impressively high.

Once at the open house, I was strolling about with other guests in the retrofitted dog water therapy room (a converted two-car garage now rubber-lined and complete with the large therapy unit and an impressive reverse osmosis water system), and an acquaintance called to me. It was Oni Butterfly, a member of the Grand Valley Citizens Alliance. Oni used to work for the Environmental Protection Agency before she moved to her ranch in Garfield County.

"Where were you Tuesday night? I didn't see you at the meeting!" said Oni.

"I couldn't come; I've had too much going on lately. I wanted to be sure to come to Sue's open house, and I couldn't do two night things this week..." I confessed rather guiltily.

"Did you see David Frey's article in the Aspen Daily?" I asked.

She hadn't, and I invited her to come out to my truck where I happened to have a copy. Oni and I continued to discuss recent media interest in what was happening to people's health that lived near the gas wells. I told her about my conversations with Judith Kohler from the Associated Press, and Jonathan Vigliati's Channel 8 piece that would begin airing this week on May 15th and run for three nights consecutively in a three-part series.

Oni coughed repeatedly in a dry, constant hack as she told me about the well explosion which happened up Dry Hollow earlier this week.

"I have this damn cough now from that mess. It was horrible; there were huge plumes of chemicals burning, huge black clouds that settled over our homes....Jim Rada from the County Health Department came, but he didn't even bring a grab canister to take an air sample. I just yelled at him and said 'Jim, how can you not have a canister at an explosion like this?' I was furious!" Oni exclaimed.

As we continued to talk by the truck, the party swirled around us. Small groups of people clustered around the impressive water therapy unit as the wet horses stepped off the treadmill and came down the ramp into the corral. It felt strange to be discussing illnesses from natural gas development at this festive celebration with visitors and horse specialists from across the country attending.

Oni shook her head.

"I need to get out of here! This is crazy; it's like Love Canal! We're all dying!" she shook her head again as she coughed.

"I need to think about this...how can I sell my ranch if I talk about it? I just want to get out! My place is already listed, but I'll have to lower the price, even by $50,000—I just have to do it, I need to get out of here!"

Oni looked depressed as she continued. "I'm going to Durango tomorrow to get out of here; I'm trying to spend as much time away as

COLLATERAL DAMAGE

I can. It's not healthy to stay here, look at me now, I can't even breathe from that explosion this week, I was breathing all that crap that was in the air around my house!" she paused again.

"I don't know, Tara. I know I should say something, but I just want to sell my place! I'll have to think about this, I'll call you when I get back…"

We drifted back to the party, where people were beginning to get serious about the huge buffet-style spread Sue had set up with the help of her friends. There were loads of all kinds of delicious foods: appetizers, salads, barbeque chicken, spice-rubbed brisket, homemade bread and baked beans, and coolers of beverages. Plates were loaded up and people claimed seats at the tables set outside under the huge elm trees.

The setting golden sun illuminated the beautiful rural panorama of meadow, mountains, ranches and hay fields, and long shadows dappled the light. It was a perfect evening for an outdoor party.

Off in the distance, the huge eerie orange flame from the flaring well trembled as it leaped into the Colorado dusk. And the strange towers of white lights illuminating the dozens of visible rigs began to shine, like huge, misplaced Christmas displays south of the Colorado River. My heart anticipates the pain I will feel when I see the first rigs arrive here on what is now the beautiful unspoiled Silt Mesa. Sadly, it won't be that way for long as the area is slated for upcoming drilling. The extent of the drilling will depend on how well or how poorly the gas wells produce, if they hit big they will go for it, if they don't hit big then it may not be too bad. Only time will tell…

Throughout the festive evening as horses and dogs were paraded through the water units and the outdoor barbeque was served in the balmy spring sunset evening under the rising full moon, I kept glancing over at leaping flame from the gas well fire. It continued to burn the entire evening into the dark hours when I left for home and drove over the mountain ridge and out of sight.

It doesn't seem fair that the people who live around natural gas development have the tranquility of their lives all but ruined. Your only choice is to live with it or go away. We live in the country to enjoy and cherish nature, away from cities, crowds, and pollution. We live here to raise hay, cattle, and horses or just to enjoy the peaceful natural surroundings. Then in comes natural gas development to steal it all away. No one believes the severity of what can happen until it happens to them, or to someone close to them. No one believes it, because it inherently just doesn't seem right or fair.

Learning about the impacts of natural gas development is not a hobby regular folks take up in their free time. Trying to fathom how to deal with a huge, multibillion-dollar seemingly unregulated (for all practical purposes) industry on your doorstep is a mammoth and impossible project to tackle. Years of nights and days spent at endless dry meetings and hearings discussing and strategizing with local groups and officials. All this time is taken away from your limited private life after working a full-time job.

I go through highs and lows, sometimes excited by good press releases, or contact with big and hopefully bigger media and then other times get as low as low can feel over no visible or tangible progress against this dangerous and polluting industry. At least that's the way it feels, too often.

Sitting in the 60-year-old barn here, on my plastic chair I pulled from the garage yesterday with my laptop perched upon my "new" two bale high computer desk, I type away on this fancy Toshiba laptop the Mobaldis brought me. It is a gorgeous day, the scrub oak trees on the mountain behind me are leafing out, and the elm trees around the house are just beginning to unfurl their leaves too. The cats are curled in hay nests beside the haystack, and the llamas are chewing their cud here in the barn. My big bay and chestnut geldings are dozing in the shade canopy of the scrub oak clump, relief from the intense Colorado sun. Dozens of birds sing their hearts out in the trees and down in the creek bed. Spring is here full force.

As I dragged the two bales on top of each other to improve my "office furniture" and hopefully reduce bad posture backaches from holding the laptop on my knees as I did yesterday, I noticed a dusty spiral notebook tucked between bales of hay. It was the latest notebook from the historical novel I have been working on at a slow but steady pace since we moved here four years ago.

It will wait; history will not run away where I cannot retrieve it again. Besides, I am nearly finished with the first draft. I will save it, like dessert, to work on after this gas book is written.

Anyhow, I am in no rush. I made a vow after my first book was published to write, as John Steinbeck said: for the joy of it. When asked at the completion of one of his novels if he was happy, he replied that no, he was saddened. It was as if one of his friends had died, and he missed them.

That is the view I take on creative writing: it won't pay the bills unless you are someone like Steven King or Larry McMurtry. You do it for the love of the craft and the joy it brings into your life.

This gas project is different; it cannot wait, not long if it is going to be of any benefit at all to those who live in the looming shadow of the booming natural gas industry in Garfield County. They say this is only the tip of the iceberg; God forbid what is to come for what I call "Gas Field County" and its unknowing and trusting residents. Trusting that is, until the rigs set up shop in their yards and the party begins.

Andrew Nikiforuk, the author of Saboteurs, an excellent book about gas development and its impacts in Canada, said that the gas industry people call it "making a play" when they go into an area and develop for natural gas.

That is exactly what it seems like; it is a play.

The gas executives read their lines and tell everyone everything is great, and everything is going to be fine. Everyone will make money or benefit in some way, and there are no problems, no health concerns, none whatsoever. The politicians—local, state, and federal, whom in some way must benefit or thrive from siding with the industry—allow unregulated practices in this dangerous and toxic development right in landowners' yards. The few who voice the public's health and other concerns are steam rollered under by the power of lobbyists and corruption.

The supposed overseers of public health are uninformed or rendered useless by the current administration. Weston Wilson from the Environmental Protection Agency is an excellent example of that. His whistleblower action regarding the dangers of the carcinogenic chemicals that are injected by the tens of thousands of barrel loads beneath Coloradoan's and Westerners' homes was ignored by the powers that be in the EPA in 2005. I'd like to see the day when George Bush, Dick Cheney, and the head of the EPA allow that deadly stuff to be injected below their homes. And the heads of Encana, Barrett, and Williams gas companies to name a few. Let's not forget Governor Owens who is pro-gas, as are two of our local county commissioners: Larry McCown and John Martin. Tresi Houpt, the third county commissioner, stands alone championing the concerns of Garfield County citizens in regards to gas development. Added to this list of people who might learn something from having their homes drilled on and fracked would be the members of Colorado Oil and Gas Association and the Colorado Oil and Gas Conservation Commission, who more often than not turn deaf ears on citizen complaints. Strangely, once a landowner makes enough noise with the press and hires a competent and lawyer and puts forward a case against the gas company for health or other damages, the companies time and time again pay up. Yes pay up, settle out of court, and have gag orders signed by the plaintiff. Very strange, since these same complaints are not found viable initially by the regulatory entities such as the COGCC and COGA.

Got off the phone with Steve about an hour ago. He left a message last night that the Channel 8 story began a day earlier than scheduled initially. I called Channel 8 today and got the air times: 6 a.m., 5:30 and 10:00 p.m. It will run for a three-day continuous series, different footage every evening. Unfortunately, we don't get that local channel with our dish satellite. Steve said the station was making them a tape.

"Did you see it tonight?" I excitedly asked Steve. "What did they say?"

"We were in it, but it was mostly the blacked out Encana guy, talking about the pollution and the toxins and the leaks. And his concerns for people's health…and the spraying of the fracking liquids from the evaporation pits," Steve said.

He continued to describe the scope of the informant's undercover exposé of what was really going on in the gas fields of western Colorado.

"They showed film of ripped pit liners, and the guy talked about how one day they would be up at a pit, and the next day they would go back and the level would be really low, obviously things were leaking right into the ground that shouldn't be; and about the spraying of the fluids into the air, and just spreading them everywhere."

"That will be the most important part of the whole thing!" I said.

Having an employee from the industry come out undercover to the media to disclose improper and flagrantly irresponsible and illegal practices is huge.

"I just got off the phone with Jonathan Vigliati, he said that two people called him today from Rifle who saw the news cast and they are sick too. One of them, her name is Nancy, has the same kind of things that Chris does," said Steve.

"What, the same thing? Foreign Accent Syndrome?" I asked.

Steve wasn't sure.

"Then Lance Astrella called," he said.

"Oh really! A big afternoon!" I said.

Lance Astrella was one of the attorneys representing the Mobaldis in their case against the gas companies. He is a nationally renowned attorney, for that matter. The firm he works with was involved in the case that resulted in the Erin Brockovich movie.

"What did he say?" I continued.

"I can't talk about anything that happens anymore…" said Steve.

"You've never said that to me before!" I exclaimed, quite surprised.

"No, I can't talk about anything that happens from here on out," he said.

I paused for a minute, trying to digest the information.

"Well, I guess that's good news, I guess things are rolling," I said.

"Yes, and the Associated Press reporter is coming Thursday at 6:00. Hey! I have to tell you what we did—we went for a detoxification. Dr. Gerdes told us to, and we found a lady in Palisade who does a foot treatment detoxification. Guess what happened?" he asked excitedly.

"What happened?" I asked.

"Well, they did Chris first, then they did me. Then they ran the water through a reverse osmosis thing, and mine turned green," he said.

"Oh," I said. I had no idea what a chemical detoxification process was comprised of.

"Then they did Chris's, and it turned black, and had crap floating around in it!" he said.

"What?" I exclaimed.

"Black and crap floating around in it!"

"Did you tell Lance? What did he say?" I asked.

"He couldn't believe it, and he said we needed to get a Western medicine Doctor on it," said Steve.

COLLATERAL DAMAGE

"That's what I told all you people before! That's what you need for court! No one's going to listen to an alternative doctor, not in court," I said.

I filled Steve in on a few things that had happened last week, and Karen Trulove's contact with Kemper Will, the Denver attorney that worked with Laura Amos who recently settled quietly with Encana Gas Company (only after waging a very unquiet media barrage). Karen was gathering together groups of people from south of Silt with similar health complaints to her own, which were severe and extensive. The forty-acre ranch and house where she lived with her husband was now up for sale.

When I talked to Karen last, she had been disappointed to hear that Jim Rada, the Garfield County health inspector, had failed to get a grab-air sample at the site of the well explosion last week.

"Do you have a grab-sample canister?" I asked her, knowing that a neighbor had helped previously in getting a sample from the Trulove's property half an hour after an "odor event." This is what the locals and officials around here refer to it when the stench in the air near the wells is overpowering.

"No, I don't, and I don't know if Jim would give me one after I questioned his information at that meeting..." said Karen.

"He has to give you one if members of the public are allowed to have them, he probably just didn't know about the toxicity levels..." I said in his defense.

"But he's the Public Health guy for the county! It's his job to know this stuff, you can find it out just by going online!" she said incredulously.

Karen proceeded to describe how she had called her son who worked for a governmental agency related to environmental issues. He told her what website to go to where one can get information about the toxins associated with natural gas production. She followed his directions, and soon had the information in hand as to the allowable levels and the danger levels of specific chemicals associated natural gas extraction. At the April Grand Valley Citizens Alliance meeting when Jim Rada gave his presentation and showed on an overhead screen graphs of data indicating the levels of some of the chemicals including benzene and toluene that were tracked in the air monitoring study, Karen questioned his assertion that the levels shown were below the allowable level and not considered dangerous.

"We took a grab sample at our place half an hour after the smell came on, and those levels are over the allowable limits," Karen said in front of the large group gathered at the meeting that night.

"It's easy to find out that information, just go on line," she said.

Jim Rada didn't have too much to say in answer to her questioning.

A week ago when I was headed to Rifle I called Jim's office and asked if I could pick up a grab-sample canister. He said sure, no problem, and we set a loose time for me to come by.

At the end of the afternoon I found the small, stark waiting room for the Garfield County Department of Health. Its bare, light-colored walls reminded me of a Social Security office, or some other generic, windowless governmental office. A few families sat in the plastic chairs waiting for their turns to be seen. One family was Anglo and the other was Hispanic. I told the girl through the square window in the wall that I had called Jim, and that he knew I was coming to pick something up.

Jim came out to meet me, carrying the good-sized cardboard box holding the grab-sample canister. He opened the box and explained to me how to operate the unit.

"Get a wrench and loosen this nut first, then tighten it with your fingers so you are ready when you want to take the sample. You don't want to have to waste time looking for a wrench and opening it, it should be ready to go. Then, loosen this one over here, and it takes the sample automatically. Then just bring it on in and we'll get the reading done on it. What well are you close to?" he asked.

"Actually, I am picking it up for Karen Trulove—she's sick a lot and has trouble getting to town some days," I said.

"She has a lot of wells around her place…" Jim said. As if by way of explaining her poor state of health, or understanding it.

I nodded in agreement.

"So what happened at that well explosion last week? It sounded pretty bad…" I asked.

"I was at the Encana meeting this morning, and they said it was caused by friction from a tarp that sparked the volatiles in the condensation fluids. It happens sometimes, and the air is so dry here it causes a lot of static electricity," he said.

"That's so dangerous!" I exclaimed. "These things can just blow up from a tarp!"

"Well, Encana said they're not going to cover them with tarps anymore; they have containers with covers in the condensate tanks." said Jim.

We talked for a bit about the air impacts from that explosion, and Jim said frankly that no one really knew what it had created in terms of dangerous toxins in the air.

I left the new brick county building carrying the large box with the silver metal grab-sample canister. It made me feel good to be a tiny part of a study on the impacts to our air from natural gas development, but it also seemed sad that the study was being done after people had been breathing the air near the wells for over a decade in some instances. Like in Chris Mobaldi's case. And look at her today…

Last night Steve told me that Judith Kohler from the Associated Press came down for her interview with them and she stayed three hours.

"How did it go?" I asked with curiosity.

"We just talked," said Steve.

"I really liked her, that's the way it was when I talked to her. She's not like some reporters who make you feel like you are wasting their time; she makes it very easy to talk to her," I said.

Her article should be released on the Associated Press wire any time now, because Judith said two weeks ago that she wanted to get it out in a few weeks before someone from the L.A. Times beat her to it. She promised she would try to track who picks it up, and let me know. I am keeping my fingers crossed, not only for that but also for Joe Brown's documentary with Fractured Films. He said he would call me back again around the twentieth. Today is May twentieth, so I will hear from him soon.

As Steve and I chatted and caught up on recent events, it didn't sound like there was too much going on in the past few days. He said that they will meet with the attorney Lance Astrella sometime this week. He will be going over to Grand Junction or Rifle, and they will meet up there. The Astrella and Rice office is in Denver, where Chris and Steve have traveled a few times for meetings. Apparently Judith Kohler, the Associated Press reporter, is meeting with Lance also in the near future. That should make a huge contribution to Judith's article, because from what I understand Lance really knows his stuff when it comes to the gas industry. Sounds like the article is a week or two down the road now before it hits the AP wire.

Steve said that today Chris is completely incomprehensible.

"She hasn't been this bad for a long time," he said.

"How many times has she been this way?" I asked.

"Oh, maybe twice; no more than that—maybe five or six times that I couldn't understand her for this long…actually, every day there is about five or ten minutes when I can't understand her, but then it stops. This has been really long," he said.

"When did it start?" I asked.

"Well, she went to bed around eleven, then she got up about eight and had breakfast. I couldn't understand a thing she was saying. I wanted to go out to breakfast but she couldn't—her balance was off too much. She was walking into walls," Steve said.

"So what did she do?" I asked.

"She got dressed, and she made the bed OK, then she kind of tidied up around the house some, and did laundry. She can function; I just can't understand a word she is saying. Then we went out to get lunch, we got Subway and brought it home. Then she got really tired and went to bed, around noon. She is still out," he said.

"What's she talking like, is it that Chinese or Japanese sounding accent?" I asked.

"No, not really, it sounds more like Russian or Polish, maybe. It just doesn't usually go on this long, it's been a long time since it has," he said with a concerned tone in his voice.

"How has she been lately?" I asked.

"Oh, she's been pretty good since the foot detox, when we were both sick for a while afterwards. She went out some this week, she went for a manicure and a pedicure, and went shopping and stuff. That was this week," he said.

"Do you think she'll sleep the rest of the day?" I asked.

"Yeah, probably. Right now she's up, and she's drinking one of those Ensure things. She's talking to the can of Ensure," he said.

"Great!" I said sarcastically. Chris, my friend, unable to communicate intelligibly with those around her in a language anyone else could understand, was having a conversation in her mind with a can of liquid nutritional supplement. How depressing. How distressing.

Before I hung up the phone, we talked a little more about the Channel 8 three-night series that I have yet to see.

"Jonathan has had a bunch more people call up, but I don't know too many details," said Steve, referring to the people who had seen the telecast and drawn correlations between their own health concerns and Chris Mobaldi's health issues.

"He said that he and Josh took a drive higher up in the mountains, and they saw a whole lot of pits with stuff in them, and the stuff was fuming; they could see it in the air. You know how you can see chemical fumes in the air? Josh was disgusted, and they couldn't believe how much they saw. Pretty much the last night of the TV thing was me and Chris talking, and the blacked out Encana guy talking about how out of control everything is around here," he said.

Then Steve laughed, "There was an Encana spokesperson who said how they got on top of every problem right away, and how they were doing a very safe job with everything." His laugh was hollow.

"Well, call me when you hear something, I'm starting to take notes when something happens if I don't have time to do the computer right away. That way, I can go back to notes later and catch up," I said.

We said goodbye, and I have been typing here in the new and improved barn office. I changed out my table of two hay bales stacked vertically for a rickety folding table my husband dragged out of the basement. He used to use it for river trips before he got a better rolling table with detachable legs. It is hot, high eighties and the sheep in the pasture below are baaing their chorus of high and low baas as the ewes relocate their lambs. The Harleys occasionally roar up and down the narrow valley road that amplifies sound so that neighbor privacy is a mute point. (If you and your spouse are going to argue outside, the entire valley neighborhood of five houses will know everything you yell. Fortunately Al and I rarely argue.) The horses are dozing under the scrub oak that continues to leaf out its bright green leaves. Soon the trees will cast effective shade to cool the big animals. By that time it will hit the nineties and push one hundred degrees. The black flies are buzzing in the fresh flytraps I hung out midday. Such is life in the barn office. I like to write outside where I can observe nature and watch my animals and whatever wildlife is flying or walking by.

Making Movies

May 28, 2006

Another week has gone by, and thankfully it is a three-day weekend. A much needed break from the seemingly constant pressures of work. Also, a chance for alone time as Al is gone on a rafting trip on the Arkansas River and to stay at Dad's ranch property up in Fairplay where we used to live. This is the highest water they have had in years, and I am waiting for the phone call that I was promised to let me know that they got safely off the river. So I have had the house to the cats and myself and yesterday I took full advantage and spent hours on the phone making what I have come to call "gas-related phone calls."

I got a call from Joe Brown of Fractured Films. We made plans for a get together either here or maybe going over to Grand Junction together when they will film with the Mobaldis.

"Sorry I haven't called sooner, but I just got back from my trip and now I'm applying for graduate school at the same time we're trying to finish this film," said Joe.

"Don't worry about it; I've been busy with work anyways," I said. "Steve told me you're coming down this next weekend, is that right?"

"Well, we're hoping to get in on that meeting with Ken Salazar, but I'm waiting to hear back some details," said Joe.

"You mean the meeting on oil shale development, the one with the other senators? I have an invitation to that right here." I picked up the letter from western Colorado Congress I just got several days ago and began reading the information to Joe. I gave him the address of the Grand Junction City Hall, the meeting time and date, and Matt Sura's number. Matt is an organizer with Western Colorado Congress.

COLLATERAL DAMAGE

"Call Matt, he'll help you out," I suggested.

"Are you going to the hearing?" he asked.

"I don't know, I probably won't because I would have to miss work. I've missed a lot of work from going to gas things. I went to Denver to lobby with the senators for the Surface Owner's Rights Bill and the press conference, and I've taken a couple days recently for the television interviews. It begins to add up..." I confessed.

The truth of the matter is that I have spent good portions of my coveted vacation days on work for natural-gas-related issues, plus a bunch more last winter and fall for animal rescue work from Hurricane Katrina. I needed to save up enough days so Al and I could take a vacation or two together this year. I've got to go on at least one long raft trip every summer, and we have our week planned for Mexico in November in the Yucatan. Last year the hurricanes in Mexico changed our trip plans. I am anxious to get back and see that beautiful turquoise water again.

Joe and I discussed the names and information I had e-mailed him earlier.

"Do these people know I'm going to maybe call them?" he asked.

"Most of them do, but I've lost touch with a few of them. But all of them have spoken publicly already about their health problems and concerns, so I don't think anyone would mind getting a call. Most of them are so mad that no one cares or does something when they call with complaints, they would be happy to talk to someone who will tell their story," I said.

We talked about the people who had already moved out-of-state or away from the area, and the people who had been bought out by the companies. They were limited in terms of what they could say because of the waivers they signed with the gas companies upon reaching their settlements. No future complaints could be voiced about the gas company, and no specific information in regards to the issues that were settled upon. Nonetheless, there was a lot of information available in the press already about some of these cases, and it was accessible.

"There are some people here who would be really good for you to talk to in person, if you want I can try to line some of them up in advance," I offered.

"That would be great, I really appreciate what you are doing to help," said Joe.

"It's my pleasure, I am happy to do it. I just want to see the story get out about what is really happening. It's so frustrating to watch the people who are the sickest leaving and getting bought out so it is all covered up. The gas companies are willing to pay a lot of money to the people who have evidence and lawyers, but the story of what's really happening never gets out because they sign the gag orders. They say it's the golden rule of the gas companies: Deny, Deny, Deny. Never to admit that anything they do is harmful to people, never!" I said in frustration.

"Something has to change, someone has to start to pay attention to this," said Joe with a firmness in his voice.

I hoped to God that his efforts with this documentary, and Judith Kohler's soon to be released story on the Associated Press Wire, plus David Frey's extensive layout in the Aspen Daily News Mountain Journal, and the Channel 8 coverage it all might lead to something bigger.

"I am just hoping it will get pushed to a higher media level. That, plus more lawsuits are the only thing that can help," I said.

We said goodbye and set a time to get in touch in a few days.

After hanging up, I called a few people and finally got hold of Nancy Jacobson from Silt. I told her about the film crew coming up, and that I was trying to find out more about the recent explosion.

"Oh, you mean the condensate tank that blew up? The one that Encana said tarp friction started?" She sounded skeptical.

"Yes, the one that just happened a couple of weeks ago. When I went to pick up a grab-sample canister (for taking air samples during odor events) from Jim Rada, he told me that's what Encana said that day at a meeting. That the tarp blowing around that covered the tank was what caused a spark that ignited the explosion. He says it happens," I said.

"I can't believe he didn't take an air sample, it's ridiculous. Apparently he said that because of the black smoke it was obvious there were a lot of carbons in the air," said Nancy.

"I asked him if he knew what was in the smoke, and he said no one really knew…he was pretty honest about it. I read an article about it, but I didn't keep it. Do you know whose place it was at?" I asked.

"I keep all the articles, I have it here in my files. Hold on a minute." Nancy paused as she went through her papers.

"Here it is, the Post Independent write-up says flames shot 200 feet into the air, and it was a condensate tank that caught on fire. It doesn't say whose place it was at, it just says Cedar Springs Ranch. I think it is at the bottom of Grass Mesa, near Hunter Mesa. You know, if you called the reporter, I bet he could tell you. It was written by Bobby McGill—he's written a few articles on these things," said Nancy.

"Do you know the lady who complained when the pits were lit off this winter over there? I know Duke knows, and there were articles about that, too. I was up on Silt Mesa hauling hay and I saw the huge plumes of smoke. I knew something was going on bad over there. I knew there must be people furious who lived around it. The smoke just billowed; it went straight up and was black. I could see it from way across on the north side of the river," I said.

"You mean Beth Dardinsky? She was the one who had the pits get burned off on Christmas Eve; they cut corners and did a cheap job on the pits, so the paraffin build-up couldn't be removed like it should be. Then they have to get rid of the stuff somehow, and it's too hard to suck out with the rest of the stuff. So they got a permit issued from the Health Department to burn it off. They didn't tell any of the landowners before they did it, wasn't that nice of them? And on Christmas Eve, when people have friends and family over and everything!" she exclaimed.

"Happy Holidays! That was so thoughtful of them.... I heard that the gas company told a reporter who came out that it was a mistake, then Duke called the reporter the next day and said actually they had been issued a permit. Someone at the gas company was either lying or didn't even know what was going on. What a surprise," I said.

"Actually, COGCC (Colorado Oil and Gas Commission) filed a NOAV on Barrett for that," Nancy said.

"A what? What's that?" I asked.

"A notification of alleged violation. They originally had nine pit burn permits, but then some of them got pulled. That's what I was told," Nancy said.

"Yeah, I heard that too. You know the time I saw the black plumes going up wasn't on Christmas Eve, it was more like February when I was running out of hay. So it must have happened other times," I said.

"I'm sure it happens a lot, and no one knows what's in that smoke they burn off. You know it's toxic…" said Nancy. "Anyways, back to why you called?"

"Well, I am trying to get hold of a few people to talk to Joe Brown and his film crew. He's doing a documentary and there are some people who he has lined up to see, but I told him I'd try to get a few more—people who have had some bad stuff happen right at their places. It would be so much more effective to get footage out where this stuff is happening, instead of a coffee shop or someone's living room. Not that a well will be blowing up or anything…" I said.

"You know Tara, I'm done with it. I'm done with the meetings, with GVCA, with everything. I need to focus on my life now. Do you know how much money my husband and I have lost? About three hundred thousand dollars. That's what we've lost in land values. We didn't want to stay here; we were planning on selling this place and retiring in Oregon or Washington. Now we can't sell this place for what it would have been worth, and it's our big investment. We're stuck, and I need to focus on my life. I might even go back to work..." Nancy said.

"I am hoping that with this recent media stuff going on, and with the Associated Press article coming out, that something will happen. And this documentary..." I trailed off.

"I've done all that. I've been on National Public Radio, I've been part of a documentary, I've been quoted in the New York Post and the L.A. Times, and I've had French reporters up here. In fact, I'm not allowed up on the well pads any more since I showed up with the French reporters. I have done it all and it does nothing." Her voice was flat, and even a bit angry.

"I heard you on Aspen Public Radio, you sounded great! You always do, when you're quoted in the paper too. I heard you, and Duke Cox, and Lance Astrella all in the same segment when I was driving to work one morning," I said.

"Really? I never even heard it," she said.

"Hey, did you read that letter to the editor last week, that one about the lady who said that oil and gas is their bread and butter? I got so mad when I read that, I started swearing in the office at work! I was so mad I wanted to call that lady up and chew her out, but I knew it wouldn't do any good," I said.

"Didn't you see my letter in reply, last Friday? I've had people call and tell me it was my best letter yet. That lady wasn't in the book; people like her who write letters and are unlisted make me sick. Too chicken to face it," she said angrily.

"There were a bunch of letters, I was really happy the next day to see a really good one in response. I wish I had seen yours..." I said.

"I can e-mail it to you; there were about five letters in response that I know of. What's your e-mail address?" she asked.

We exchanged e-mail addresses, and talked more about lawsuits, settlements, the several failed attempts by some neighbors to do group lawsuits, and the few who had settled independently. I could tell that Nancy was pretty angry with some of the lawyers who she said had "cherry-picked" the most prominently impacted neighbors out of the group.

"I had a lawyer and wasted $2,000 on fees. She did absolutely nothing for us. I am so disgusted.... I am happy for Laura Amos, she deserves it. She worked really hard for what she got, and there was a NOAV filed by the COGCC on Encana over what they did to her water. Encana really got pressure to settle with her," said Nancy.

The Amoses' well had been damaged during the fracturing process, and their water had been very fouled and contaminated. The adrenal tumor that Laura developed was most likely directly linked to that carcinogenic chemical exposure. That's what Laura's stance on the issue was, although the gas company publicly denied any wrongdoing.

"That's funny, when I called the COGCC the day after the Divide Creek seep hearing last summer a lady told me that Amoses' well was not contaminated," I said.

"You mean the hearing that they moved and re-scheduled to Denver, so none of us could come? Like they always do..." she said sarcastically.

"Yes, I had just gotten back from Boston at three in the morning, and I called Laura to find out if day two of the Divide Creek seep moratorium hearings was happening, because I was going to go. I wrote my speech at the Logan Airport in between attempted phone calls to Colorado to see if the hearing was going to take place, as I waited for our plane to leave. If the hearing was going to take place as scheduled, I would have to set the alarm and be ready to go Rifle after a very few hours of sleep. That would mean I should try very hard to sleep on the four-hour drive home from the airport as Al drove.

"When I got hold of Laura the next morning that was the first time I ever knew that Chris's illness was likely associated with gas, when Laura asked me if I knew Elizabeth Mobaldi, and I said no. I know her as Chris, so I didn't make the connection. Not until Laura started describing her. Then I was absolutely stunned!

"I said: 'Chris, Chris Mobaldi? Is that who it is? I was her boss and her friend, you know, three years ago. I was in her hospital room a few days after she had her head sawed open for the second pituitary brain tumor surgery. She looked like Frankenstein; she had metal staples across her forehead holding her skull together. She got really sick after the surgery, she about died."

I still vividly remember the strange phone calls I had had with Chris during her long and bumpy recovery at Saint Mary's Hospital in Grand Junction. She has little to no memories of those calls; she was so drugged up most of the time. Several times she was put on ventilators and she almost died of pneumonia. Her voice would sound over the phone in a barely audible whisper as I told her it was me.

"When I found out that Chris was sick from gas development, I called the attorneys Astrella and Rice because Laura had told me that they said they needed more than one client for a suit. I thought that perhaps with Laura and Chris they could make a go of it. Lance Astrella called and left a message a few days later, and then I got hold of him one day from work. I told him about Chris and her illnesses and the recurrent tumor. He said they would call the Mobaldis but in all likelihood it would take quite a few other parties to merit a lawsuit.

"Much to my surprise Steve Mobaldi called a few days later and said that Astrella and Rice were interested in taking on their suit. Alone, with no other parties involved. It seemed that Chris's medical information was substantial enough to merit such a move."

Nancy and I talked about the Mobaldi's pending suit, and Nancy said she hoped the lawyers would be successful for Chris. We said goodbye, and I looked up Beth Dardinsky's number in the local phone book. I dialed the number listed under Dardinsky, and by luck got Beth who happened to be at her brother's house that afternoon.

I told Beth about Joe Brown and his documentary, and she was excited to hear about it. We talked about the pit fires on Christmas Eve, and also about the condensate tank explosion. She had driven over to see it, along with a neighbor of hers Sarah Wussow. I told Beth the two dates when Joe planned to be this way with his film crew, and she said she would be happy to talk with them. We discussed the recent and ongoing attention from the various media outlets, and of the discouragement of some of the locals over no results—or seemingly no results from the coverage.

"Joe said that he could protect people's identities if they wanted to, if they had concerns about coming forward," I said.

"No, I don't care about that; I want people to know that I am angry. I don't care if people think I just want my name in the paper; that's not it. Someone has to do something about what's going on, and people have to know about it. When it's all said and done, I want to know that I did everything I could, even if it doesn't help," said Beth.

"Yes, lots of people just feel like it's too big to fight, so why bother? I think my husband feels that way, but he doesn't really mind me doing it. He just won't get involved," I said.

"That's the way my husband is too! But like I said, I want to know later on down the road that I did everything I could," said Beth.

We agreed to talk in the near future to pin down a date for Joe's visit. Beth was hosting an open house for her neighbor's to meet Tresi Houpt, the one and only (out of three) Garfield commissioner who listened and responded to community members' concerns over health and other impacts from the gas industry's development in our county.

I was quite sure I had just made a new friend whom I would soon meet.

COLLATERAL DAMAGE

As others seem to be fading away from involvement with this issue, either from exhaustion, disgust, or fear of not being able to sell their homes and get out, it was a shot in the arm to talk to someone as concerned and motivated as Beth Dardinsky.

I am out in the barn again today, with computer freshly charged up. Yesterday I had been on a roll and had a mission to complete writing for several hours when I was happily interrupted by a neighbor inviting me to a post-wedding party at the ranch next door. Their daughter had just gotten married, and the huge white tent would be there all night—so they invited all the neighbors to come down for an extended party—complete with a barbeque and live music. I turned off the laptop, fed the horses and llamas, and rode my bike down the county road to the neighbor's ranch.

It nagged me throughout the evening that I had not finished what I intended to write. When I got home, I finished up writing. So, I am done for today, and hope that the work I have done on this Memorial Day of 2006—plus the other days I have put into this project and will continue in the future—are not in vain.

May 30, 2006

At work today I got a call from Steve Mobaldi.

"Just calling to let you know that the photographer from the Associated Press went up to the Trulove's yesterday to get some pictures for the story; I knew you were wondering about Karen," he said.

And I had been, as I had been calling Karen for two weeks straight to no avail. I was glad to hear that she was back from wherever she had been. I thought maybe she had thrown the towel in...

I called Karen and she said she had been out of town for a while, and had not been feeling good either.

"I called you for a few weeks, but when I couldn't find you I gave the air grab-sample canister to Dee Hoffmeister. I didn't know if you left and went away, or what. I'm glad that you're OK," I said.

And I meant it too; Karen was a kindred spirit to me. She was an avid horsewoman, and also brave enough to come forward in spite of her illness to talk at the state Capitol for an entire day about what had happened to her from living near the gas wells. And how it had devastated her and her husband's life, and their home and ranch.

"I'm not OK, I'm feeling sick..." she reiterated.

"I know, I mean I'm glad you're not gone; like to Texas or somewhere," I said.

Gone like Susan Haire was, who finally had enough trying to live on her Morseana Mesa retirement ranch unable to deal with the toxicity. Trying to do outdoor chores and enjoy her life in the county wearing a respirator and with three heavy-duty air scrubbers in her house so she could function had become too much. The day after lobbying in Denver with the senators for the Surface Owner's Rights Bill and speaking at a press conference she fled the state and had her ranch up for sale. Running for her life and leaving it all behind — everything that she had worked and invested and dreamed for.

"I was afraid that you didn't want to deal with any of this anymore, and I felt kind of guilty because I called you so many times. But then I remembered that you went to Denver and testified, and no one pushed you into it," I said.

"After the last interview, my husband said: 'This is my last interview,'" said Karen.

"I don't know how to say this right, but I need to ask you. You know, sometimes when a husband and a wife don't do the same thing—like my husband, he went to one, the first gas meeting here — then he never went to another one. And he won't, he won't deal with it unless it happens to him. But he doesn't get mad that I do what I do, he just doesn't want to do it too. What I'm trying to say is that I don't want to ask you anything that would cause a problem, but if you are willing can Joe Brown and his film crew contact you for this documentary?" I asked.

"Let me know if you are able to do it." I followed, not wanting to commit Karen to a decision at that moment.

"I had so many messages when I got back, and I just didn't want to deal with it. Then I had the problems with the horses," she said.

We talked about the activity with some of the other neighbors who wanted to do a group lawsuit, and where that was going. Karen at this point in time was the one who spearheaded that, after two others from the area before her had also tried. Both of those parties had since settled individually with the gas companies that developed their properties, as the group initiatives had failed to gain the necessary momentum to take off. It is a hard task to motivate enough people to raise the $100,000 necessary to enter into legal negotiations with the gas company. That was the amount that Kemper Will, the attorney from the Denver Firm of Figa, Kemp, and Will had proposed.

Then Karen told me about her horses. One of their mares foaled last week and something was wrong with the foal. Its legs weren't right, there was some kind of deformity in the tendons; they were too short. There were other problems, so the Trulove's had a vet called and the vet came out to the ranch to attend to the foal. The next day the foal was having trouble standing so they drove it into Glenwood Veterinarian Clinic in Glenwood Springs. The vet worked on it, and wrapped its legs to give it more support. Then Karen said something that made me sit up and really listen.

"It was the vet who mentioned the effect of toxins on the tendons, that's where they will gather and do damage. I didn't bring that up," she said.

"I wonder who knows more about how the toxins can cause these things. It's pretty well known that abortions and birth defects are caused by exposure to these chemicals. I wonder if the vets know some specialists, or the lawyers?" I said.

"I have another mare who's about to foal, I have to go because she's pacing by the gate right now," said Karen.

"I hope your foal is alright..." I said. I could not imagine, as a horse owner myself, the pain that Karen must be feeling over the deformed foal that may not make it. That, on top of her own health problems that were significant, was adding undue insult to injury.

"This just isn't right, this shouldn't be happening..." I said from my heart. I cannot stand bad things happening to animals. Somehow they affect me more deeply than people, because they are so often at the mercy of humans.

"Good luck with the mare, I'll talk to you soon," I said.

I sent a few e-mails to Joe Brown about the itinerary for the next weeks when they would be filming out here. I feel confident today that there are some good people for him to talk to.

Nonetheless, I must confess that this is all very strange and depressing.

I write historical fiction, and typically for the last ten years I have enjoyed the time I spend writing as an escape into a type of refuge of the imagination, with a firm basis in history. Specifically, the history of the settlement of the Colorado Rocky Mountains during the Gold Rush, which was a very exciting, dangerous and colorful period of human history in the United States.

This story today, of a different type of mineral rush, the rush to extract natural gas from beneath the earth, is a sad story to tell. It lacks the energy and excitement of the Gold Rush.

For some people, they have gained lots and lots of money.

But for others, those who live in the shadow of natural gas development and breathe the air and drink the toxic water of the contaminated wells by their homes; they are falling seriously ill.

And their horses are throwing deformed colts, and their goatherds have increased stillborn numbers.

COLLATERAL DAMAGE

I wish I were writing about something else; something more inspiring.

Today Karen said "We are at war, but what about us? We are dying here from what they are doing to us…"

June 10th, 2006

Today did not turn out as I had planned; I supposed I would have been down in Grand Junction doing yet another interview with the Mobaldis—this time with the film crew working on the documentary. A mid-week phone call changed that. It was Joe Brown of Fractured Films, calling from Denver. He was returning my latest call, asking if I could hook up with them and catch a ride and back to Junction for the interview so I didn't have to boil in my non-air conditioned vehicle. It has been up to one hundred degrees recently, and it would be awful. Besides, I could save gas and we could carpool. And talk, undoubtedly, on the way.

"Tara, sorry for not getting back to you earlier, it's been really busy for me," said Joe.

"No problem, I'm just checking in on the plans for the weekend. I'd like to go down with you to the Mobaldi's, if it will work out. It's at ten on Saturday, right?" I asked.

"We've had a change of plans," said Joe. "The Mobaldis can't do it then."

"What? Why not?" I asked, confused to say the least. Steve and Chris had always welcomed any opportunity to deal with the press, and this documentary was certainly a good opportunity potentially on a scale above just another paper article. Not that the articles haven't been increasingly better, and we have yet to see where the latest Associated Press article will go.

"Are you still coming up here then?" I asked.

"Yes, we're going to meet with Wesley Kent and go film at the nuclear blast site in Rulison. Then we have an interview at four with the Utesch's," he said.

I knew Peggy and Bob Utesch quite well from Grand Valley Citizens Alliance. They were some of the very most involved and influential citizens voicing concerns over natural gas development in our area. And they had paid a painful price for their knowledge of the industry; they had sold and left their dream country home in Silt when their area was infiltrated by natural gas development after Peggy developed skin rashes and even more problems followed. After selling their home they relocated to a cookie cutter subdivision in New Castle, but remained fiercely dedicated to the fight against reckless natural gas development.

Joe and I discussed the future plans to interview impacted landowners south of Silt in the area thick with so many problems from gas development. We pinpointed the twenty-fourth of June as a possible date, and I agreed to begin working on spreading the word and arranging plans with interested parties.

I called Steve that evening. It turns out that the lawyers wanted to hold off on the documentary film interview Saturday to do an interview with Rebecca Clarren of Sojourner Magazine. The attorneys wanted to do the interviews first with Rebecca Clarren the freelance writer, and Judith Kohler of the Associated Press, who were known entities to them, before the Mobaldis did any more.

So today, Saturday, instead of a trip to Grand Junction I am again sitting in the barn writing on the computer under the stinky, buzzing flytrap that is now ready to be tossed and replaced. It is hot, and I am already overly sunburned from previous days outside recently, so I don't miss the long four-hour round trip drive in the scorching heat past the endless clumps of gas rigs decorating Interstate 70 along the banks of the Colorado River. I am curious how the Mobaldi interview went, and wonder how Joe Brown's crew did up at the nuclear blast site. Time will tell—in a few weeks we should be doing a similar event here south of Silt. All I have to do is spread the word and commit people to come. It feels at times like herding rabbits; scared and timid and tired-out rabbits who are sick of everything and just want out.

June 13, 2006

It is late, almost 11 p.m. I should be going to sleep, but I just got back from a gas meeting. Grand Valley Citizens Alliance, that is—not a gas company meeting as people usually assume. And as usual, I am too wound up after the rare-enough late night event to go right to sleep. Out of restrained courtesy towards my husband, I am not turning on the television to wind down. Instead I sat on the front porch, (which is actually just a poured slab of pebbled concrete, more of a step than a porch) and drank a can of cold beer while visiting with two of the barn cats. It is a beautiful dark night, and the front door light cast just enough light to see the cats but not enough to block out the blankets of stars that sparkled in the black sky. In the distance across the creek I could see the neighbor rancher's headlight glowing and moving through the large pasture as he irrigated. I had noticed the truck parked by the shoulder of the road as I drove home, and I spotted the glowing light in the field. A raccoon wandered nearby, nose in the air, as she went on her evening perusal.

So, it is a good time to pull out the laptop. I wish I felt like I had exciting and hopeful things to say; yet again it seems I am coming from an extremely informative meeting with increasingly depressing news. That is not the view of everyone else there, and I don't want to give up hope nor will I allow myself too, but some of the information dispensed tonight about what the gas industry is pulling off and what a sorry job any entity is doing of regulating or caring about the effects of this industry is completely unfathomable. Completely.

Let's see, what other bad news did I hear tonight? Oh yes, the proposed development of a series of gravel pits along the Colorado River from Rifle to Silt to New Castle. I'm sure it is already well paved out by the local powers that be that stand to profit and their gas and oil buddies who will use the product then leave the communities with wonderful "parks" and three "lakes" when the pits have been scraped thoroughly. The trees next to where the bald eagles have been nesting for three years have reportedly been cut down, and those trees contained large nests too. People at the meeting put their faces into their hands and groaned. The next gravel pit is already being prepared for.

Someone asked that it be proposed that prior to a new pit being started, the last pit must be completely reclaimed. Oh no, the response came. They already said that couldn't be done…they need the dirt from the second one to reclaim the first one. The questioned was asked by one member as a joke of sorts:

"What about the last pit?" Another member jokingly retorted, "Oh, they'll make it a lake!" the GVCA members chuckled at that.

The answer was not a joke; as we shortly found out. A member spoke up in a somber tone,

"That's what they're going to do—they are going to leave three lakes."

The room was quiet; it was all too familiar. What seems preposterous and ridiculous is being done on a daily basis in Colorado by the gas industry. With the help of an administration and a local government that certainly seems to supports the industry all the way. Although some members of the local government voice opposition to certain projects and practices, in the end the gas companies and Colorado Oil and Gas Association seem to get their way.

The low point of the evening was hearing about the new statewide regulations being proposed by the Colorado Air Quality Control Commission for tougher regulations regarding air pollution from the oil and gas industry. The new regulations would decrease the amount of emissions, but areas outside of metro Denver would have twice that amount. The reason being for this was that metropolitan Denver is one of the fourteen areas around the country that don't meet ozone standards.

As a result, Colorado is at risk of federal sanctions that would result in the loss of highway funds. So, only in the metro Denver area would the emission levels from condensate tanks be limited to eleven tons per year. The rest of the state would have almost twice that amount of emissions allowed: twenty tons of emissions from each condensate tank per year.

How could it possibly be fair for the Denver metro area to have lower emission levels than the rest of the state, while the rest of us got two times the emissions? It feels like a slap in the face...

The rationale given was that Denver was already so polluted they were being forced to clean up the air by the EPA, but the rest of Colorado wasn't at that level yet. Yet. Also, it was mentioned that the Western Slope lacked adequate ozone monitoring data.

The one issue that was discussed at length at the start of the meeting was the study recently conducted about property values being affected by the gas industry. Again, the study being conducted seemed to be heavily influenced by the hand of the power of the industry. More on that later...

Also discussion was had on the current totally inadequate taxation of the gas industry in Colorado. Why are we so stupid here? Why is the State of Colorado being given away to this industry when other Western States are not? Matt Sura of the Western Colorado Congress raised that issue tonight, noting the huge amount of money other states are collecting from the profits of the industry while Colorado fails to.

Someone at the meeting mentioned the Natural Resources Defense Council's *OnEarth* quarterly magazine article about Silt and the impacts of the gas industry. Strangely enough, I found it in a packet of today's mail at the dining room table when I got home tonight. Chris and Steve Mobaldi had been interviewed for this; I recall them telling me about it numerous times. I had no idea it was affiliated with NRDC, which I am a member of.

I honestly don't like these late meeting nights, coming home unable to fall asleep, and needing to tiptoe as quietly about the house as possible to not wake Al. He gets up at 5:30 and needs his sleep. He's a light sleeper to boot. I haven't had a bath; a washcloth will have to do. Dinner was somewhat lacking—a rolled up bunch of turkey cold cuts grabbed as quietly as possible from the refrigerator at 11:00 p.m. in the dark kitchen.

I will say, however, that I feel good when I go to these meetings. It is better than ignoring it all, and pretending nothing is happening and that it isn't going to get worse. Much worse if no one cares how the industry goes about its business. The dedication of so many of these people in Grand Valley Citizens Alliance is amazing.

I have my cat Mister purring beside me on the bed, and I cannot wait to crack into the unopened *OnEarth* magazine. This national magazine has on the cover a highlighted article entitled "Wrecking the Rockies" and says in its subtitle:

"Under the Drill: Is Halliburton Hazardous to Your Health?"

Inside the front cover is a double-paged color picture of a blue and white gas rig towering over a huge black rubber-lined pit containing liquid of some type. All the ugly industrial equipment spread across the area around the rig: pallets of bags of mud mix, large tanks of various colors and sizes, the familiar dark blue rectangular container trucks and even a small brown cabin-type house trailer on wheels line the shores of the pit. If it were not for the gas rig and pit and all the industrial mess that accompanied it, that cabin house could have been someone's idyllic mountain camping retreat in the Rockies. Just visible over the bulldozed piles behind the well pad you could see the high mountain peaks rising like gray ocean waves to the north. The blurb in the table of contents reads:

"**Wrecking the Rockies** by Michelle Nijhuis

The nation's energy crisis wears many faces, but few are as ugly as the devastation of tranquil places like Garfield County, Colorado where high natural-gas prices and powerful drilling technologies are tearing apart lives and landscapes."

It is eerie to know that my own close friends are going to be in this article, in this national magazine. It is eerie because of the reason why: deadly tumors and illnesses likely caused by natural gas development in Rifle, Colorado.

COLLATERAL DAMAGE

Reading the gas article in *OnEarth* magazine will be my dessert tonight, and then hopefully I will go to sleep. It is getting too late.

Saturday June 17, 2006

Thank goodness the weekend has arrived. After going to the GVCA meeting on Tuesday, then staying up way to late writing on the laptop and after that reading the *OnEarth* article, it made for a tired few days at the end of the week. Some people do well incorporating meetings into their weeknights; I am not one of them.

Going back to the *OnEarth* story, it was excellent. The subtitle said it exactly like it is: Wrecking of the Rockies. On Tuesday night, actually early Wednesday a.m. after one o'clock I poured over the magazine. The multi-page layout had incredible photographs and the story was so familiar to me. I was thrilled that the coverage of Laura Amos's situation was so well done, from descriptions of the entire drilling progress on her Silt ranch to the day when the Amoses' water well blew sky high due to the work and hydraulic fracturing being done on the neighbor's gas well.

And again Weston Wilson spoke bluntly out about the EPA's total lack of recognition of any harmful effects of the carcinogens dumped by the ton beneath the earth's crust here in western Colorado to loosen the 'tight sands' during the fracking process to aid in the extraction of natural gas.

The image that stays with me more than any other is the photograph of Laura Amos holding her daughter close to hers, their so-similar faces cheek to cheek. Lauren's beautiful blue eyes mirrored her mother's, but the dark circles under Laura's eyes told a different story. A story of her illness, the fury she harbored against a huge and powerful and intimidating industry, and fear for the health of herself and her family. Most important to Laura was the health of her precious daughter. Laura told me many times of Lauren's continual nosebleeds, and of her concerns that her daughter may have been affected from the tainted water in their well after the fracking job caused benzene and toluene to contaminate it. Although the Amoses contended this to be so, both the COGCC and Encana disputed it along the way.

Finally, however a cash settlement had been reached in out-of-court negotiations.

"If something happens to my daughter, they better be afraid of me…" was a quote Laura made in an interview with the local paper.

Few people have the drive to put the kind of time and effort into their fight against the natural gas industry that Laura Amos did, and what I commend her for the most was her interest in including others in her effort. She tried valiantly to enlist a group effort to mount a legal action against Encana, or perhaps more accurately put, to engage Encana in a dialog with the affected landowners through an attorney. She contacted an experienced attorney who had once worked for the EPA. His name is Kemper Will, and he is out of Denver and practices with the Firm of Burns, Figa and Will.

I met Laura and Kemper last summer at the Silt Café for a burger dinner in the small, smoky establishment before we set out to introduce Kemper to more of Amoses' affected neighbors. We took the 15-minute drive up to Divide Creek and stopped at Laura's next-door neighbor's ranch. I stayed in the car when Kemper and Laura talked to the people. Laura had explained on the way up to the ranch that these folks' water well had run completely dry during the fracking of the wells. Encana denied any possible correlation between the natural gas well activities and the sudden dry-up of the water well. A large plastic cistern erected in the yard held the water that was trucked in for household use.

We drove back down to the Silt Fire Station for a meeting that Laura had arranged with neighbors to meet with Kemper. There were about fifteen people there, and after Kemper pitched his proposal to the group and fielded some questions, he asked if people would be willing to put up $1,000 each to begin the process of retaining him to represent their concerns to Encana. Not everyone raised their hands, but a fair percentage of those in attendance did. After some discussion of the amount realistically needed to launch such an effort with any hopes of success, it was decided to propose the idea to the Grand Valley Citizens Alliance, and from there approach the Western Colorado Congress. If we needed to raise $50,000 then we would obviously need help.

COLLATERAL DAMAGE

Sadly, when Kemper Will launched his proposal to the GVCA, it was decided that it duplicated the efforts that the Calgary Group was already doing in private meetings with Encana. The Calgary Group was small core group of GVCA officers were who were having continued meetings with higher-ups at Encana. The activities of this group were never made privy to me, and this winter in 2006, right about the time of the second failure of the Surface Owner's Rights Bill in Colorado, the Calgary Group's project was declared pointless. They felt that Encana was merely placating them, and not seriously interested in making any meaningful concessions to the impacted landowners' requests.

I remember leaving the GVCA meeting that night I heard of the end of the Calgary Group. As I walked to my truck under the black Colorado night sky, I was feeling doubly depressed that both of these hugely time consuming projects had failed. The ballot initiative "eyesore"—acronym ISOR' that would take the place of a Surface Owner's Rights Bill was announced to be an even more monumental undertaking. Great—another huge project with even less hope of succeeding. I must confess that my heart was not full of cheer as I drove home that night, the lights of the rigs glowing eerily and brightly from the distance.

When I closed the *OnEarth* magazine in the after midnight hours I was proud of Laura Amos and the efforts she has made to bring her story out in the public eye. Even now that she and her family have sold their ranch and settled with Encana quietly, then moved away from Silt, she still deeply cares about the fate of those who live near these wells who are likely being exposed to serious dangers from the toxins. She has from the beginning of her fight.

I was puzzled not to find any mention of the Mobaldis in the article. I was quite certain that they had mentioned numerous times that they had been interviewed by *OnEarth*.

I called Laura the next morning and told her how impressed I was with the NRDC *OnEarth* magazine story. We promised to get together the next time she came to town.

Later that day at a Democrat's lunch meeting in Glenwood Springs, I pulled my friend New Castle City Council member and attorney Greg Russi aside. I had some questions for him about the DOLA grant that Garfield County commissioner Tresi Houpt had argued for, to fund a study of air quality in Garfield County. The grant money was made available from taxation of the gas industries operations here, but it was up to a board to decide who got the funds.

Greg clarified a few issues for me, and let me know it was Ken Wonstolen from Colorado Oil and Gas Association, (not Ken Salazar our senator) who had said to Tresi at that meeting:

"Can't you see they don't want the study done?" or something very close to that effect. I put my hand to my forehead in relief when he said this, and replied,

"Oh, I am so glad to hear that! I could not believe it when Tresi said, 'Ken doesn't like me very much', thinking she meant Senator Salazar. I was shocked to think that the senator would be against a study of air quality in Garfield County because 'They don't want to know what they would find out'…"

I breathed out a sigh of relief.

We talked about a few more issues I wanted clarification on, the most important on what I learned at the meeting of lower-populated areas outside of metro Denver having two times the allowable VOC counts in air quality.

In layman's terms this basically means: if you live in a rural area, what industry releases into the air as a result of their gas extraction process doesn't matter as much as it would if more people lived there. Even more bluntly put, if you live in the country near gas extraction, you are more expendable than a city dweller. I guess they think our lives and health are worth less because we are rural people. I have news for everyone: those of us who have worked all our lives to live in the country do not agree. Our health and welfare are not worth less than those of city dwellers. What is democracy supposed to be about anyways? Isn't it about common human decency and equality for all?

It is hot this afternoon as I write in the barn. The horses are pestered by flies, the two female llamas Shanty and Minnie just got into a rare and noisy fight in the oak brush behind the barn. As they lunge and bite at each other they cry out in their strange llama screams. These fights seem to be territorial or dominance-oriented, and they don't seriously injure each other but they keep at it for a good while until a spitting standoff is reached. The charge on this laptop will go out soon, and I have so much more to write.

Yesterday County Commissioner Tresi Houpt returned my call. We talked at length. When I said that some regulatory entities needed to be held accountable for the lack of enforcement of this industry, she agreed. I mentioned the EPA and Colorado Oil and Gas Conservation Commission. She brought up the point that even on a county level we were not being accountable. Just a few days ago she was outvoted by County Commissioners John Martin and Larry McCowen over the issue of permitting of compressor stations. COGCC cannot monitor stations for emissions and violations if they don't know where they are because they are not permitted; yet the other two commissioners voted down the need to enforce the permitting of gas compressor stations. I am going to find out more about this, along with the proposed air quality standard differences for rural and urban areas.

Joe Brown of Fractured Films left a message yesterday. I got his e-mail earlier in the week and was happy to learn that Karen Trulove and several others had committed to doing interviews next weekend with his film crew. When I read that e-mail I threw my hands into the air in my office and said,

"Yes! Yes!"

It had been frustratingly hard to get impacted owners to say they would talk to press or film crews. So many un-returned phone calls…

Joe and I talked at length. He had his first time set with Pepi Langegger, the landowner of the property where the highly publicized (and largest finable violation for the gas industry in Colorado to date) Divide Creek seep had occurred in 2004.

I told Joe that Matt Sura of Western Colorado Congress had a discussion with Duke Cox, (President of GVCA) about Erin Brockovich's current involvement with a group of people suing a gas company in California over their children's illness which they believe were attributed to toxins from gas wells by their school. Another lead to investigate… I had contacted the law firm that Erin Brockovich works for some time ago, but after the Mobaldi's signed with Astrella and Rice I didn't pursue it. California is too far away, and you can only do so much at one time.

But maybe now it is time to think about connecting the dots and drawing parallels from areas far away from here. We know from reading <u>Saboteurs</u> of the similar and horrible woes of those living near natural gas development in Canada. California is even closer to home.

As I said to Joe Brown this morning over the phone,

"It's ridiculous, but once something becomes the stuff of a movie or you mention 'Erin Brockovich,' then people pay attention. You have to make a Hollywood movie out of something before the American people pay attention and care…"

The computer battery is beeping, so I am turning the computer off for the afternoon.

June 18, 2006

Today is Father's Day. I talked to Dad this afternoon for several hours, and I told him of some of the recent developments with natural gas issues here in Garfield County. One of the subjects we touched upon was the idea of suing one of the regulatory agencies that were doing little to nothing to enforce regulation of both air and water pollutants caused by natural gas development. I told him about how frustrated I felt after last week's GVCA meeting when I learned that not only did DOLA refuse to fund an air quality study of Garfield County with tax money available from industry for that specific use (surprise,

surprise—industry players were part of the decision-making group as to where those funds would be allocated), in addition, less populated areas stood to have higher allowable VOC counts in air quality, according to the proposed state regulation; twice as high the amount of Volatile Organic Compounds as the Denver metro area. Basically, this means that no governmental agency would enforce the same air quality standards in rural area of Garfield County that would need to be enforced in large population densities such as Denver.

It is hot today, and after my call with Dad I went to unload the truckload of hay. This morning I went below Harvey Gap Reservoir to a hay farmer's ranch where I have bought from before. Don has a nice place, and a beautiful log home with deep, cool porches. I met him in the field where he was driving the tractor, and then he asked me to take him up to the house so he could get his respirator. He suffers from serious asthma and never loads hay without wearing the respiratory. Don and his wife are avid lifelong hunters, and the interior of their immaculate home displays the trophy animals in museum style setting. Although I am not a hunter by any means, I could appreciate the pride the couple took in their display. They graciously offered me an iced beverage and we chatted casually before setting out to load the hay in the hot Colorado midday sun. Charlotte brought out photo albums of their last litter of fine pure-bred Australian Shepherds, one of whom was now full-grown and happily sniffing at my horse-hair covered pants. I had ridden my horse Doc Holiday bareback this morning, and kept the dirty pants on to load hay. After discussing a variety of topics leisurely, it was mentioned that Charlotte worked full-time. I was surprised, as Don was retired and they seemed well set up.

"Where do you work?" I asked.

"Williams Production, in Parachute," she answered.

I got the strange, sinking, and now all too familiar feeling growing inside me. She worked for the gas industry. At least I hadn't gotten into any kind of gas discussions yet.

We continued our pleasant chatting, and then went out to load the hay. I told Don that I would like to come back with the double-axle flatbed and get a big load, but I needed to get Al to pull the trailer with his bigger truck. I have never practiced or mastered trailer pulling, although I know I should. My truck can't pull that kind of weight.

"Next weekend I am going to be busy; on Saturday a film crew is coming up here to do some work," I said.

"Oh, I remember you telling me about that before," said Don, referring to a conversation several years ago when I had told him of my filming of old-timers telling their history stories. That is when my writing was focused on recording Colorado's past history, not the present oil and gas boom. "Book Two," as I call it, is stacked in piles of spiral notebooks, and will be ignored until whenever this current gas-development writing endeavor is done. Past history will always be there, but what is happening here every day is current and needs to be recorded now.

"Well, I'm doing something different now. It's about natural gas," I said.

I didn't think too much before I said that, I just said it honestly. We were neighbors of sorts, and his wife happened to be working for a gas company. And I happened to be planning on touring impacted properties that had gas production activities with a film crew next weekend. Without going into many specifics, we both commented on the high level of gas activity going on here.

"Yep, there's lots of things going on," said Don.

I figured he was referring to all the disturbing articles we read on an all too frequent basis reporting on explosions, leaks, fires, gas truck rollovers, cancers, and air and water pollution reports. I waved goodbye, said I would call when Al could pull the trailer up for a big hay load. I like Don and his wife, and we had a nice visit.

When Dad and I were talking this afternoon I mentioned my idea of holding some entity responsible legally for lack of enforcement of public health issues.

COLLATERAL DAMAGE

"You know, you don't have to focus on winning when you sue an agency, you can get something done even without winning. If you have a good lawyer…" said Dad.

"Yes," I agreed. "You can get them to do what they're not doing by bringing attention to the issue. Nobody is enforcing anything down here. The EPA has waived all water quality standards for the gas industry, and they are injecting all the fracking fluids, which are undisclosed and include carcinogens, down into the ground. Now the air quality standards are lowered and we can't get grant money available to do an air quality study done."

Dad and I said goodbye.

"You better go unload that truck of hay," he said.

As I unloaded the bright green fresh bales onto the wood pallets in the back pasture I sweated and thought about the day.

I thought about my encounter with a local rancher; yet another one who now has a family member working for the industry. I have quite a few friends that either work for the gas industry or have spouses that do, so many than I can count on all my fingers. I like and respect and even love some of these people dearly as close friends, but out paths cross in an awkward ways when it comes to natural gas development.

It is a path you need to tread on with care, but both sides can share the same road if it is done right. I believe that.

Take One—Film Crew Arrives

Saturday night June 24, 2006

Last night Joe Brown and his crew arrived, a little later than we anticipated, but they got here before dark. Even with the poorly marked and confusing county roads and no cell phone service in the area, they made it to the juncture just below the house about nine p.m. and sat pulled off the road as many other semi-lost motorists do. I went down the gravel drive and waved, and they came up and pulled in and we met in person for the first time after many phone call meetings.

We chatted for several hours as evening descended, talked about the plans for the following day, and generally got acquainted. My numerous cats provided entertainment, and about 10:30 the visitors went outside to pitch their tent on our front lawn below the elm trees. Warned about impending visits from the local raccoon with his nightly route by the deck, the outside cats, and other unknown travelers (from skunks to mountain lions), all settled in for a well needed rest before the upcoming big day. I told them if anything happened they were free to come in and sleep in the living room, but unfortunately we have no spare rooms in our small farmhouse here.

As usual, my inability to drift off quickly to sleep after a night event occurred. It was after midnight when I put my reading book down and clicked of the brass bedside lamp. I finally forced myself to go to sleep, surrounded by purring cats and my mind full of thoughts of tomorrow's interviews. The alarm was set for 7:00 a.m.

The alarm went off, I sprang out of bed and put on a big pot of coffee, and got out the fast breakfast no cook items I bought the night before. It would be a long day, but we needed to get moving quickly so I had bought bagels and cream cheese, a bunch of yogurts, a coffee cake, and fruit. We ate, and then Joe's girl friend Marissa and I hayed the horses and llamas. She threw the armloads of hay flakes over the top of the fence into the wind, and received back a face full of hay.

"I'm not much of a ranch person," she laughingly said as she brushed the hay of her attractive face. Marissa was a quiet, intelligent and very strikingly beautiful woman. She looked a lot like Sandra Bullock, the famous movie actress. That's what I thought, anyhow.

The visitors packed their camping gear, and Joe and I took off with a cooler in the back of my truck full of lunch and a lot of beverages. Marissa and Pete followed with the substantial load of film gear in another car. It was going to be a hot and dry Colorado day, I knew. Ample food, hydration, and salt were going to be important. There was a twelve pack of green iced tea, two twelve-packs of sparkling water, four quarts of Gatorade, and some plain water. In addition to the large quantity of beverages, I also had bags of potato chips for salt. That, and the bottle of 50-SPF sunscreen from Mexico that I had in the truck should help us survive the heat of the day.

Joe and I drove in my truck, and we went up the longer way past Harvey Gap Reservoir and down south through Silt Mesa toward Divide Creek. I explained what I knew of the latest projected development plans for the area, which today is unspoiled by gas development. Unfortunately, that is likely not for long. The first group of rigs is already up on the north side of the Colorado River near Rifle, and development is predicted to spread east and north from there up into Silt Mesa. Only time will tell how the semi-rural neighborhoods and ranches will look with gas development.

The first stop was at nine a.m. at the site of the Divide Creek seep, on Pepi Langegger's large elk ranch. We headed north from the Colorado River around the area of a large and ugly gravel pit, past its dirt lot full of gas trucks. Divide Creek Road wound up into lush valleys with shade trees protecting from the glaring sun.

It was an idyllic country scene, just the type of country any one would fantasize about living in. Six miles up we spotted the sign for the ranch, and drove down into the large farmyard by the old ranch house and huge barns. It was obviously a lovely property, and high elk fence ran around the pastures as far as the eye could see. Pepi drove up to meet us on an ATV, his young blond grandson sitting in his lap. We split into two groups, and Pepi drove with Joe on the ATV and the rest of us followed in my truck. It would be steep going, but Pepi said a two-wheel drive truck could do it. We bumped along the dirt road after the ATV and climbed and descended frighteningly steep hills. I had visions of not making it back up those steep inclines, and needing to tow the vehicle out. That would absolutely ruin the day, and with those fears in mind I declined to drive down the last steep pitch, but instead left the truck parked at the top of the hill with the emergency brake firmly set and the truck in reverse gear. We carried the heavy film equipment down the hill to where the seep site was.

Divide Creek wound through this beautiful hidden oasis below steep canyon walls. It was a lovely, remote site. The hungry populations of brown-and-yellow-striped deer flies greeted us enthusiastically and immediately began their attack. We began slapping at our arms and legs, and jumping at the painful bites. Bug spray was one thing we had forgotten.

Pepi walked us over a large, new green metal bridge to the seep location. A small, wooden cabin the size of a tuff shed sat in a clearing. Electrical type equipment was visible through the windows, and a series of pipes, boxes, monitors, and wires were hooked up into some complex system. Over at the creek large metal pipes were submerged just below the water, and bubbles blew forth from the water. Pepi explained that the air was being circulated to help clean the water, and he pointed to the large, round area of trampled dirt where two water wells were drilled. The area was contaminated, he explained. There were pipes buried underground that served to contain the contamination. The two test wells were still contaminated. A number of mostly dead small shrubs dotted the area, each root area surrounded by a protective landscape cloth square. Pepi told us that Encana had spent $40,000 to landscape the area, and had irrigation hoses run on top of the ground to water the shrubs. Unfortunately, in spite of the money spent and the efforts made, most of the plantings were dry sticks.

As the camera rolled, Pepi continue to explain the uses for the equipment erected. A camera mounted on what looked to be a large lamppost monitored the area. Pepi said that the company monitoring the clean-up watched from Denver via the Internet which was hooked to that camera. It was strange to think that as we stood in this bug-infested remote valley far from any visible civilization, we were apparently being filmed and viewed on the Internet by an employee from Cordiara, the testing company which does work for Encana.

The seep itself was caused from problems with the Schwartz gas well on a neighboring property. The cement job was faulty, and a large amount of gas escaped through fissures: 115.5 million estimated cubic feet of natural gas escaped and bubbled up through Divide Creek. Pepi's neighbor discovered the gas bubbling from the creek when he came down on his livestock access area of the creek.

"At first they thought it was naturally occurring gas, sometimes that happens…but it wasn't. You could light the gas with a match, and when you put a can over it, it kept burning…" explained Pepi in his thick, German accent.

Pepi said that the company had gone to good efforts to deal with the problem, but the larger and more disturbing issue was the lack of regulation of the industry.

"Sure, we found the gas seep here, but how many other problems are there that no one has found?" said Pepi. "The government has made it so they can do anything they want, and nobody cares what happens. It doesn't matter if it's Democrats or Republicans in charge; it is the big money and the energy companies that are in charge."

We discussed the recent passage of the Energy Bill, (which as of last winter now exempts all fracturing chemicals such as 2-BE and other carcinogens often used in the secret "frac" mixes) and how our state senators and representatives had endorsed the bill in spite of this exemption. Allowing for the unregulated usage of these toxic chemicals was a trade-off for the so-called gains to be had from the Energy Bill. Try selling that argument to the people who live on top of the watersheds where these dangerous toxins are being pumped in by the hundreds of truckloads per day. That argument falls on deaf ears here south of Silt where known fractures and bad cement casing in wells have come to light. How can anyone living near a gas well have confidence that their water is safe, or the air they are breathing, for that matter?

Industry argues that the procedures are fail-safe, and nothing can or ever does go wrong. Public health is never in jeopardy, and never will be. What a ridiculous claim to make, and how can the Colorado Oil and Gas Commission with its eight members ever pretend to adequately oversee and monitor the estimated astounding numbers of gas wells in Colorado?

As we left the site of the most infamous natural gas violation in western Colorado, the Divide Creek seep, I walked through the pretty, cold and clear water as it tumbled over rocks in my Teva rafting sandals. I remembered years back to my childhood, when finding such a spot as this in the woods was pure joy. A small cabin in the forest, a meandering cold stream, all enclosed in a secluded canyon thick with grasses, wildlife and trees.

This however, was no idyllic spot. This was for all practical purposes a hazardous waste clean- up site from a huge natural-gas well failure. The charming cabin contained expensive monitoring equipment and pollution detectors, pipes both visible and buried wound through the ground and creek, and an eerie out of place camera on a pole in the forest recorded the goings on around the seep sight, easily found as the large area planted with dead and dying shrubs was devoid of even weeds. It was a barren patch of dry earth. How many years would it be, if ever, before this spot would be pollution-free?

On the way back to the ranch Pepi pulled over the ATV to feed some of his elk. He held his hand out, full of feed that he dipped from his bucket, and the beautiful, huge glossy brown animals ate from his palm. They ran to meet him, obviously they were used to being fed. Pepi raised them for breeding, I was told. He certainly had some magnificent creatures in his herd. As we drove by they stared at us, docile as horses with large dark eyes and impressive antlers covered with brown fuzz.

We thanked Pepi, then drove away from the 500-acre or so beautiful ranch to destination number two, Karen and Tim Trulove's. Pepi had given us some directions heading to the south, which would be a shorter route than retracing our steps back toward the Colorado River. Within a few miles we were at road junctures we were unsure of, and I turned the truck toward what I thought was the best route as the crow flies to Dry Hollow. After a few broken up cell phone calls in an attempt to get directions, we soon made our way to Dry Hollow, which I recognized. Without too much more trouble, we found Owens Drive and located Trulove's property by recognizing the for-sale sign and the horses and foals and corrals. And there was Karen, riding with another woman in the round pen. We made it.

Karen looked great, so much better than the last time I had seen her late this winter down in Denver. At that time we were there with a group to lobby senators for the Surface Owner's Rights Bill, and to conduct a press conference. Then Karen had appeared haggard and pale, and devoid of hope or energy. There were dark circles under her eyes, and lines on the papery skin of her otherwise attractive face. As the group of landowners from Garfield County and the organizers from Western Colorado Congress met with the Environmental Coalition lobbyists in the large Capitol cafeteria, we each were invited to introduce ourselves and briefly describe what we wanted to speak about.

When her turn came, Karen, dressed in a dark suit with a pin on her shoulder, stepped forward hesitantly. In a shaky voice choked with emotion she described her fragile health condition and the realities of living surrounded by gas rigs and the accompanying fumes at her Silt horse and cattle ranch.

"I just don't plan my days anymore; I just try to get through them. It's like having a terrible hangover that never goes away…and I used to be very healthy, very active, and never sick…I just don't have the energy to do anything anymore…"

Her eyes shimmered with tears as she spoke. During the long day at the Capitol, Karen hit her own stride. She became a passionate representative voicing the concerns of impacted landowners from the invasion of the oil and gas industry as she and her group went from each senator's office to the next one. We had split into two groups and saw many senators.

Today June 24, 2006, when Karen spoke to the film crew from her own Silt ranch, she was a stronger and more hopeful woman than she had been 4 months before. Her newly tinted hair glistened gold and brown in the Colorado sun, and her tanned face was set in a determined manner. As the crew filmed, and Karen described the defects of one of their foals who had been born with deformed limbs, and the unusual lingering illnesses exhibited by the broodmares (whose crusty nostrils still displayed the evidence thereof), she never wavered.

"The vet asked if we had presented the bill for the foal to Encana; that's why he thought you went to film the foal," she said.

Two weeks earlier the film crew had come to the Glenwood Vet Clinic to film the deformed colt with stunted legs and strangely short neck that was given a fifty-fifty chance of survival. He stayed at the veterinarian's clinic for some time before being deemed healthy enough to return home. Karen talked about the vet's referencing of goats whose tendons had collected toxins and developed resulting deformities. Apparently toxins accumulated in connective tissue, and Karen described her own leg pains in the same muscle areas.

Then she spoke about the goats; a neighbor's goats. The goats had delivered sacks of water, not baby goats. All heads turned in disbelief at this information.

"Sacks of water? What?" asked one of the film crew.

We discussed the connection between exposure to natural gas development and birth defects. In <u>Saboteurs</u> it was mentioned numerous times that both animal and human birth defects of frightening natures had occurred near exposure to gas. Stillbirths, malformed fetuses, and more… One Canadian cattle rancher had a herd of cows that gave birth to calves whose legs were so deformed they walked on their ankles. Another rancher committed suicide when he woke one day to discover that his herd was suffering from horrific deformities.

COLLATERAL DAMAGE

After filming by the horses and foals, the crew went up to the house where the well pads and new rig were visible across the property line. Karen described the odors she and her husband had noticed just days before, and she talked about the instantaneous headaches that she got when she smelled the gas.

"The wind was blowing right from the rig; I looked up at the direction the wind was blowing and it was from the rig," Karen continued, and talked more about what it has been like to live with numerous wells and their fumes surrounding her home.

Here we stood, in the lovely garden. It was a huge raised bed with neat rows of lettuce, tomatoes, chili peppers, and more. Below us on the hillside was the large and tasteful stucco- covered tan home, with inviting decks and lovely landscaping. The horses trailed nose to tail through the fields, moving downwards toward the corrals where the mares and foals were. Karen's husband Tim was down in the lower pasture putting up hay. The grinding of the baling machinery sounded in the distance, and the yellow and red bale wagon lumbered slowly through the field as it picked up and stacked the fresh, rectangular green bales one by one. It should have been an idyllic scene, and it was - except for the gas wells and rigs surrounding the ranch.

After some trouble locating the right spot to plant the camera and get in both Karen and her property and the wells, filming began again. The interviewer asked the questions, and Karen answered in a confident and determined manner. She rested her arm against the fence, and silhouetted with gas wells and rigs behind her, talked about having put the ranch on the market, and her family's decision to be gone from the ranch by October. They had bought other acreage in the area, and this was their last summer here. There was a determination in her voice and strength showed on her face. A far cry from the defeated woman I had met some months ago; Karen now had hope. She and her husband were leaving the scene of what had brought on her illness, and that reality had brought her hope.

As the film crew stood wilting in the unrelenting sun during the fifth hour of interviews in high and dry Colorado midday heat, Karen didn't seem to be affected, even in her thick shirt and jeans and boots.

She talked about her symptoms, their patterns, and her family's decision to leave their home and ranch behind. Some neighbors had come by recently, she said, and complained of the illnesses plaguing their families.

Next we went to Dee Hoffmeister's house. Karen pointed out the roof of what she thought was Dee's home in a cluster of ranch properties set amongst dry knolls to the east. Dry Hollow wasn't called dry for nothing...

After getting a bit lost a few times, we finally found Dee's place. By this time we were thoroughly dehydrated, and a bit overcome by the sun. Dee invited us onto her shaded porch atop the small hill. The well pad was clearly visible up the hill to the south about 700 feet from the house. The nearest well pad, that is. There were plenty more visible in the area, and towering rigs where they were drilling new wells now.

"Can we bring our cooler up here and make our lunch?" I asked. "We are hungry and we need to eat. I guess we are the only film crew who comes to your house to interview you and asks to make sandwiches at the same time."

"Of course!" said Dee, with a big smile on her friendly face.

Ever since the Amoses had received what was believed to be a sizeable settlement from Encana, others had begun to speak more openly about their own issues. The Amoses' old home, now sold during a buy-out with Encana, was visible just up a rolling grassy hill to the south. It was reported that their water well had been contaminated after the neighbor's well had been hydraulically fractured improperly. Of course what really happened depends upon who is speculating; the Amoses settled out of court with Encana, and no public admission of wrongdoing ever was made public. In spite of Weston Wilson's and other EPA officials' request to launch an investigation of the Amos case, their supervisors told them to stop—that it was not going to be pursued... Nonetheless, Encana reached a settlement with the Amoses, who are now relocated in another part of Colorado and away from the oil and gas wells of Silt. I did speak to a woman from Dry Hollow this weekend who has a taped conversation of Encana personnel admitting that they "paid off" several different land owners, including the Amoses.

When Laura Amos was told that it was said in a local meeting that she had settled for three million, she laughed heartily.

"So now they're spreading their own rumors…" she said.

After fueling up on sandwiches and drinks, the filming with Dee began. As she sat on her covered front porch on the knoll, Dee relayed her story. She described arriving home with her husband after a trip away for their fiftieth wedding anniversary to find a towering gas rig by their house. After running through a visible cloud of gray fumes into the house, she fell unconscious. Then her family carried her out of the house and took her to her daughter's home in Glenwood twenty miles to the east to escape the fumes. Dee's voice became choked several times as she continued to describe her health problems, and her inability to breathe the air at her home. It was so bad she could not live there for eight months; every time she tried to come back she couldn't stand the fumes and had to leave again.

By the time the filming was done, we were all exhausted. We thanked Dee, packed up the cameras and gear and wearily drove to the ranch. After the crew dropped me off and we transferred gear from my truck to Joe's car they prepared to return to Denver. We said our goodbyes, made plans for another filming day soon, and I stood at the end of the drive and waved goodbye to the first film crew I was to encounter. In spite of numerous applications of sunscreen, and the quarts of water that I drank that day, I felt definitely overdone by the sun and thoroughly exhausted. Even so, I was thrilled to be involved in helping with making a movie about what was happening to the landowners who live near the gas wells. It seemed so much more tangible a way to convey this important issue than my endless days and hours of typing on the laptop computer.

On Sunday Judith Kohler called to wrap up a few details for her Associated Press release. We talked for quite a while, and I told her of the filming the prior day.

"Karen looked great, and I don't know why, but when she was talking I began to cry. I didn't mean to or want to — tears just leaked out. I had to wipe them away. I remember her talking down at the Capitol, and she looked so bad then. That was the first time I saw her, in a cafeteria room before the press release, the day when we met with the senators to lobby for the Surface Owner's Rights Bill. She seemed so sick, so pale and drawn, and so nervous. She was almost crying when she talked about what happened to her health…" I said.

"They have to leave their home and ranch; it is very sad…" said Judith gently, as if in explanation.

There have been moments in the days of working on these issues when things like that happen. After a lot of hard work, after what should be important and uplifting interviews with larger than local press, depression strikes. One struggles to be heard, to have the story of those suffering the negative effects of industrial development in their yards here in the countryside of western Colorado heard.

Strangely, several times it has been just at the time when the best opportunity for exposure comes along in weeks, it hits you in the heart. The reality is that as the story is being told in the moment of the suffering and hardships and illness brought on by the gas wells it should be enough to sway anyone's mind and bring a stop to the industrial practices causing it. Yet, on the heels of excitement over attention from some form of big press, comes sadness.

As Bill Solinger, one of the landowners from Silt who went down to the Capitol to tell the legislators about the health problems he was having said,

"Why did I go to Denver? They listened but they didn't do anything, and they are the people who can. I want some payback for my efforts, and there are none…"

Oh how true that is. One begins to rate the importance and effectiveness of the meetings one will attend, and weigh the value of the time spent. How will it assist the cause? What phone calls are most important, and what meetings? What press will really get coverage?

COLLATERAL DAMAGE

What lawyers are the best, where in the end will it really make a difference and protect the people of western Colorado? No one knows—I certainly don't.

That day in March of 2006 at the Capitol the landowner's group waited patiently for hours outside each senator's office for the prearranged meeting. The typical visit would last about five or a few more minutes, and the senators truly seemed to care about our issues and impacts and health concerns regarding natural gas development in Colorado. The majority, if not all of them, also seemed to have no idea that any Coloradoans were being negatively impacted in any way by natural gas development.

"You all need to come here more often, we need to hear from the people!" said one senator with a cheerful smile as we left his office.

I wish I could spend more time doing that, but the reality is that we need to earn our livings too. No stipends have been made available to the impacted citizens to replace lost wages when we take days off to work on gas issues. Our free time is ours to portion out, a few days and evenings here and there on a regular basis for gas-related work. Weekend hours spent on the phone, researching and writing. One's ability to effectively do good work without damaging one's own quality of life is an art, or an unpredictable science. The bottom line is, however, that we need to keep our moneymaking jobs as we do our nonprofit work. No matter how tired I have ever been, I have never doubted the value of the efforts that so many make. I am merely a tiny drop in the bucket of the people who work indefatigably to shed light on the impact issues of the natural gas boom in the West.

The day when Karen Trulove spoke to the documentary film crew, perched against the fence above her ranch, gesturing toward the wells that continue to exude the fumes that bring her illness, tears crept out of my eyes behind my sunglasses. She was not crying, she was leaving and she was telling her story to the public as well. As I told Judith Kohler, the Associated Press reporter,

"She walked the walk and talked the talk."

But it still made me shed tears.

And she, like others, was leaving because of gas development...

June 28, 2006

Today I attended the Wednesday weekly meeting of the Garfield County Democratic Party. One of the subjects brought up was the ISOR ballot. We are still waiting for the final approval from the State Supreme Court. It seems that there are some stall-and-delay tactics going on somewhere; the OK of the final version was promised some time ago. Some weeks if not months ago... Now it is going to be down to the wire to collect the 65,000 signatures needed to get the initiative on the ballot for November.

The health study being conducted by the Saccommano Institute came up also. The fact that those conducting the study were not allowed to actively pursue cases, but had to wait for people to come and report to them was mentioned.

"Well, can't we ask the people to talk to the research group?" one member asked.

"I already have. They are either sick, sick of it, or just don't do it. You can lead a horse to water, but you can't throw him into the trough," I said sarcastically.

In today's Denver Post there was an article entitled "Anti-drilling backlash could bubble up, says Sen. Salazar."

In the article the water pollution issues are mentioned, and the lack of recognition of public input on drilling permits issued to the BLM, and more.

If people knew what was being pumped beneath the Colorado earth's crust; the carcinogens, the unknown cocktails of toxic chemicals, and where they end up—not to mention the air-borne pollution—they would or should be horrified.

COLLATERAL DAMAGE

A co-worker Tracy, who was chatting with me yesterday, was horrified when I told her of fracturing practices. We initiated the conversation as we chatted about our weekends, and I told her the documentary film crew had come to film natural gas development impacts.

When she heard about the dangerous and undisclosed chemicals being dumped under ground then spread with explosives to expedite the extraction of natural gas, she stared back at me with a look of nothing other than disbelief.

"That can't be! They can't let that happen!" she said in a stupefied voice.

"I don't really like to talk about it, but it is happening and it is real…" I said.

"But how can they do that?" she asked, disbelief clear in her voice.

"They just can, because they make money, and politics is on their side. This administration has thrown the doors open wide to energy development, and the people it kills don't matter. The people it is slowly poisoning don't matter. The energy companies are in charge of the people who set the rules, and they get what they want even if they are killing people and wrecking the land," I said.

"I have a friend who works for Encana, and he tells me about stuff. He says that it's really hard to go up to these beautiful places—places way up in the mountains with waterfalls—and then wreck them. He says they just rip them up and leave them a mess. Most of the guys he works with are meth-heads. They probably don't care; they just want the money to get high," said Tracy.

"I was up at the Red Stone Inn near Aspen this winter, and a group of energy executives were there. They started hitting on us in the bar, and they were asking where they could find some people to hire to work in the gas fields that could pass a drug test. They said that was their big problem, finding people who could clear a drug screen," I said.

After Tracy left her shift, she called me back an hour or so later when she got home.

"I checked out those websites, and I called my friend. You need to call his brother; he works for a law firm that works with this stuff. Maybe you could help each other…"

She gave me the names and numbers, and I told her I would call soon.

Sunday July 2, 2006

On Wednesday I called the law firm and asked for Don. He was out, but the woman who answered the phone asked what the nature of the call was and I told her it was in relation to legal matters involving natural gas development. We got into a short discussion, and as it turns out Kauffman and Kauffman had done legal work to assist in the ISOR ballot. I told her that I was working with the group also, and that I was a friend of John Gorman who was doing the bulk of the work on the initiative. Small world, or small county, that might be better said.

We discussed the status of the bill that had survived attempts by the gas and oil powers that be to destroy the initiative. The State Supreme Court was still holding the initiative, although John Gorman believed it would be in his hands in the final approved form any day, or any hour for that matter. He was on the phone daily with the State Supreme Court tracking the ballot initiative's progress. The hard part was the time crunch that loomed, less than a month before over 65,000 signatures could be collected to get the initiative on the ballot.

Friday night I called John and asked him where things stood. He said he was still waiting, and he believed in good faith that the initiative was about to be released, and that stall tactics were not at play. There has been speculation on that issue, as each week and day that slips by gives less valuable time to collect the huge number of needed signatures. Hopefully by tomorrow I will get an e-mail with the good news that the petition is out.

"John, have you talked about using Grand Valley Citizens Alliance as a contact group for ISOR signatures? There are over two thousand members right there, and also there is Western Colorado Congress. They are state-wide and even bigger," I said.

"I was just on the phone with Matt Sura, as a matter of fact," said John.

"Yes, Western Colorado Congress and GVCA will both be great to have," said John.

"You know, there is OGAP (Oil and Gas Accountability Project) out of Durango too. Do you know them?" I asked.

He did not.

"Well, I know them and they are great; they are an oil and gas watchdog group. They have some great people; Lesa Sumi just e-mailed me this week asking about health related issues that I knew about. They have a great lawyer too, Carolyn Lamb. I've met her a few times, and she knows her stuff."

I remembered being so impressed with the pretty young blond woman who flew in to Denver from Durango to testify at the Capitol two winters ago on the first Surface Owner's Rights Bill, sponsored by Kathleen Curry, the Democratic representative from Gunnison. Carolyn Lamb unfortunately was not able to testify before the Colorado State House of Representatives as the session was abruptly cut short. Many of us who had traveled long distances at great expense to testify had done it for nothing. We got to hear the testimonies of others of course, but our own speeches remained unheard that day. Surprise, surprise, the bill failed anyways.

I remember that trip vividly...that was the first time I met Duke Cox, President of Grand Valley Citizens Alliance; and Orlyn Bell, the extremely knowledgeable landowner from Dry Hollow (once State Water Engineer); Mary Ellen Denomy, the certified accountant who represented the Savage Land Company's interest in the successful suit they mounted against a gas company for failure to pay royalties due to the mineral rights holder; and a passel of other committed and knowledgeable others.

Two years ago I had made the sudden decision, the morning before the testimony session for the Surface Owner's Rights Bill at the state Capitol, to think about going. I called someone from Grand Valley Citizens Alliance and they secured me a hotel room and gave me a phone number for one of the cars driving down to Denver so I could carpool. My wonderful boss gave me clearance to go at short notice, and I made my plans hastily. I told Al what I was going to do.

"You're going to Denver?" he asked, obviously surprised that his stick-in-the-mud wife who never left the ranch except to go to work or to Mexico or rafting would all of the sudden up and go to Denver.

Yes, I explained, I was going to Denver with people from western Colorado Congress to testify at the Capitol on surface owner's rights issues. I would be staying at the Ramada Inn in the Capitol District and would drive with Carol and Orlyn Bell.

We drove in a raging blizzard from Garfield County over the Continental Divide and down into Denver in the Bell's big pickup truck. Once or twice we came mighty close to slamming into the vehicle in front of us on the ice-and-snow-packed roads. Thankfully, we made it safely.

The group met at the Ramada, and we went to dinner at the Mexican restaurant in the hotel. Oh, what a mistake that was…

The group was hungry and tired, the restaurant was almost empty, the service was dismal and there was no salsa for the corn chips or the entrees. A Mexican restaurant with no salsa?

I was seated at the end of a long rectangular table, and once the food finally came out of the kitchen everyone began to focus on their slow-in-coming dinners. Out of the corner of my eye and off to my right, I noticed a tall man dressed in dark clothes wearing a ski mask going over to the cash register podium. He leaned over the machine, and began pulling wildly at the cash drawer. I saw the receipt tape unrolling, and soon began to realize that he was robbing the restaurant. Fear immediately overcame me, and the man ran from the room. As no one else seemed to have seen him, I rose from the table as if in autopilot and walked as quickly toward an exit to another hall as I could. I said to someone at the table as I walked by,

"This place just got robbed!"

I had nothing else on my mind other than leaving that room before bullets might begin to fly. I pushed open the door into the industrial and brightly lit white kitchen, and the Hispanic employees in white aprons with their hair in nets didn't seem to comprehend or care too much when I said that a robbery had just occurred.

Carol Bell went to the Ramada lobby to report the robbery. She returned to the table and said,

"They said they didn't own the restaurant, so they weren't involved."

The police, eventually arrived and the officers requested that we stay to give statements. When I was being questioned I asked the female officer how many robberies or shootings had occurred at that hotel.

She looked at me squarely as if I should have known.

"Don't you realize that you are in the middle of the drug and prostitution zone of Denver?" she said.

No, I hadn't known that and apparently Deana from Western Colorado Congress hadn't known either. She had booked online and gone for a bargain price.

"I was just trying to save the nonprofit money..." said Deana.

Before we went upstairs to our rooms, Mary Ellen Denomy announced to the group that she thought she saw the robber going up the stairs; she was pretty sure it was him...

That night, behind double-locked and dead-bolted doors I sat in the double bed propped against pillows and wrote a rough copy of a speech for the next day. I always did better speaking with something written to glance at. Outside the second-story window, beyond the security grating, the non-stop lights of Denver at night flickered. I missed being at the ranch, even for one night. Cities never have done much for me, and after the robbery tonight they did even less. I propped some pillows against the bed headboard, and finished my speech on hotel paper and finally fell into a not too restful sleep.

Early the next morning the alarm rang and I got up and dressed in respectful clothes: black pants, a white and blue pin striped shirt, nice boots, and a tweed coat. I tried to make my hair look nice and put on mascara — a rare event.

We met for a briefing in the old cafeteria in the basement of the Capitol building. Strangely dingy for the Capitol, I thought. We had coffee and talked about the timeline for the testimony.

One thing I learned that day was to restrain my internal negative emotions towards the industry people there. It was obvious that we were on opposing sides of the issue, and that was that. I personally believe that those who advocate for the industry over landowner's rights or concerns are lacking as human beings in a vital area. They place money over fairness, and believe that the pursuit of money gotten at any means entitles them to ruin other people's lives and properties, and in some cases blatantly poison them to death. How do they sleep at night?

It has become far easier to not look at them as people, to avert my gaze and ignore them. They will never listen to our concerns and issues as landowners, and they will never care. Not even when some one as seriously ill as Chris Mobaldi stands before a room and attempts to relay in the horrors which have befallen her as a result of gas development by her Rulison home.

Today's Glenwood Springs Post Independent has an article entitled:

"State may tighten regs on oil and gas air emissions" followed by a header:

"Proposal would have significant impact on operations in county, officials say."

Well, it would be nice if anyone gave a damn about what is dumping out into our air, let alone what is dumped into the ground. It would be nice but time will tell if this will fly.

Last week after spending all day up Dry Hollow and Divide Creek doing the interviews with the film crew, I immediately developed a dry, hacking cough. I had numerous nosebleeds, one or two quite severe and hard to stop. The cough was irritating for days, and made me choke to the point my eyes burned red and overflowed with tears. I can't say it was from exposure to the air up there by the gas wells, but then again I can't say it wasn't. I have been fatigued all week, and still feel kind of draggy.

We hauled hay yesterday, and at home re-stacked one hundred and twenty six bales in the barn. I am still tired from the workweek, and from the physical work yesterday.

Time to turn off this computer and go outside. Judith Kohler said a few days ago she thought the Associated Press article would go out this week. Someone from COGA (Colorado Oil and Gas Association) said to her something to the effect when she called them for comment:

"What, you're doing an article on health effects? That's old news, it's been done…"

July 3, 2006

Today Judith Kohler sent an e-mail back to me saying she got my message about Donna Gray's article about air impacts from oil and gas development. She thought the article was good, and gave a good view of the localized situation here in Garfield County. She also sent along a copy of an article she wrote on the same subject, but on a more statewide scope. I just scanned it; I will have to read it later.

Joe Brown called, and apologized for not getting back to me. He said he took a break for a while; he needed to. I told him I understood, and that I had kind of hit a wall at the end of last week in between work obligations, writing about the gas issues, and having done a full weekend prior doing interviews in the (hot) gas fields of Divide Creek and Dry Hollow then writing all the next day.

It didn't leave much time for R and R, let alone house cleaning or laundry. Can't complain, but the urge that pressed me to write the details about the day of filming each night after work and after ranch chores and feedings were done had left me pretty wiped out by the end of the week.

Actually, there have been times when I have been grateful for the distraction of the gas issues from steering me away from focusing too heavily on work. It is nice to be able to let one thing go completely from one's mind. Definitely after some of the meetings I have come home with the mental slate wiped clean from the small nagging issues of the nine to five days. But of course, then one dwells on the issues of natural gas development instead.

Somehow I must believe that all of the work that people are doing to protect the health concerns, the landowner issues, and the "big fight" are doing something. That it will bear some fruit.

Joe e-mailed me a schedule for their next visits up here to do more filming for the documentary. They'll be here in about a week. We discussed the best landowner prospects, and the timing of the calls to those individuals. He also hopes to interview the county commissioners. That, if successful, will certainly be interesting.

July 6, 2006

I called John Gorman a few times since Friday to check on the status of the ballot. After not hearing from him, I became a bit concerned as last Friday just a handful of days ago he was pretty sure the State Supreme Court had approved the final version.

COLLATERAL DAMAGE

On the way home from a long day at work, I called Duke, the President of Grand Valley Citizens Alliance. I filled him in on my communications and no recent callbacks from John.

"Has there been a problem? What is it?" I asked.

"He got the final version approved today," said Duke.

"It's good to go? Do we have enough time?" I asked. Sixty daunting thousand signatures were necessary to get the on the ballot for November.

"Yes, we just need one hundred thousand dollars to do it. Our grass roots can do good work, but we need to raise the money to do the rest of the votes. Do you know anyone?" Duke asked.

"Well, there's the Pillsbury heiress on Silt Mesa with the big high-end horse place," I suggested.

"Who can we get to talk to her?" asked Duke.

"I don't know, I just know people who work for her. We need some Aspen people, some big money people. That's who we need," I said

Duke and I talked for a bit more, discussing possible leads.

I got home and the red light on my machine was flashing. There was a message. I grabbed ten minutes of relaxation time watching the tail-end of the evening news before my husband came home. We seemed to be for some reason in unusually bad moods. I do not like bad moods at all.

I listened to the message: it was John saying that the initiative went through.

Shortly, I called him back. Kim was at his house stapling petitions together and numbering them. He was momentarily on his way to Grand Junction. I would pick up some petitions soon, but I knew the only hope for success in even getting the initiative on the ballot was getting a large donation to assist in that effort. Or a number of them…

I wish I had confidence in the local and statewide people's interest in this issue, but it seems that if a well is not plunked onto your land and doesn't ruin your life or future or compromise yours or your family's health, the regular Colorado citizen doesn't need to think or care about it. That is a shame, because what happens to landowners whom the gas companies roost on top of or next door to certainly bear the brunt. And the majority of the most impacted run for their lives…

July 8, 2006

Last night I went by John Gorman's house on my way home to pick up petitions. He had just gotten back from Grand Junction where he delivered a number of petitions to Western Colorado Congress.

It was busy inside the Gorman's house, relatives were there and the family was preparing for a dinner with guests. We introduced ourselves and talked about the ballot initiative as family members drank tea and prepared a side of ribs and the rest of the meal. People came and went from the kitchen, and the graying black lab rested on the kitchen floor by the table.

John's wife Susan asked me after a while of chatting, point blank,

"Do you think it can work?"

She was referring to the impossible goal of getting 60,000-plus approved signatures to allow the initiative to even get on the ballot.

I was honest in my response.

"I don't know… Last night when I knew the petitions were here, I didn't know what to think. It may be doable, but it is a huge task… I don't think we can do it without the help of some really big organized groups. There is just not enough time," I said.

COLLATERAL DAMAGE

When I got home I thought of a few ideas of larger groups statewide: the Sierra Club, and the League of Women Voters. They would be great contacts to make with a broad base.

This morning I left and went to town to get my hair cut at the salon near the grocery store. As usual, we talked about gas issues as the friendly beautician cut my hair. She now lived up Dry Hollow, basically ground zero for gas development. In fact, she and her husband had bought Bob and Peggy Utesch's place when they fled the gas patch. Simultaneously as I was thinking about asking to leave a petition with 100 slots to be signed with her, she asked me if I would give her one.

I went from there to the nearby farm stand where drilling was slated to come in the near future. I waved at the woman who I have come to know over several years from my trips there to buy plant, vegetables and melons. She was out in the field tending to the rows of plants. She and her small dog came to greet me.

We talked about gas-related issues for a long time, and I bought some hot pepper plants and flowers. She was more than happy to take a petition to have people sign that came in to do business at their farm stand. She also said they had been having a lot of odor issues lately.

"Really chemical smells blowing in over here," she said.

Duke called, and we talked about the ballot initiative, in realistic terms. And we made battle plans.

Shortly after that, Joe Brown from the documentary film crew called. We discussed the recent goings-on, and he was getting ready to set up a new computer system so we cut it off.

I told him that it seems like sticking a fork into an iceberg, and I am thinking about what each of the prongs is.

The documentary being done is very important, the fact that what is going on here now will be recorded is probably going to be more valuable than anything we can try to do on the ground here today.

Because, with the current local and federal political situation it seems we are doomed. It is open doors no-holds-barred for gas and oil development. And at any cost; to both the environment and the population.

On that cheerful note, I turn off the computer. Unfortunately, I believe it to be true. And we are sitting here in the middle of it.

July 9, 2007

Tonight when I got home from work, there were two messages blinking on the answering machine. They were both about the ballot initiative.

I sat down on the sofa to catch the last five minutes of the national evening news, and the phone rang. It was Matt Sura from the Western Colorado Congress. He wanted to discuss the status of the ballot initiative. We had a lengthy and realistic talk not only about the initiative, but also about the Colorado Oil and Gas Associations role in delaying the proceedings by throwing up legal roadblocks. We would touch base later on other issues, such as holding the county liable for not requiring permitting of compressor stations, a major source of air contamination from gas development. The current Garfield County commissioners' majority vote did not require the permitting. It was widespread and local knowledge that two of the three county commissioners — John Martin and Larry McCowen— sided almost unanimously with the gas industry in any decision regarding the ease of their ability to develop with no thought to public health. Anything that could help them expedite the process most cheaply…

I went out and fed the horses and then Laura Amos called. I had not heard from her in a while; I was pleasantly surprised. We had had many a lengthy conversation regarding health issues and the gas industries doings here in Garfield County.

"We finally close today!" said Laura, a jubilant tone in her voice.

"It took us this long!" she continued.

"How long has it been? Since March?" I asked.

"No, since January. That's when we moved," said Laura.

Carefully, I constructed my next question. I knew that the Amoses had been in legal settlement proceedings with Encana. Serious and final ones, and that they were not at liberty to talk openly about it all...

"So, Encana closed on your place? Or was it the settlement?" I asked, referring to any damages clauses that might have been simultaneous to the buy-out.

"It was the whole thing..." said Laura.

Tuesday July 11, 2006

Just got home from a Grand Valley Citizens Alliance meeting in Rifle. It is almost 11 p.m. and I need to get to bed, I am tired from a round trip drive to Grand Junction for business training, then immediately after that had a suddenly planned meeting with another documentary film crew from Boulder: Zach Fink, the owner of Soul Films. He had called me earlier in the day, and I had given him information about several impacted landowners he might want to meet. Like Joe Brown, he also was making a film on the impacts of gas development in this area.

I got to the Mexican restaurant early, and just planned to sit and do paperwork for an hour before they were appointed to meet me at dinnertime. I saw a tall, athletic looking middle-aged man in a ball cap talking on a cell phone and reading from a notepad as he paced back and forth in the parking lot. His non-cowboy and non-oil field worker look, plus the fact he was near a Subaru with sports racks made me guess he might be with the film crew.

Inside the restaurant I told the fellow seating me that I was waiting for two others. The Hispanic gentleman gestured toward a dark-haired younger fellow sitting alone. He looked well-educated and well-traveled, if one can surmise those things about someone at first glance. I raised my hand in a wave and said with a slight question in my voice,

"Zach?"

"Yes, you must be Tara?" he asked.

We shook hands and exchanged greetings.

"How did you know it was me?" he asked, somewhat surprised.

"Well, you didn't look like a rancher or a gas field worker. And actually, I told the waiter I was meeting two people." Then I glanced around the large dining area, taking in the other diners assembled there.

"You know, I think I could have picked you out anyhow pretty easily. You just don't look like the local type. And, I saw the bike racks on the car…"

As though bike racks were a dead-giveaway of documentary filmmakers…

For the next hour-and-a-half we talked non-stop about the local health impacts of natural gas development that were known to me. I told them of the booming development and my concerns for not only the environmental and personal impacts, but also the more frightening and all too real health issues.

We discussed individual cases, including my own friend Chris Mobaldi's heartbreakingly debilitating illness after her recurrent tumor, and her mystifying acquired foreign accent syndrome. We discussed legal cases that settled out of court, and how the worst situations seemed to fade from any public notice once the out-of-court settlements were reached.

The film crew had hooked up with Dee Hoffmeister and spent three hours that afternoon interviewing with her up Dry Hollow, at her home just 700-some feet from the well pad that had caused all of her illnesses and problems.

"What is your role in all of this?" Zach asked me.

I stared back at him and pondered the question.

"My role? I don't really have a role, I am just scared of what is happening here. It is out of control and they say this is just the tip of the iceberg. I want to leave, but I can't. Maybe if I was by myself I would, but my husband has a good job here, and he likes it. I like my job, too. We didn't know about gas development before we moved here…" I answered.

"No one does, until it happens to them," said Zach. Obviously in researching for his documentary, Zach had done his homework. That was exactly true. Until natural gas development is on top of you, it is far away and someone else's complicated problem.

The majority of the people in Colorado and in fact in Garfield County have no clue, no inkling of an idea of what injustices occur to the property owners in the vicinity of gas wells.

We agreed to set another time soon, to go out into some of the nearby impacted areas and do some filming. I knew they had taken in the information seriously, and I noted the way their faces took on that strange shocked dumbfounded look when they heard the most bizarre stories of health impacts.

"I know that a year from now, I will be looking through my notes and see this quote:

"Goats gave birth to sacks of water…and I will wonder what that is about…" said the cameraman.

July 13, 2006

Last night I went online and checked out *The Inconvenient Truth* website, and I wrote a 'contact us' letter about gas development in

Garfield County. I can't say I have much hope that I will get answered, but that is what I chose to do last night.

Feeling the crunch for the ballot initiative. I called John Gorman and his wife before and after work, and stopped by on the way home to pick up more signature petitions. I was headed to Rising Hearts horse ranch in Silt, a huge breeding and training farm in the area of potential future slated gas development. They had graciously agreed to take one of the signature sheets for the ballot initiative.

I talked for a while with the friendly staff in the lovely redone two-story log ranch house, one of the many buildings on the vast and expensive-looking property. We discussed the petition in the spacious western and rustic décor office and the staff said that many of their clients would be eager to sign the petition, as the majority of locals here who are in the know are not at all happy with the prospect of seeing gas development on and near their own properties. Let alone the fact that there is no compensation of any sort required currently by Colorado law if you do not have mineral rights, which the majority of the landowners lack. The gas companies can set up shop 150 feet from the building where you or other humans dwell, and erect drill rigs. And more, including fracking equipment, multi-acre well pads, and evaporation pits filled with carcinogenic chemicals. Just what we all want to see in our country backyards… Apparently there is a subdivision in Rifle where they are going to put gas wells where the properties are less than two acres, and another area close to the Rifle MacDonald's toward the west where it is in the heart of a sub-development in close proximity to homes and children riding their tricycles through the streets.

Up in Carbondale the movie *The Inconvenient Truth* has been drawing large and long lines. John and Susan Gorman went up there tonight with petitions. There are more contacts from that area that have been called, and hopefully there will be some assistance in this daunting process of getting 65,000 signatures in less than one month's time.

I went to the store in Silt to get beer after going to the horse farm, and got a few spontaneous signatures there. I ended up coincidentally

COLLATERAL DAMAGE

meeting the sister of Bill Ritter, the man who is going to run as the sole Democratic candidate for governor of Colorado. We talked a long while about gas issues in the parking lot after she signed the petition.

She was yet another example of a person who lives in the heart of areas of future development who didn't know many of the facts. There are so many people who live here in the looming shadow of gas development who have no idea that their own property can be drilled upon, and that they will get no compensation. Let alone the dangers healthwise of living close to the fumes and the chemicals used in the process.

Garfield County needs a big wake-up call, and very fast. And very, very, fast... As Bill Solinger said,

"We are collateral damage..."

Take Two—The Pit At Mamm Creek

Saturday July 15, 2006

What a day! Joe Brown and his crew were down in Rifle last night at the bars, hoping to get some contact with the gas workers. The day and night before they were up filming on the Roan Plateau — as Joe said,

"To get it before it got wrecked by drilling."

He said it was absolutely gorgeous, and they camped at the edge of the cliffs and had great views of the impacted areas glowing with bright rigs light up in the night to the South across the Colorado River... if that can be considered a great view. They also visited a well-known beautiful waterfall on the Roan Plateau. He insisted that I get up there and see it, and soon. It makes me sad to think that I have to go up and see places quickly, before they are ruined. Is it better to never see them first, so that later when the area is strewn with well pads, compressor stations, pits and miles of roads and the pollution that comes with the whole mess, you feel less sad? I don't know.

I met Joe after 8 a.m. and we went on a drive up Dry Hollow and over to Mamm Creek. I am glad I wasn't alone; I think it helped temper my reaction to the whole astounding mess that unfolded before me across the once pristine rolling mountain valleys. It looks like something from *Star Wars* up there now, in between the few ranch homes and hay barns. As far as the eye can see, gas infrastructure and more rigs poking up here and there amongst all the finished-off sites. We drove up into the southern section of Mamm Creek and turned through a large gate surrounded by a number of signs, one entitled "Rules of Entry."

COLLATERAL DAMAGE

We carefully read the rules. No alcohol, low speed limits and more. There was nothing stating that access was prohibited, so we drove on through. After about five minutes of driving uphill toward the rig that toward on the hill, a pretty red-haired girl in her twenties came driving up beside us on an ATV. She was smiling, and asked us if we needed anything. Her hands were covered with green paint—she must have been doing some ranch chores on equipment.

"We're doing a documentary film on natural gas development, and we were wondering where we could get some good shots of rigs," explained Joe.

"Well, they're everywhere. You could go up that road, or over there. They are really all over the place," she said.

"Is this private property here? We read the sign and it didn't say anything about that, is it alright if we are here?" I asked. I certainly had the sense that she was checking on what we were doing, although she was nothing but friendly.

"Yes, it's our private ranch, and usually there is a guard at the gate, but he's not here today. Most people just know it's private," she said.

We apologized for not knowing, and then we gave her our phone number and let her know that a group from Colorado College wanted to do a tour of rigs and impacted areas next week. I asked if they would be willing to let us tour on their ranch, and she said she would ask her brothers. Given the fact that it is a huge privately owned ranch that I understood willingly leased their minerals to the gas companies, my guess is that they will not want to invite the college group out for the tour, even though she was wearing a local college t-shirt. It's just a hunch.

We headed back down toward Rifle on Mamm Creek Road. Gas trucks and water trucks barreled uphill past us, taking up a good amount of the road. Several times I cautioned Joe to slow down, or watch for the trucks. There have been collisions on these gas roads with those big trucks as they come around the curves; a few years ago a woman was killed up in the Dry Hollow neighborhood.

There was a big pit off to the right side of the road, with a ripped black rubber lining. It was flagged with a streamer of colored plastic triangles, like you would see at a fair. In the bottom of the pit lay a stagnant substance of a brownish color. Around the wellhead concrete rubble was strewn; it was not a very good-looking well site to say the least.

Joe slowed down and was looking at the pit, obviously interested in it.

"Do you want to stop and check it out?" I asked.

"Yes, I do," said Joe as he backed up to the entrance of the well pad then turned in off Mamm Creek Road.

Across the street was a modest small modular house, and up the road a short distance was a fenced of driveway with some RV's and other equipment stored. The dry, scrub-covered hillside climbed fairly steeply away from the road to the west.

Joe got out the big camera, and set the tripod up on the shoulder of the road where he had a good view of the pit. Industry trucks continued to drive up and down the road, and Joe politely waved at each one. The huge semis carrying water, gas, or chemicals roared along the county road, and kicked up a good amount of dust. I turned away from the trucks and walked away from the street when they barreled by. One typical industry large white pickup with the a tank in the back slowed and checked us out as it drove by, as though the driver might be considering stopping. But he didn't, and I was relieved not to have to talk at that time to an industry person. While some are friendly, some are not when it comes to people being around the gas wells.

Before Joe had a chance to roll any film, a small dark older sedan drove up the road and stopped. A friendly young man in his thirties wearing a car racing ball cap and blue jeans rolled down the window. He asked us what we were doing, and indicated that the house across the street was his.

"We're a documentary film crew, and we were wanting to get footage of this pit. Is that all right with you? I'm from Denver, and my name is Joe," Joe said as he reached out and shook hands with the man who introduced himself as Jeff.

"Hey, not a problem. I'm glad that someone is interested in what's going on here, because it's out of control. This pit was so stinky it made us sick, we had headaches and we were nauseous. I called and complained to Encana a few weeks ago, when it got hot and there was no wind blowing. Man, it was terrible. The heat made the stuff in the pit evaporate, and you couldn't be near it without getting sick. We couldn't even turn on the swamp cooler at night because it was sucking in the smell. I called them up and yelled at the guy. I said,

'You have to do something and get that stuff out of here, you're killing us!'"

Joe and I shook our heads as we listened.

"We've met lots of other people like you who say the same thing; you're not alone. They smell the fumes, they get headaches and feel sick," said Joe.

"Some of the people have actually been knocked unconscious by the gas, one lady Susan Haire was holding onto her truck door after the gas hit her. She barely got out of there and then when she got home she was violently ill all night. She lived up on Morseanna Mesa, but she got so sick from that one exposure that she couldn't go outside without a respirator and had to have three air scrubbers inside her house. Her horse died with some strange symptoms, and she finally couldn't take it anymore. She left, I think she's in Texas now and she put her ranch up for sale," I said.

"Yah, the neighbors with the ranch up the road own a lot of property, and they have a really nice cabin way up there but the fumes were so bad they couldn't go there. They complained to Encana about that too. I think they came and pumped the pit out, just like they did to this one. You know, my son likes to come out and jump on his trampoline by the house. And all those fumes are right here, and he's breathing them in. I'm not against gas development, but this just isn't right. They shouldn't being doing it like this. You know, I was raised by Democratic parents, and I'm a Republican now. But I'm not voting for George Bush again. It's not about politics, I don't care if someone is a Democrat or a Republican, but this is just wrong," said Jeff, his face creased with anger beneath the brim of his ball cap. It was a dark-colored cap, with flames along the sides; the kind that racecar crews and Nascar fans wear.

The sun beat down relentlessly on the dry, desert-like landscape of rocky hillside and hard- baked reddish earth. The heat radiated of the road, and just across the street was the house with the trampoline out front. That is where Jeff's son played in the unpopulated area where no friends lived near by; he jumped on the trampoline breathing in the fumes from the frac pits when they were full and stinky with toxins.

We discussed the Divide Creek seep issue, and talked about the geological fault that ran through the area, and about all the unknown things that were damaging the environment and causing health concerns as a result of gas development. Then Jeff started to talk about the pits that were above his house.

"There were some pits above the house that they didn't have any liners in, and there were no fences around them. The wildlife ran through them and some of them got stuck. The sagebrush around the pits was coated with this oily stuff, and you know the deer were eating that in the winter. And those pits stunk too, just like this one did. Right now there's a breeze, so you can't really smell anything. But if there is no wind, you can still smell it if it's not windy," said Jeff as he sniffed at the air being moved by a slight breeze.

A small car drove down the road past us, going way too fast and driving in the middle of the road as it screeched around the turn in front of the evaporation pit.

"The speed limit's thirty-five here, he must have been going fifty-five! They do it all the time, he probably just got off his shift and he wants to go have fun. You should see the trash they throw out, it's disgusting. Vodka bottles, Red Bull, fast food trash. Some of these gas workers look horrible, they don't have many teeth, and the shit they eat and drink. They don't seem healthy. Encana used to hire Mexicans to walk the road and pick it up, but they stopped. So now I walk the road every week and pick it up. And no one regulates what they do. You know, I'm a trucker and I get my truck weighed and checked all the time. No one ever weighs or checks these trucks up here. And they make a mess on the road and they never clean it up. Look at the road there, see all that mud, and that's nothing. After they fracked, they dropped clods of mud all over the street and they never came back and cleaned up the road. If I had a skid steer I'd do it myself, but I don't have one. So we have to breathe all the dust from that," said Jeff.

On the road were large circles of ground in dirt, and I wondered was chemicals might be in that dirt if it fell off a fracking truck. If that mud had contact with the fracking chemicals, then it would very well be carcinogens in the mud mix. But I had no idea for sure.

"I can't imagine that some of these workers are very happy, I met a guy last night, he was an engineer we met at the bar in Rifle. He said he made over $140,000 this year working for the gas industry, but he didn't seem very happy. He was pouring down shots one after another, and he was alone," said Joe.

The image of a man alone in a bar pouring down shots in rapid succession, far from home was depressing. Regardless of how much money he was making, is that a quality life? But who are we to judge? Maybe he had a fabulous time when he wasn't at work; maybe he had a wonderful supportive network of friends or family, interesting hobbies and a deeply satisfying life. But possibly, he did not.

"How could these guys from far away be happy, they are away from their families or girlfriends, and they are living in motels. So all there is to do is go to the bar, or wait to go home. Apparently there's a lot of problems with drug use. We met a gas executive one night at a hotel near Aspen, and he point blank asked us how they could get workers who could pass a drug test," I said.

Other friends who know people working in the industry have mentioned the high use of methamphetamines, a dangerously addictive drug, now known to be worse even than cocaine.

We talked more about the pits above the house.

"They never drained those pits, they just covered them up, so all that crap is still in the ground. It's all contaminated up there, all that ground is. It must be seeping underground too, so we don't drink the water here," said Jeff.

At this point I was beginning to feel somewhat stunned with all this information. How could the gas company have been so flagrantly negligent? Filling in evaporation pits with no liners, and no fences? Animals getting trapped in pits full of toxins? His son jumping on the trampoline a ball's throw from the stinking pit full of hazardous chemicals, and breathing in the fumes? Leaving underground dumps of carcinogenic stews above people's water systems? I had trouble believing that the company would actually do that, but I knew it was true. How did those people sleep at night, I have often wondered.

"They never could get away with this stuff over there, around Dry Hollow and Divide Creek; at least not near people's homes. There's too many people who know what's going on and they call in stuff like this. They don't have to leave the stuff in the pits; they don't even need to have pits at all. They can truck that stuff out and it never hits the surface. They do the pits because it's the cheapest way, and they don't care about what damage happens." I became angry as I said this; it is unconscionable that no oversight or monitoring to any effect seems to be done with this industry here in Garfield County.

"Boy, you should have seen those pits up there. I wish I had a sample of what was in them, they were awful. I have pictures of all of it, though," said Jeff.

Joe and Jeff exchanged numbers on scraps of paper, and again Jeff thanked us.

"I'm glad that somebody cares about what's going on, I just never called the papers or anything. I guess I could but I didn't," he said.

We agreed to hook up in the near future, then Joe and I headed back down toward Silt so he could meet up with his crew and we could all head to Burning Mountain Days in New Castle to circulate the petition for the Ballot Initiative and meet up with John Gorman.

Five minutes or so downhill from the pit we saw a water truck pulled up by the intersection of the frontage road and Mamm Creek Road. I-70 was in sight. A very obese short man was standing by the truck, and he was holding a large hose down in the irrigation ditch, parallel to the Colorado River.

"Hey, is he sucking up water like Beth said they do?" asked Joe.

"Yes he is, that's exactly what he's doing," I answered.

We turned onto the freeway and headed east.

"How do they do that? How can they get that water? How does anyone know how much they take if there's no one watching? What about the ranchers and their irrigation water?" Joe asked in a puzzled voice.

"They can do what ever they want, they are the gas industry," I replied in a flat tone.

I had no idea how the water situation with the gas industry was set up, yet another question to puzzle over. What was clear as day is that no one was monitoring it, how many truck loads of water were being sucked from the ditch every day or every hour. Were they allowed free range of all the ditch water they wanted? If so, could I have some too for my pasture? Could I drive a water truck down there and suck water out of the ditch and fill a big truck tank up? I'm sure the answer is no.

I could not resist a sarcastic comment.

"Don't you know the gas companies are saving us from the Taliban? And from that guy, you know, that guy who's hiding out in the mountains of Afghanistan: Bin Ladin. They can take all the water they want and wreck our land and air because it's saving us." The terrorist's name had escaped me for a moment.

Joe looked straight ahead at the highway traffic as he drove toward Silt on Interstate 70.

"Oh, yes, I forgot that,", he said without cracking a smile.

I dropped Joe off at the motel parking lot in Silt, which was filled with the signature large white industry pickups with diesel tanks behind the cab. I noted the out-of-state plates of the truck behind mine. Wyoming. Lots of them were from Wyoming. People were coming from all over to work in the booming gas fields of Garfield County and elsewhere in the West. I heard a newscast recently talking about training veterans from Iraq to work in the energy industry. It was good money for a recently returned vet, and they went through fast programs then out onto rigs.

In New Castle the Burning Mountain Days parade was in full swing. Incidentally, the festival has that name because of the fact that when the coal seam caught fire back in _____ the fire took off underneath the mountain and has never been extinguished. Companies and specialists from as far away as Japan have come in and tried various techniques, to no avail. Across the county road from my kitchen window I can see the plumes of steam rising from the ground when the temperatures dip below freezing; the heat being released from one of many vents from the coal seam fire burning underground.

I left my truck on the outskirts of town where the police directed people into the middle school parking lot. With the ballot initiative petition clipped to the clipboard, I proceeded down the shadiest path toward town. People lined the streets for the parade. I had no trouble getting signatures for the initiative; I just had trouble trying to stay out of the sun. I would not ask for signatures unless there was enough shade to stand in. Nonetheless, the sun was brutal as temperatures climbed above 100 degrees. I do not handle sun well at all. Once I hit fifty signatures, I began to slow down. My ability to get the words out effectively diminished, and I needed a break.

I went back to the Grand Valley Citizens Alliance booth where others were stationed under a white shade canopy with petitions and pamphlets of information laid out on tables. It was good to meet with my friends, and then I ducked into the Mexican grocery and fast food restaurant and collapsed in a booth. I was fried and dizzy, and my stomach hurt. Too much sun, definitely.

One barbacoa taco heaped with jalapeno salsa and an ice tea later, I rallied and returned outside onto the broiling sidewalk. I said goodbye to the GVCA friends, covered my head with the clipboard as a make shift parasol, and trudged with measured steps toward the car which was five minutes away. I felt woozy from the heat, and hoped I would make it home OK.

Back at home I hid inside the wonderfully cool house shaded by the huge protective elm trees that hung over the roof. Although I had put on sunscreen (rather spottily) that morning, I felt like my skin was quite burned. I worked on the computer, sending e-mails for the ballot initiative.

That evening I spoke to my parents back East. My father and I often have long conversations about the environmental issues we are both involved in. I got him up to speed on the latest goings on here. I heard myself telling him that I wasn't sure I wanted to stay here, at the rate that things were going.

"I'm not talking about right away, but some years down the road. If it gets really wrecked, I don't think I can stay here. I won't leave Al, I love him and he has a good job. We like it here, it's a nice community, but I just don't know."

Dad and I discussed the family property up in the high country of Colorado, the small ranch up there that Dad had bought some years ago. Our house had been just down the road, and I had kept my horses and llamas there and looked after the house and barns. It was adjacent to a large Nature Conservancy piece of land, where the endangered mountain plover lived, plus a few species of plants and a rare bog of ancient peat that was as thick and rich and black as oil.

I am sad to even think about moving again, even many years from now. All the work we have done here, all the thousands of dollars of improvements that have been made here, all the hours of work on the ranch to fix it up faster than it can fall down. I honestly don't think I will, but in the back of mind I worry when people keep saying about the gas boom here "This is just the tip of the ice berg…"

If this is just the tip, then God forbid what the berg itself will look like.

Sunday July 23, 2006

Yesterday I talked to Chris Mobaldi, I called to apologize for not having stopped in for a visit when I was in Grand Junction earlier this week on business. It had been blisteringly hot, and I did not easily have time to spare.

"Haaalloooo? Taaraaa?" she said in her strange accent when I called.

"Hoooo nace toe hey fom yew…" she said in her new strange dialect. Or one of them, I should more correctly say.

I was relieved. This time I could understand just about every word. It was not always that way when we talked. Sometimes, I had difficulty making out even a portion of the conversation, other times, Chris was either too ill or too incoherent to even speak on the phone.

"How have you been doing?" I asked.

"Ah hev been do-hing gooood. Ah yuh co-mang toe sae mey?" she asked.

I told Chris that I had hoped to stop in that week, but would have to do it later. When it was cooler. We talked for quite a while, and we discussed the status of the lawsuit.

"Ah cahnt tall yuh what is goh-ing on rut noh…" she said reluctantly.

"I know that, and that is good news; that is good news for you. It means the lawyers are doing their work. That is what is best for you," I said.

COLLATERAL DAMAGE

"Si, si" she responded. In Spanish. Once I asked her if she spoke Spanish and she said no.

We talked about the press, and the articles that were being held back from print due to the legal case.

It is near impossible to describe what it is like to talk to a friend of yours who has gone into another state both physically and at times mentally, and who has had their ability to function and communicate so heavily damaged. There is that strangeness of a distance that illness can bring; severe illness—but this is different.

Chris is good on some days and horrible on others. Yet through the layers and veils of illness and strange tongues, I know my friend is in there. And I can tell from the joy in her voice on the good days that she is happy if not overjoyed to hear from me. It is like communicating by shouting to someone down through a pipe, or a well. This time, ironically enough, I guess it is because of a gas well.

Theo Colburn is a world-renowned specialist from Colorado working on the effects of the toxins that the industry is using here. She knows of Chris Mobaldi and Laura Amos and their tumors.

She is working with a team of four PHDs studying the issues. They have presented their findings to officials up in Gunnison, Colorado.

"We are pushing to get full discloser of all the chemicals used…" said Theo.

Currently, the chemicals used in the hydraulic fracturing process are exempted from disclosure; they are "secret" proprietary industry information.

July 26, 2006

Today in downtown Glenwood Laura Amos's truck was parked at an intersection as I was crossing on busy Grand Avenue. She rolled down her window, smiling and waving back at me as I waited to cross the four-lane street.

I ran over to the passenger side, and Lauren her beautiful daughter stuck her head out.

"Do you want to see my dog?" she asked, a huge smile covering her face.

"You are bigger than you were in the magazine picture, you are getting big!" I said.

"I am five!" she exclaimed with pride, as she held forward for me to see the small white and black wiggling cock-o-pooh.

I stood in the street between the lanes for five minutes or so, right at rush hour catching up with Laura. They were on their way up to camp, at the outfitter's station.

We talked about the recent goings-on, and Laura said we should go boating together on their new pontoon boat.

"It's been real busy here, with the documentary film crews, meetings, the press; yet nothing seems to happen…" I said with depressed resignation.

"It's like Bill Solinger said: we're collateral damage. And one of the industry workers said this area is a national sacrifice zone," I said, thinking of the pitiful ranches trying to maintain their way of life as wells and pits encroached on every side.

"Encana's lawyer's in the court house, we just saw them there," said Laura.

They waved goodbye, and we said we would catch up soon. The familiar truck with the company logo and trademark mountain lion picture on the back window merged onto the highway and faded out of sight into the rush hour traffic. I had the feeling of wishing that I was going with them — not up to the outfitters camp (I am not a hunter and would not enjoy that environment) — but to another country town away from the drilling. Away from Garfield County

(Gasfield) which the media continually refers to as "The Saudi Arabia of the West".

The Amoses were one of the very lucky few families who had reached a substantial enough settlement for damages done by the gas industry to afford to relocate to a safer area. Most people either hunker down and deal with it; fumes, pits, impacts and all, or just abandon their properties and flee and hope that their homes will sell. The very sad fact of the matter is that most people out here do not have the money to flee. It takes a padded bank account to do that, and Silt and Parachute and Rifle Colorado are not Aspen. Not by a long shot.

Who wants to live in the Colorado equivalent of Saudi Arabia's oil fields? I don't. We moved here to be in the country, surrounded by nature, and to work hard on creating the home for the rest of our lives on this small ranch. Why do we have to fight so hard on every level from local town and county governments to state and federal agencies to protect our health and homes? Why?

I walked to my truck and had the feeling that I was happy for the Amoses; it was like seeing old compadres. Yet, as I drove by the courthouse in Glenwood Springs it made me angry to think of the Encana lawyers inside. What violations were they perhaps settling on this time? Which side would win this one? Of course we would never hear about it in the papers; everyone settles out of court because it is less expensive. Yet, why were they in the courthouse now? Who knows...

I knew that this was going to be an endless process. The lawyers, in their expensive suits, would file in and out of the Colorado courts and argue cases one by one for violations, health impacts, and worse. Some people would get money for their problems, and some would fail against the well-heeled gas lawyers. One thing they had plenty of was money. Of that there is no doubt.

July 30, 2006

In late July the temperatures have spiked to an almost unbearable level; over one hundred in the sun every day for a long stretch now. It is hard to be productive at anything. Hopefully August will not get too much worse. Motivation is a hard thing to come by in this heat; just getting to and from work and doing a regular workweek is about all that one can manage. Especially without an air-conditioned vehicle... how I regret that I passed on the air conditioner option when I purchased the truck years ago. But then, living at 9,000 feet in Summit County, being too hot was never, ever an issue. Live and learn...

Farmers on Silt Mesa, Grass Mesa, and Dry Hollow cut and bale and load the hay. The fields are striped with even rows of fresh cut hay, and the scent of it baking in the sun floats in through the truck window as you drive by. Up by Harvey Gap Reservoir a young boy does his job and kicks over a bale to line it up properly for pick-up by the big stacker truck following him. How those tough ranchers deal with the heat is unbelievable. Perhaps he is thinking about how good a plunge in the nearby icy reservoir will feel after hours of sweaty work in the hay fields.

This is what Colorado must not lose to the energy industry as they do their pillaging of the land; we must not lose the farm-based economy. Yes, the suburbs are encroaching as they always do from cities. But, Colorado and other Western States deserve the right to keep a piece of what we are about. About what this place has been about for over one hundred years? It is ranching and farming. And, before that, the Native Americans and the wild game roamed these mountains and valleys.

The gas industry is here—it will not go away. But why can't they do it better with concern for the people's health and property? Why are Colorado landowners not compensated when their homes and properties are taken over by the industry? Why, when it is so inherently unfair?

At work last week I was talking with a fairly new employee with whom I share an office. He is from the East, and worked in Manhattan for years, and witnessed the whole 9-11 disaster and aftermath. He and his wife chose to move to Colorado, an area they had visited in the past and found beautiful.

For some reason we got into a discussion about gas development impacts. I had pointedly avoided that subject once I learned that he had bought a home on Silt Mesa, an area slated for the next expansion of gas development.

We got into a discussion of the fracking operation and the injection of the chemical mix underground. I don't recall how we got onto that topic, perhaps from a local newspaper article. Gas related articles are common in all the Colorado papers. This is a huge industry operating here and the impacts are also huge.

"What, they blow explosives underground and inject chemicals? I didn't know that!" said my co-worker with great surprise in his voice.

I could not and will not tell him of the impending development of Silt Mesa, where he now lives with his wife. It is not my job; I don't have to do it. And it does no good to do so; it just gets people upset.

Just like I am upset that I and my husband moved here with no education about the gas development in the outlying areas. We didn't know about it, it seemed like a good leap to make, so we did. And we have thrown a lot of money into the move. It is the major investment of our future, just like homes and property here is for most folks. I can't say that if we even had a hint of what was to come with gas that we wouldn't have come. I just know that we had no idea, absolutely none at all. And most people don't.

As Teri MacGuire from Dry Hollow says;

"People don't care about it until it happens to them…"

Terrie and her husband have a ranch where gas wells were forced upon them unwillingly.

"How are things up there now?" I asked her this morning.

"It's really slowed down a lot, but we got gassed out a few times last week," she said.

They got gassed out. That means the fumes were so strong that they couldn't bear the air outside. And they are breathing it in, and their horses and dogs and cats are. Other people don't care about horses and dogs and cats when it comes to death, mostly they respond to human issues.

The bottom line is that all living creatures, human and animal, are being exposed to unknown and un-researched and unmonitored dangerous chemical carcinogenic cocktails where we live. Why does no one care? Why does no one do anything? Is it fair the Chris Mobaldi's life has been ruined? And countless others? Why does the industry not use best practices when they know of them? Money, money, money…that's my guess.

July 31, 2006

Talked to the Mobaldis tonight.

"Hi Chris, how are you?" I asked.

"Ae um fine Tada, hoe ah yu?" she asked.

I could tell a distinct difference in her accent from the last time we spoke about a week ago. Her accent sound more clipped, with a faster pronunciation of words. It sounded like one of the Eastern European accents, unlike the slow and drawn out speech she often has, or the fast, chattery and indecipherable Asian type accent.

We discussed the heat wave, the recent rain and relief. When I asked how Chris had been feeling lately, Chris said that she was not vomiting regularly any more, which was incredibly good news.

"Are you able to stay awake through the day?" I asked.

"Yas, yas ah cahn stey a-wahk no," she said.

For so long Chris has had so many bad stretches of severe ill health, with her continued weight loss, nausea, vomiting, and inability to function physically or cognitively for any length of time.

"Whey ah saying a bedy gode doh-coh nohw," she said.

"Not Dr. Gerdes?" I asked.

He was the specialist in Denver that the neighbor had gone to for diagnosis and treatment prior to filing a lawsuit against the gas company for some of the same health issues that Chris had. That neighbor ended up settling with the company for damages out of court. He and his family have since moved to Oklahoma.

"Who is the new doctor? Did the lawyers find him for you?" I asked.

"Yas, deh deed. Hay ees ay beeg spa-sholisht..." she said.

"He must be, if he is bigger than Dr. Gerdes..." I replied. This was heartening news; not only had Chris finally made it to the level of being treated by Dr. Gerdes in Denver, she was now at a different and presumably even higher level of treatment!

It was monsoon season early here in western Colorado, and as we spoke I looked out the window to the east and watched the sheets of rain pounding down endlessly. The ranch animals huddled in the open sided barn on the hill, taking shelter. Hopefully they had eaten a good enough amount of evening hay before it was to be ground into the mud.

Thank God for the rain; with this all-time history-breaking record heat, we were ripe for a terrifying fire season this summer. Something we can all live without.

Just outside the window I noticed a large, bulbous spider hanging in a web woven under the eaves. It was large and black and I had little doubt that it was a black widow. As Chris and I talked, I watched the grape-like spider go up and down her web, until finally she crawled into the high and dry area closest to the roof and out of sight.

After chatting for a bit, I asked her,

"Chris, you sound different tonight than last time we talked. Can you hear what your accent sounds like when you talk, or does it all sound like regular English to you?"

At first, she had a bit of trouble understanding what I was trying to say. The she got it. By this point her speech was slowing down more.

"Ah jest hae Onglish wan eh spek…" she said. She was getting tired, so she passed the phone to her husband.

We talked bout the lawyers, and the general progress of the case unfolding without me being able to really hear anything substantive, which is fine.

There are two law firms in Denver working on this case, and Lance Astrella has an affiliate in L.A.

"Why so many lawyers?" I asked.

"Because it's a huge deal…" said Steve.

I hope and pray that is true.

August 2, 2006

The Colorado sun is slanting its way thought the narrow valley, shadows deepening on the hillside as the minutes pass. The earth, recently saturated from several days of good downpours, is still wet and emits a fresh and cooling scent in the air. The blades of tall grass clumps illuminated golden by the setting sun as they ascend the steep mountainside are stunning; as beautiful as any Los Angeles or New York interior designer could ever re-create in a sterile setting.

The evening bugs are thick, and clouds hover as numbers of salmon-bellied swallows gracefully soar, glide, then descend for their evening meal. Groups of swallows collect on the power lines along the road like evening drinkers lined up at a local bar. Between their hunts for bugs the swallows come to rest and survey from the high wires.

In the garden, the sunflowers are illuminated by the light like stained-glass sculptures of oranges, browns, and yellows.

COLLATERAL DAMAGE

Cucumbers and tomatoes are coming on, the arugula is bolting but still is edible, the chives need cutting, and the California poppies are curling their golden petals tightly shut for the night. One of the special large poppies has bloomed; it is tissue paper white, with pink edges.

The llamas sink into the hay piles on the hill to feed, and the birds carry on. This is why life here is wonderful…and we want it to stay that way for a long time.

August 5, 2006

This week the local papers were full of front page articles about gas development, articles on the gravel pits, controversy about the roadless areas, Grand Junction and Palisade's Ballot Initiative to stop drilling from an area that supplies the cities' drinking water, and yesterday's article covering the meeting of the mayors from Aspen to Grand Junction as they did a flyby of areas impacted by gas development.

It is typical for the papers to have plenty of gas related articles, but this week's articles took on a distinctly political bent.

Silt Gravel Pit:

The owner of the 181-acre property for the proposed pit, Scott Balcom, pulled the application with the town of Silt on July 27th, 2006. According to Silt Mayor Dave Moore, a letter was sent by Balcom stating that the permit was pulled based on the applicant's inability to reach an agreement with the town and that further negotiations would be fruitless. Moore states that the town had hoped the owner would negotiate with them on issues such sales tax, royalties, traffic, road maintenance, and reclamation of the pit.

"We're the ones being impacted and we want them back to give us control over things such as dust, noise, vibration, and unsightly pollution," Moore said. "If they go to the county, it will minimize our ability to control it."

(Quotation taken from the front-page article in the Glenwood Springs Post Independent on Thursday August 3, 2006.)

Moore understands that the town of Silt has no control of the permit being sent to the county, but he hopes the county will in deny it and send it back to the town.

Roadless Area Issues:

Governor Bill Owens appointed a Task Force to travel across Colorado to gather public input to utilize in making a final decision on how to manage the State's roadless areas. In Glenwood Springs the June 21st meeting at the Hotel Colorado was packed with citizens, the majority of whom supported preserving roadless areas.

As usual, the Garfield County Commissioners were divided two to one on their stand on the issue. Democrat Tresi Houpt supported preserving the roadless areas, while Republicans Larry McCowen and John Martin did not.

Martin questioned White River National Forest Supervisor Mary Beth Gustafson about a particular area already designated as a gas and oil lease. This lease is on a 125,000 acre property South of Carbondale known as the Thompson Creek Area. Discussion occurred over the ability to create roads for existing gas and oil leases in area that would be designated roadless. Gustafson explained that under certain conditions roads could be constructed in the roadless area, and that being deemed a roadless area did not cancel out the opportunity for gas and oil leases.

Further discussion continued questioning the Forest Service supervisor on plans to be flexible with roadless areas to deal with the large problem of beetle-killed forest. Martin and McCowen supported a position that would allow roads to be put in to access those beetle-kill areas.

The concern for the future would be then, how would those roads be restricted for gas or oil development? Constructing road for one purpose is an initial foot in the door for further use by multiple industries. "Healthy forests" can also be interpreted in another context as logging, and that can go hand-in-hand with gas and oil development.

The public locally has voiced their majority opinion, which is to preserve roadless areas. The public also voiced their majority opinion to preserve the Roan Plateau and protect it from gas and oil development.

That majority opinion did little to stop energy development. The first rigs are up there, looming on the edge of the Roan Plateau Cliffs. Joe Brown and his film crew are up there somewhere this weekend camping and filming, taking in the beauty of the area before it is destroyed.

Meeting of the Mayors:

On Thursday August 3, 2006, mayors from Aspen to Grand Junction met to discuss area issues such as gas development and its impacts, and to do a flyover of the Colorado and Roaring Fork Rivers. Silt Mayor Dave Moore arranged the meeting, and was impressed by the response. The group discussed roadless area policies, gravel pit reclamation, and how to hold the gas and oil companies more accountable.

Rifle Mayor Steve Lambert was pleased with the group's progress. He stated that there was a good potential of productive outcomes to be had in the future from this group continuing to work together. Collectively, the mayors represent a lot of people who have similar concerns. The group plans to meet again next month in Aspen to continue their discussions.

August 5th, 2006

This week, while driving East out of Rifle on Highway 6 and 24, which parallels the Colorado River, I turned my head in surprise as I went by the new rigs that hug the roadway. In addition to the towering and foreboding structures were the countless blue rectangular container trucks set side-by-side, the stacks of mud mix, and whole industrial mess. But what turned my head was a huge area of bulldozed earth between two of the rigs where a large hill of freshly excavated earth had been created. What was that for, I wondered? Due to the dense brush and trees and bushes it was impossible to see much behind there. In the short stretch of less than a mile (if that) were now three new rigs, and yes, another few popped their metal heads up across the river, close to the shore.

Where quiet riverside dirt drives with mailboxes had led toward farm houses and barns, now large wide roads busy with industrial traffic turned off Highway 6 and 24 where the metal signs announced the rig number, with a picture of a logo and a company name. There were several on the right side of the highway, and I noticed with a sadness the first sign on the north side of the river, by a road going into Silt Mesa.

One was with the Grey Wolf Company, and a nice picture of a wolf howling at the moon adorned the sign. Somehow, the wolf just made it seem friendlier, like gas development was a wildlife kind of thing… it's like those subdivisions popping up everywhere on what used to be ranching property; they have names like "Lakoda Canyon," "Warm Springs Ranch" or "Elk Creek Meadows," politely naming themselves after the ranches that were destroyed and divided into suburban lots. If you give something a cozy name that sounds outdoorsy and western, then you are helping keep a piece of it. Even though it's just a name…. As I drove pass the Grey Wolf rig area I didn't notice any wolves lurking, but I could smell a deer carcass rotting in the ditch off the road. On the north side of the road I drove by the farmhouse where the family lived who reportedly couldn't sleep at night due to the sounds of the rigs operating. Duke Cox, President of GVCA who lives miles away higher up on Silt Mesa says he can hear the rigs all the way up at his place, grinding away in the night.

Sunday August 6, 2006

I have been re-reading parts of this manuscript today, as I have no new information to insert. There is plenty of writing that waits, but I have not broken the book into sections yet.

In the midst or reading through one of the sections, I became inspired to pull out the Natural Resources Defense Council magazine *OnEarth* and re-read the article "Wrecking the Rockies."

COLLATERAL DAMAGE

I pored over the article for a long while as the laptop hummed in front of me on the kitchen table. I wanted to write in depth about everything covered in that so very well-written piece; Michelle Nijhuis did an excellent job collecting and reporting upon her information.

There are so many pieces to the complex whole of the issues involving human, animal, and environmental impacts from the gas industry. I continue to dwell on one of the most disturbing events that take place, (to me, anyhow): the injecting of undisclosed carcinogenic chemicals deep underground during the fracturing process.

Whistleblowing Weston Wilson Of The EPA

Hydraulic Fracturing Chemical Exemptions

The EPA allows the industry to inject any chemicals underground in the fracturing process; that is, except in Alabama due to a challenge in the mid-nineties when LEAF, the Legal Environmental Assistance Foundation based in Florida went to court on the issue.

The case was successful, but it held only in Alabama where the case was filed. So, unless every other state takes the issue to court, the EPA will continue to allow the gas industry to dump whatever type of toxic and carcinogenic chemical underground that it wants. Wonderful...

"Though the EPA could have chosen to enforce the ruling nationwide," says LEAF attorney David Ludder, it decided not to. "The agency took the attitude, 'If it's going to be applied elsewhere, they're going to have to sue us.' Alabama, the only one of the three states where hydraulic fracturing is used in coal-bed methane wells, remains the only one required by the EPA to permit and oversee the injection of hydraulic fracturing fluids under the Safe Drinking Water Act."

I guess the idea I had the morning after the last Grand Valley Citizens Alliance meeting was not so bad; sue the EPA. I am waiting for the promised phone call back from the group, I already talked to the attorney who agreed to rally the forces and discuss the issues of taking one of the regulatory entities to court who are not doing their job of enforcing our health.

One local attorney I spoke with was more interested in doing a legal action at the state level; he thought the EPA was too big to deal with. But, here it is in print, friends. If they were successful in Alabama, why can't we do it here? How many cancer cases and illnesses do we have to lay out for evidence before there is any discussion of a health concern? How many?

Re-reading this *OnEarth* article is bringing many memories crowding at once into my mind. The majority of the people in this article were people with whom I have had numerous phone calls, mostly regarding the exemption of hydraulic fracturing from the Energy Bill. I lobbied long and hard to stop the passage of that exemption, unfortunately to no avail.

I learned how to make calls from Colorado to Washington D.C. at 7 a.m. when it was nine there. For an hour before leaving for work I could lobby the key senators' and representatives' offices as I made coffee and brushed my hair with one hand, and held the phone with the other. (Getting dressing with only one hand is a bit trickier.) By calling the switchboard, at times I was successful in getting transferred back to the operator to go to the next politician's call. Otherwise, it was just a matter of hitting redial to ring again at the main switchboard at the Capitol.

I faxed packets of the best newspaper articles with worst health concerns and ugly pictures of industrial rigs and pits by people's home here in Silt. I spent many hours, and hours, and hours.

There were phone calls with Weston Wilson of the EPA, (who I consider a true hero), pestering phone calls to the Inspector General's Office (who did not follow through on Wilson's whistle-blowing action involving hydraulic fracturing chemicals proposed exemptions from the Safe Drinking Water Act), calls to Udall's office to put pressure on the passage of Senator Jeffords's Bill to undo the de-regulation of fracturing chemicals—calls, calls, and more calls.

How strange it is, to read all these names in this article, and to watch the path the whole process took from a state to a federal level. And how sneakily the game was played…

Who cares if some people die or get sick from this stuff, anyways? There is so much money to be made and so many oil people from

Cheney to Bush in Washington to Governor Owens here in Colorado and even locally. Why do County Commissioners McCowen and Martin push for gas development everywhere with no holds barred? What are they getting?

Don't they care if their families and neighbors and grandkids are being exposed to toxins? Don't they know? How could they not?

Just because an industry is making multibillion-dollar profits doesn't mean they can't be held accountable for killing people. Even if in the scheme of things "It's not that many... ." The industry could spend more money and do it better, and there would be fewer problems. For one, they could use non-carcinogenic fracturing mix ingredients.

I am too angry to type anymore today. Yesterday I wrote a letter to the editor.

August 8, 2006

Wish I could say I was in a better mood, but due to numerous circumstances things are not very cheerful tonight.

Driving into work today I head on Aspen Public Radio the announcement about the ballot initiative failing to meet its mark of 65,000-plus signatures. Then I heard my friend John Gorman speaking about the issue.

The front page of the Glenwood Springs Post has an article talking about Kathleen Curry's commitment to change the make-up of Colorado Oil and Gas Conservation Commission and create a more balanced commission.

I called the NRDC Office in San Francisco and talked to them about potential support for mounting a similar case against the EPA as Alabama did regarding fracturing chemicals. I talked to Duke Cox of GVCA and told him what I wanted to do.

Yesterday, I had a long telephone call with Keith Lambert the Rifle mayor about passing information to the mayors group. As I detailed

COLLATERAL DAMAGE

for him the stories and facts of some of the illnesses and the lawsuits dealing with exposure to toxins used in natural gas extraction, he said at the end of the call somberly:

"Everything you said is true."

The farrier came to trim the horses' hooves and my lame horse did not do well. His injured front foot can no longer bear weight to allow him to be trimmed on his other front hoof. Buddy had always been a very gentle horse to trim; he never put up a fuss as his feet were being worked on in the past. Tonight however, he reared up in pain when the weight fell on his bad front leg. We had to give up trying to make him bear weight on that foot; he was going to injure the farrier by coming down on top of him, or he was going to cause more injury to his bad foreleg.

That is what is the worst news of all. How was I going to keep this horse going if he couldn't be trimmed any more? Tears loomed beneath my eyelids and threatened to spill over as we did our best to trim three out of Buddy's four feet, and as quickly as possible the farrier finally had to saw the tip of Buddy's last hoof off, in a triangle form. We pried the horse's leg up an inch or two off the ground, and shoved a flat board underneath the overgrown hoof. Then the farrier sawed away at the hoof tip with a short saw. This was merely a quick fix, and not an ideal one.

Another problem to deal with…

While out in the barn, when I was putting out hay for the animals, the red horse Doc swung his head suddenly around to bite at a fly. I caught it square on the right topside of my skull, like a baseball bat had hit me. I never knew how much power a horse's head had…ow!

I put my hand up to check for blood, and staggered out of the barn hoping not to go unconscious from the blow. Luckily, I did not.

Later, with ice packs on my head, I set to work on the computer and was promptly interrupted by an emergency phone call from work. Tonight was just not meant to be a productive evening. I went to bed with a blue vet ice-pack from the freezer under my head, hoping for a better day tomorrow.

August 9, 2006

Today is a better day, in spite of a good-sized headache that came on this afternoon and will not leave. The aftermath of the head hit from the horse. I am lucky I didn't wake up feeling worse...

After a typical busy day, I was headed home and got a call from Lance Astrella, the attorney from Astrella and Rice law offices in Denver. I had called and left a message yesterday inquiring into the Alabama lawsuit with the EPA regarding hydraulic fracturing chemicals.

I was on the phone when he called, but I retrieved his message and quickly called him back. We talked at length.

"I'm calling because of an idea I had after re-reading the NRDC's *OnEarth* article. Did you read it? It's called 'Wrecking the Rockies'."

"No, I haven't," said Lance.

"Well, it is excellent, and it goes into great depth about the political reasons why the Energy Bill allowed for the fracturing chemicals to be exempted from the Safe Drinking Water Act; how it was all set up in time to stop Weston Wilson's whistle-blower action against the exemption."

"Can you e-mail or fax it to me?" asked Lance.

"Yes, I can fax it later tomorrow," I said.

We continued to talk. Lance said that the only way to effectively fight the industry would be with litigation; no legislation will ever get through due to the amount of money that is being spent to lobby on behalf of the gas and oil industry.

I described to him the people who I had met, some on the trip down to the Capitol with Chris Mobaldi and others to testify before senators.

COLLATERAL DAMAGE

"I was shocked by what I heard on that trip," I said

"After I got back I called the press, from the local paper to the Associated Press. David Frey from the Aspen Daily did a great article in the Mountain Journal, and Judith Kohler of the Associated Press is holding hers…"

"We have a big case going on right now about a fracturing issue," said Lance.

"Is it the Mobaldi's?" I asked.

"Yes," he said.

"I am the one who called you about them…" I said.

We continued to talk, and Lance asked questions about the Grand Junction water issues. Just today the leases were approved for gas development in the area of a water supply for Grand Junction, but a moratorium was put on development for a year until a study could been done.

I asked questions about the future development here, and got the sobering answers I expected. I inwardly cringed as I asked specific things, and said,

"I might be up at 2 a.m. thinking about your answer to this one."

Lance indicated that the industry can do things much better; they can reduce the pollution and emissions both in the air and the water, but need to be legally forced to due so.

"In Denver it's happening, they are tackling air emissions, and it can happen up there too," he said.

"You have a great group up there, the group you are with; they are not radical, and they get things done," he said in a most complimentary way. I knew he was referring to Grand Valley Citizens Alliance.

I explained that I was fairly new to the mix, and that the first gas meeting I went to was the one where he was a major speaker. I said that I had been very impressed with his information, which I had been. I remember lying in bed that night late for a long time that night, rehashing what I had heard at the meeting, and specifically thinking about what Lance Astrella had said. That was a year-and-a-half ago by now.

Thank God I brought those papers home that were spread out across the folding tables in the Rifle middle school gym at the gas information meeting that night. Papers stacked in neat piles, all with different information relating to gas issues and contacts. I picked up business cards too, from the two lawyers who spoke at the meeting to the 200-plus audience in the gymnasium; rows of people all nervously sitting on the folding metal chairs waiting to hear what was going to happen to their homes.

Tonight there's a Community Development meeting to discuss the plans for gas development on Silt Mesa. I'm not going to this one; you have to pick and choose your meetings since you could go to one just about every night.

"These meetings are very sad, because people go hear if their properties are going to get wrecked. I know the Community Development Plan is a great thing, but it is still sad," I said to Lance.

I told him about another meeting I went to where they put up a big map. The blue line across Silt Mesa indicated where the drilling was predicted to expand to first, and then peter off due to the geology closer to the hogback mountain range.

"The people sitting next to me who are hay ranchers I buy hay from that live on the blue line; they went rigid in their chairs when it was announced that the company expected to drill successfully for gas right up to their property line…" I said.

"The man from the gas company said 'You people who live above the blue line can all go home…' and he laughed in a friendly way as he said it, meaning those folks didn't have to worry about getting gas wells on their places, so they could leave," I continued.

Lance extended his offer to be a resource for people up here in Garfield County, and we wrapped up our conversation and said goodbye.

August 12, 2006

Today I spent all day today at a Pet Vaccination and Spay and Neuter Clinic in Parachute. It was a good day; I got up early and worked hard all day with dedicated people. We gave out vouchers for free spays and neuters to a lot of people, and dogs and cats and even one ferret got their needed shots from the vet who donated his time to sit under the white tent on a hot Saturday.

Last night, after a long workweek with a lot of gas and oil business crammed in between, I was fried. During the week as I was writing I ran into a computer problem. The discs were not saving consistently, and I got frustrated staring at the screen as it failed to do what I wanted.

Hopefully my last trip to the computer store fixed this, and the "memory stick" will make things better. I am not that great with high-tech computer stuff. The basics I can do, but new things are always a bit much to deal with. We'll see.

Yesterday, Friday morning, I got a call back from Kemp Will, the attorney who handled Laura Amos's case. I told him my interest in trying to duplicate what Alabama did with the hydraulic fracturing chemicals. He had not read the NRDC article, which surprised me as Laura's story was showcased there. We talked at length. I like Kemp Will.

"It is stressful to do these cases," he said.

"I know, it is depressing to work on these issues that seem to go nowhere, but we have to," I responded.

"When people hear what is really going on, like the fracturing chemicals being injected underground, they say 'That can't happen!' in great surprise. It is sad to tell them that yes, our governmental entities — local, state, and federal—allow it. All through fancy maneuverings of political manipulation, it is OK to inject carcinogenic chemicals underground that can pollute the Colorado water, and leave pits of evaporating chemical waste to pollute the air. That's only the start of the huge list of things: flaring, spills, and on and on and on…"

Kemp Will, as did Lance Astrella, agreed to be committed to helping us on the Western Slope protect ourselves from the gas industry. I don't think anyone has any delusions about just how insurmountable a task that is. I recall saying to one of the lawyers,

"I don't want to be doing this…"

As if dreaming of a better reality, I imagine having moved to a farm where one could live without the impending knowledge and fear of what was happening, and what was to come.

Today I saw on television farms in Kentucky with rolling green pastures and white fences, horses grazing. It looked so beautiful, and I felt jealous in a painful way. There were no well pads, no evaporation fits filled with stinking chemical mixes, no dug-up acres near homes with industry equipment and trash strewn all over on a horrendous scale. Why do we have to have it? Just driving to and from Parachute to do the vaccination clinic was painful; seeing all the gas infrastructure crammed in all along the way. There is no other way to describe it; it just makes me incredibly sad inside.

Matt Sura from Western Colorado Congress talked to me yesterday; we had a very good conversation. We talked about the strategy being pursued to enforce better regulatory restrictions on the industry. He e-mailed me the latest letters that were drafted to different agencies. I told him what it was like to continue to wait for some kind of result from all the work, yet all the time knowing that one would be a long time in coming, and most likely as slow as a snail's progress.

COLLATERAL DAMAGE

"It's like beating your head against a brick wall, but you have to do it," I explained.

Matt said that he had been up late the night before.

"I was reading <u>Saboteurs</u> and I couldn't put it down..." he said.

"Did you get the same feeling I did when you hit about page thirty, the second chapter? When they started talking about all the chemicals and the symptoms? That's when I knew the same thing was happening here. I will never forget that, I went cold when it sunk in. I was reading in the canyon behind the house, on a big rock in the dry wash under the juniper trees. It was then that I knew that the industry knew exactly what damages it was doing, and all the rhetoric in the papers about nothing being dangerous was a huge lie, and a deliberate and calculated lie. It was then that I realized what the danger really was," I said.

People, companies, and industry were willing to knowingly kill people with exposure to deadly toxins to get their profits. And to lie about it all the way, and cover it up with money...

I am glad I spent the day at an animal event, doing concrete real and tangible good work toward some living thing. I could hold the dogs and cats in my arms, talk to the owners and explain what services they could receive for free, such as vaccinations, and expensive spay and neuters. Doing that kind of work, and spending the day with like-minded people just makes one feel good very deep inside.

I wish working on gas-related issues were that fulfilling and rewarding, but the cold hard truth of the matter is they are not. It is depressing, time-consuming work.

But, as I told Kemp Will, every time someone like Laura Amos gets a big settlement from a gas company for damages it is a huge victory. If it is only going to be one lawsuit at a time, each one still is a victory. And all the work that everyone is doing on behalf of this cause adds to the successful potential outcome of each of these cases.

Kemp agreed.

August 14, 2006

After the busy weekend with not much time to turn on the computer or read the paper, tonight I noticed a few gas-related articles in the Glenwood Springs Post Independent from Saturday. One was entitled

" Encana pulls the plug on well in roadless area"

and it described how a low-producing pad on the Mamm Creek section of the roadless area of the national forest wasn't considered worth the effort to continue working on.

"What we found when we were drilling [two wells close by], was we weren't getting as much production [as hoped for]," the industry representative said.

Encana declined to pursue the unavailable well sites, and then began the $50,000 reclamation process, reworking the land surface and reseeding the area. Heavy equipment moved in to rip up the cobble road that led to the well pad, and ditches and logs across the road area would be put in to prevent water run off and soil erosion.

This is a rare good story of the fate of a well pad, and mainly because it was being abandoned. Hopefully the reclamation will be more successful than the attempt that cost $40,000 at the site of the Divide Creek seep, where the majority of the plantings became dead and dry brittle stumps and stems in spite of the efforts made. The abandoned pad I see from my kitchen window resembles nothing more than a huge gash cut into the side of the mountain, as if with a giant cake serving utensil that lifted a chunk off the steep hill. The large flat gash remains, and the trees planted there many years ago all died on the impossibly steep-cut hillside. (It did however provide a convenient site for a rescue helicopter to land during a 911 emergency when the steep road was too muddy for the ambulance to negotiate after a construction worker fell of a roof of a new home being built, and sustained life-threatening injuries.) Any plants that grow on that old well pad and provide a green cover of fuzz from a distance are the most hardy and undesirable species of invasive weeds. They can grow just about anywhere.

The second article in the paper is entitled:

"Grand Junction voters may consider drilling regs."

A measure qualified for the November ballot by turning in a total of 2,635 certified registered-voter signatures out of the total collected number of 4,270 signatures. Only 1,580 were needed to qualify, so they met their mark by over a thousand. The measure will address the manner in which the city might regulate energy development in its own watershed. The petition was brought forward by the group The Concerned Citizens Alliance, which is a local chapter of Western Colorado Congress. After approximately 135,000 acres were leased in an auction by the Bureau of Land Management (13,000 acres of which were in the watersheds for Grand Junction and Palisade), the grass roots group began its crusade. Because the federal government owns the mineral rights to the land, the leasing is allowed despite any impacts that might occur.

The BLM rejected formal protests of the leases by the cities, but for a year they have suspended development to try and reach a compromise and address local concerns.

"The incredible outpouring of support this issue received from the business community, churches and from citizens of every background proves that protection of our water supply is truly a matter of common sense," said Janet Magoon, local schoolteacher.

(Quoted from 8-12-2006 Glenwood Springs Post Independent) p. A4

August 15, 2006

Today I left a message for Lance Astrella at the Denver office. It was a tentative message, an issue Chris and Steve Mobaldi and I have talked about before. If and when their case settles, how public can their information go? If at all? It just seems too vastly important to get Chris's story out and show what horrific things can occur when gas drilling goes awry. I told Lance I didn't need a call back, I just was thinking about the idea of how much would be able to be said about the specifics about what caused the debilitating illness that had all but ruined her life.

I was surprised when I got a call back less than an hour later from Lance. We discussed the importance of such stories not just going underground and fading away from the public. I too agree and hope that the Mobaldis receive a fair and just monetary settlement (but how can money ever make up for what happened to Chris? It never can; it can just ease the circumstances of what is left of her life).

The Mobaldis hope, however, that her story can go forward to prevent these things from happening to scores of other individuals living in the shadows of the gas wells, and in the affected water tables .

How many times Chris has told me that in her strange halting speech,

"Ah just don wan et toe ha-pen toe ehn-ee-wahn-ales…"

I know she means it, with every ounce of strength she has left. And she always expresses particular concern over the children who are exposed to the toxins.

"Theh shod nut due eht nae thahd cheelrun, nut nae tha cheeldrun…"

Which is Chris saying, in her strange accent,

"They should not do it near the children, not near the children."

August 16, 2006

What a beautiful August Colorado day it is today, clear blue sky with a few late afternoon thunderheads gathering and casting their cooling shadows over the green pastures and mountainsides. The temperatures crept up to just below too-hot-to-be-out: somewhere in the nineties. I am in a somber mood after spending a few hours in consultation with the veterinarian about the prognosis of one of my horses who suffers from severe arthritis, more commonly known as ringbone.

Aside from administering painkillers and anti-inflammatory drugs, there seems to be little hope for the future of this beautiful horse. In the kitchen of the farmhouse we examined the last series of x-rays, the ravages of the disease evident in the raggedy edges of the coffin joint just above the hoof. Then came the discussions of insanely expensive surgeries (such as the infamous Barbaro race horse underwent recently after breaking his leg on the race course), and somewhat less expensive but still risky alternative procedures; the grim and looming future were too much to take with a stiff lip.

"He's living on borrowed time," said the vet when I asked for the lowdown.

Not wanting to hear or accept the bad news, my hand hit the door involuntarily in dismay. In spite of my efforts to maintain composure, I turned away as tears rolled unstoppably down my cheeks. This was not what I wanted to hear, but I needed to know. How does one evaluate the level of pain and quality of life of a living creature, and determine the exact "right time" to make the final decision?

"You are an animal person, you will know," said the vet as he and his assistant backed out of the dirt road by the corral in the big white vet truck with the unmistakable equipment compartments in the back.

I could spot a vet truck from across the valley as it drove down the county road. The red truck was the emergency vet, the white trucks and the tan truck and the blue truck were the regular vets.

Steve Mobaldi called me just after the vet left; I was in the house searching for a misplaced tube of Surpass, the expensive topical medication for the horse's bad leg. I had spoken to Chris earlier that morning — she had sounded pretty good and I could understand her well enough.

"Chris said she's doing good today," I said as I continued my search of the house.

"She was, but she's sick now. She threw up and had diarrhea, and I thought she was going to pass out..." said Steve.

"Oh, that's not good news," I replied. So much for her rebound, I thought. I was disappointed to hear of the returning symptoms that since the last reports, I had been hoping were diminishing. They had said that since the visits to the specialist, things had been going better, and she was gaining more weight and vomiting less, and being able to stay awake and functional longer.

Steve said they were waiting for more toxicology reports, and they were busy with the new doctors. He had recently talked to the attorney also, and we discussed our latest happenings. (The parts we could talk about, that is, with out jeopardizing their case.) Steve seemed quietly surprised that I had just spoken with Lance over the past few days. In spite of the fact that we are close friends and committed on this issue, I spoke carefully as I described my call to Lance yesterday, and his reply to my questions.

Basically, I was hoping that they would be able to let the facts of this case go public. Of course a monetary settlement is what the Mobaldis need, and nothing should come first before that objective. But these stories—or better, realities — need to go public. For the sake of everyone who lives here and continues to be affected by this industry and its toxins.

Today, Lance said, "It needs to be told, before a big event occurs."

"What?" I asked, "When the whole town falls over dead in a gas event?"

Basically, that is what he was implying. Realistically however, it will continue as a slow and lethal progressive killing.

COLLATERAL DAMAGE

August 17, 2006

It is evening, and the last rays of the sun are hitting the bottoms of the handkerchief gray clouds that hug the mountaintops, as if placed there in the otherwise clear sky just to light up a soft pink before darkness descends. The cool of fall is in the air in the evenings now, yet hummingbirds still hover about the ruby-red back-lit liquid in the feeders, calling their distinctive shrill. The hum of crickets choruses through the air, and other night sounds join in. It is too pretty to go inside to write; the beauty of the scene outside compels me to stay.

Yesterday I spoke at length with a neighbor of the Mobaldis; an old neighbor who has since moved away. We talked at length, and I knew I was talking to a knowledgeable and frustrated and savvy individual. This person could not commit to revealing his identity in a book without consulting his family.

The information I heard was horrible, astounding and frustrating.

Another settlement, huge and glaring facts collected about the toxins found in the water after the blow-out of the Goad's well in the Mobaldi's neighborhood. Huge…

I discussed with this individual the frustrating circle of pushing people to comprehend the all too real and dangerous facts. It seems people do not want to think about or accept the idea that maybe something is going on that will endanger their health and perhaps one day kill them. That is not news anyone wants to hear.

After the Goad well blew out, one of the neighbor's wells tested positive for trichlorabenzene. With urging from the Savage Land Company, a five-mile radius of water wells was tested from the blow-out sight. A home well was tested that had reportedly had huge levels over the acceptable level of what is known as "slime-forming bacteria;" an industry by-product.

Other chemicals — a potpourri of them including mercury — and other heavy metals showed up in the water tests. The mercury was at over a two hundred times the allowable level. That is as high a contamination level as the water can be tested for, so it could be even higher.

The nameless individual with whom I spoke said,

"You have to get tested, you have to get the scientific data. And you have to leave; that's what the doctors say. You have to leave.... Take a plane from Las Vegas to Grand Junction and look down; look down at the brown cloud. Guess what's in there.... There are sick people there where I lived; I can see they are very sick."

It is dark now, the crickets are in a deep rhythm of a chorus, the mosquitoes are out biting hard, and the last bit of light to make out the letters on the keyboard is fading under the darkening sky.

The Beginning Of *Split Estate*

August 19, 2006 The Picnic—New Friends and Allies

Today was the Grand Valley Citizens Alliance annual picnic; held once again at the beautiful ranch home of Orlyn and Carol Bell in the Dry Hollow area of Silt. A band played, shade canopies and umbrellas were set up, and the brunch was laid out for the numerous guests who arrived to enjoy a day of mingling, munching and taking in the incredible mountain views aside from the countless rigs and gas pad infrastructures that dotted the rolling pastoral landscape in every direction. Long industry gray tube trucks crawled up and the down the road every five or ten minutes all day long, as did the signature white industry pickups with their tanks and pumps in the truck beds.

A documentary film editor Debra Anderson from New Mexico was there, filming intermittently and jotting notes discreetly on her pad of paper. By the end of the day she had lined up several more interviews with impacted landowners with serious health issues believed to have been caused by the gas industry; interviews for the next several days that would keep her very busy.

Rick Roles of Hunter Mesa showed up, and he was a newcomer to the GVCA group but not to gas issues. He had lived surrounded by wells since they began drilling in this area of Colorado—for a good ten years—and he has the stories to prove what horrors can and do occur.

Debra Anderson is meeting with him tomorrow to film at his ranch. As Rick described what has happened to him, and as Debra heard more stories of other impacted land owners (who have had a plethora of problems from contaminated water wells to exposure to airborne toxins, being knocked unconscious from clouds of gas then developing acute and unmanageable sensitivities to toxins), her jaw literally fell open and her eyes took on a look of horrified disbelief. That look on people's faces is now recognizable to me; people cannot comprehend that things like this are happening and nothing is being done. Nothing.

Today I learned of numerous more people who have huge issues ranging from tumors to flagrant exposure to toxins to goatherds with 50% stillbirth rates and other unbelievable problems. Several of Rick's goats have large growths on their necks; one of which burst open recently and exuded a black gooey substance.

I developed a searing headache up at the Bell ranch by the afternoon. Thankfully, it has subsided now that I am at home some number of miles away. It was good to see friends, but I have a very full mind from what I learned today. More sick people; seriously sick. People whose body parts are puffing up to balloon sizes, people with tumors, animals herds with strange deaths and symptoms, and unregulated industry stories.

The people are now falling into categories: individuals who have gathered evidence from toxicologists, contacted lawyers, and settled out of court and signed gag orders so their stories can no longer be discussed.

Then, there are the people who have gathered evidence, contacted lawyers, and have cases pending, such as the Mobaldis who have now filed claims against three gas companies from what I have been told.

Next, the people who know their health is being jeopardized and are complaining, but not to the right places. The gas companies, the COGCC and the Health Department are not handling these complaints effectively. That is putting it lightly...

Lastly is the huge population of unaware and uninterested residents of western Colorado who do not think much or care about these issues. They don't have to; it is not in their backyard. They blissfully assume that everything is fine, and the powers that be are looking out for public safety.

TARA MEIXSELL

Wake up and smell the gas, friends...

August 22, 2006

After the GVCA annual picnic, several of us stayed behind on the Bell's deck to discuss issues. It made for a long day, but a good day. Spending time with the friends at a lovely Colorado ranch home for a more enjoyable span of time than at the usual monthly meeting, in a gorgeous surrounding (aside from the wells and pads looming in the surrounding landscape), and feasting on a five star buffet (mostly hand-prepared) under canopied tables was a great way to spend an August Saturday. Carol, Oni and I washed dishes and dealt with the dirty platters and leftovers while Duke and Steve serenaded us with live guitar and vocals from the living room.

I crashed more silverware into the sink of hot water and dish liquid, and Oni asked,

"Tara, do you want gloves?"

Oni held up a pair of white translucent rubber gloves, the type dentists and doctors wear.

I was momentarily confused. I thought, why would I want gloves to do dishes? Then I realized; the water. The Amoses' well that blew was just up the hill, the fracking chemicals could be in everything—and everyone knew not to drink the water up here. Ever.

"I always wear gloves," said Oni.

I have realized on more than one occasion that as time goes by, my trust and friendship for this group of people has reached a different level than I have ever found with many others. We may not always agree on nit-picky details, we are not from the same walks of life or same lifestyles, but the commitment, character, and dependability of these folks is impeccable. I hold them in the highest regard as human beings go, and it is an honor to work with and spend time with them.

COLLATERAL DAMAGE

Saturday Sept 2, 2006

Now that the raft trip is over, I am getting back to the regular activities. Yesterday I hauled hay all day in the small pickup, five loads of twenty-five bales. After too much time in the sun I retreated to the shady house in the late afternoon for a break from stacking hay by the barn. I called the Mobaldis and talked to Steve briefly. He knew about the fires west of Rifle, and they were near his place but he wasn't up on the details. I told him I had spoken to Lance a few times for business other than their case, and I think he was surprised. I told him that more documentary reporters had been up here, and let him know that some new bad things were going on. He said Chris wanted to talk to me, and handed the phone over to his wife.

"Ta-rah, hoe ah yuh?" she asked in her strange voice.

She sounded perkier than usual, and her speech wasn't as slow as it usually was. We talked for much longer than we usually did, and Chris asked me a lot of specific questions about people we used to work with. I could tell from her responses that she comprehended things well, and that she really remembered the individuals we were discussing.

"Chris, you sound really good!" I said.

"Ah doo, wall, tank yuh bery mooch!" she said proudly, as if I were complimenting her on a stunning hairdo or some upgrade to her appearance.

We talked about the activities here, as much as I could freely speak about. I told her that I had met some new people, and that they had bad things going on health wise. I told her about Rick's strange swellings that traveled over his body randomly at times.

"He has health issues that are as bad as yours almost, but not as bad…" I trailed off, thinking of the brain tumors and Chris's horrible condition. Rick had not yet approached that level.

"You know, it is like a balancing act. With the attention that this is getting from the reporters, the documentary crews, and the calls with

Lance I feel like something is going to happen; like something on the next level can happen. Then I get too excited, and I feel like I have to prepare for failure, and for nothing to happen," I said.

"Noh, dant es-spect ah-nee-thung!" said Chris with caution in her voice.

"You know Chris, it gets depressing, but then I think back to a year ago when we never could have imagined that you would be represented by the firm that worked on the Erin Brockovich case, or that documentary film crews would be all over this story. When I am in contact with those kinds of people I find it hard not to expect some results. I just want them to connect the dots, all these cases of people getting sick and settling, and then no one knows and it is like it never happened. I keep finding about more people all the time.

"I called up Lance's office one night, and left a message for him. He called me right back the next day and I told him that I really hoped that your story could go public, and that people needed to know what happened to you. Otherwise, it will just keep on happening. I feel funny telling you this..." I trailed off.

"Tah-da, yuh ah lak theh reel Ah-rin Brock-ah-veech!" she crowed.

"They say that about a lot of people, there are a lot of Erin Brockoviches running around here," I said.

Before she hung up, Chris told me that the Montel Show called her that day.

"What? The Montell Show? How can they call you?" I asked, confused. I knew she had contacted the Oprah show (or family did previously on her behalf), but Montell?

"Theh hov eh fon line, ahnd theh called meh ahn theh fon," responded Chris, with what I took for dry wit. (They had a phone line, and they called her on the phone.)

COLLATERAL DAMAGE

Called Rick briefly last night and planned to meet at a Rifle Café at 9 a.m. I was sick of talking on the phone, and not always convinced that it was the best idea either.

I met Rick inside the packed Base Camp Café on Third Street. Everyone and his brother, and sister and father and mother, for that matter, seemed to be there. Rick was sitting at a large round table that was set for eight easily, and I spotted him at the back of the restaurant and joined him.

We caught up, and I ordered hash browns and gravy while Rick just drank cup after cup of coffee. We talked about the recent goings-on, and I relayed the story of the bear on the raft trip that had swum across the Green River to join a picnic on the opposite shore. The rafters had leaped to safety on their boats, but the bear managed to acquire and consume a large powder packet of Gatorade mix. Then shortly after that, he ripped into a backpack and ate a bottle of hand sanitizer.

Rick threw his head back and laughed.

"Maybe he wanted to get the taste of the Gatorade out of his mouth!" he said.

"Well, after that, the ranger posted a No Camping Allowed sign at the campsite, and mounted a camera also. The bear came back and tore down the sign, and ripped the camera out of the tree and attempted to eat it," I relayed.

We laughed about the clever bear, foiling attempts to keep his next picnics from being delivered. (Actually, however, I am frightened of bears on the river and one night I woke the entire camp in the dead of night with an ear-splitting wolf whistle when I heard crashing noises in the brush outside our tent. To this day I continue to be teased about it on our raft trips.)

"When we took off the river at Dinosaur, Ranger Doug was there and I asked to see the camera. It was a big black box, and it was powered by a nine-volt battery. It had bear teeth marks in it. He said you could get them at Wal-Mart, or Cabellas. They last for about a week, I think," I said.

"Last night one of the wells went off, and then I lost feeling in my foot. I can't even bend it, it is like it's broken," said Rick.

"What's it like when the well goes off?" I asked curiously.

"You know, it just sounds like a jet is taking off in your back yard," said Rick matter-of-factly, as if everyone knew.

We talked about some of the others who are sick, and some of the ones who have lawyers.

"Are you going to go to a doctor and get the tests? Are you going to go to Dr. Gerdes?" I asked.

"Well, I've been tested, but I don't have the money really right now. I just try to stay away from there as much as I can, and come to town where the air is better," said Rick.

"I haven't worked in two years 'cause of it, and I don't borrow money," he continued. As he looked at me from his serious brown eyes underneath his dirt-encrusted cowboy hat, I was sure this was true.

"I've talked to a lot of people, and the lawyers, but the only thing people seem to want from me is information," he said.

I looked at him and responded, "Well, I don't know what I'm doing, but I can't stop doing it, either."

When we moved to the counter to pay our bill, I saw how Rick dragged his leg along, as if he had no muscle strength in it at all.

We walked back to Rick's truck parked in the shade and I played with his pretty black dog Rowdy. I know I will see Rick soon again.

After a trip to the dreaded Super Wal-Mart for cat food, litter, and skin lotion to sooth the burns I got on the raft trip, I drove my truck out the lower Rulison Road toward the Mobaldi's house, which now sat vacant. I talked to an ambulance driver and a sheriff's deputy at the hospital across from Wal-Mart to get an update on open access to the fire area, but neither knew. So I just went.

A few county fellows were near the Road Closed sign half a mile from Mobaldi's, and they said to just go on in when I told them my friends were out of town and their house was up there by the fire area.

As I turned off to the left into the Mobaldi's driveway, I saw the small tent city that was set up in the next neighbor's pasture, with a fleet of support vehicles. It was the fire fighter's camp.

On the large, scorched hillside above, a few stray plumes of smoke billowed, but no live flames showed orange. The area that had burned was impressive, and had spread and climbed a substantial portion of the steep mountainside there.

I got out of the truck and walked through the yard around the house. Instantly I noticed the peeling paint, curling off the sun-exposed sides of the Mobaldi's two-story home. It looked like fraying, burnt skin peeling off a badly sunburned arm; the paint fluttered and blew in the breeze. I walked around the house and noted that some sides of the house were mostly peeled and had a gray color of the metal predominantly. The areas that were shaded were still light brown, and the paint still held.

The last time I had been here was to visit Chris during her post surgery recuperation, after her head was sawed open to remove her second pituitary tumor that lay between her eyes. We had sat in the shade of the porch, and visited.

The barns and the corrals where the lamas had died from respiratory illness were now vacant, and piles of horse manure from the recently vacated renter dotted the yard and garage areas.

I turned on the front spigot and watched the water come from the spout by the front door, wondering if I could even see anything visibly that was a sign of the contamination that had so sickened Chris. The water flowed easily, and looked perfectly normal to the eye.

I straightened the doormat of the abandoned house, and looked at all the improvements that had been done to the property over the years. All the tidy outbuildings, animal pens and fenced pastures, and the lovely home which now sat vacant. Many, many hours of work and sweat had been spent to get this place the way it was, yet now the peeling siding was a visual testament to the chemical dangers that hovered nearby. After doing research, Steve Mobaldi believed that chemical fumes from the gas wells, likely including hydrogen sulfide (which has been know to be released from natural gas wells) combined with heat from the sun and condensation on the metal siding and caused those sun-facing portions of the home to peel. Or, perhaps, it was a cocktail of assorted chemicals that resembled the effects of H2S—hydrogen sulfide—that is listed right there on the warning placards on the tanks at the well site…

When I think of that place, I recall the abandoned house in Canada that Andrew Nikiforuk described in Saboteurs, where the doors swung in the wind after the occupants left due to health concerns (or even death) from the gas wells nearby. The energy company bought the house.

I drove home, taking the frontage road along the river south of Rifle and Silt. Industry trucks passed, raising their dust. Going by an area of reported illegal waste dumping, I stopped for a minute or two. It looked a quiet enough spot, but what, I wondered, really did happen there? Especially after dark…

Sept 4, 2006 Labor Day

Today I was at an impasse, I wanted to take advantage of the holiday time and get some writing in, but not sure what direction to go. I pulled out some of my notes and stared at the jumble of names and phone numbers written on the back of the crude map I drew of Red Apple subdivision where the Mobaldis and their other neighbors had lived not too long ago. Next to the names were notations of symptoms: "dizziness, bloating, bleeding, headaches," and the like.

COLLATERAL DAMAGE

One of the names caught my eye — it was a neighbor of Rick Roles. It merely said "tumor" next to his name, and the phone number was written down also. I must have looked it up it the book earlier.

I called a few other people and left messages, then I again stared at the name Neil _____ with the word "tumor" written beside it. I picked up the phone and dialed the number, and a pleasant-sounding woman answered the phone. After explaining that I was a friend of Rick's and that he had mentioned her husband's situation to me, she was very willing to talk.

"I'm getting information of health effects of the gas industry. I'm working on a project. I'm not a scientist or a reporter, I just have a good friend who's really sick from the gas wells; she's had two pituitary brain tumors and she's not doing well. So I'm just trying to get all the information I can about possible health effects from people who live near wells. This isn't my job or anything; I guess it's kind of a hobby. I'm writing something," I said.

I gave her some background information about the two people I knew who had tumors they believed to be related to the gas industry. Then I told her about some of the others who had different health effects they believed to be from exposure to toxins in the air. I described Susan Haire's experience of being hit by fumes while outside doing ranch chores, and barely holding herself up on the truck door.

"Her doctor told her that if she had hit the ground she most likely would have died, because the gas sinks. The hydrogen sulfide—or likely a chemical cocktail that gives the same effect," I said.

"That's HS2S, right?" she asked quickly.

"Yes it is," I responded.

"I called and asked if there was any HS2S and they said no, there definitely isn't!" she said angrily.

"Well, from what I have been told and what some of the lawyers say, there definitely is hydrogen sulfide. The safety placards on the well infrastructure list it right there," I responded.

"That can kill you! Just a small amount!" she said.

"Yes, it can," I agreed.

Because of the Mobaldi's pending lawsuit I didn't elaborate further with the information about the paint peeling of the Mobaldi's aluminum siding on their house. I can picture it in my mind so clearly now, the paint fraying like feathers in the breeze in the areas where the sun hit the house. I recalled the phone call I had with Steve late this winter when he was excitedly telling me about what he learned on the Internet: how with a combination of moisture, heat, and hydrogen sulfide the paint would separate from the substance the siding was made of.

"Well, Neil's tumor was a very rare kind of a tumor, a acinic cell tumor. It was on his salivary gland. It was very rare for it to be cancerous, also. We asked them to test it, and they didn't really know what to do, and we were kind of caught up in everything, so we didn't follow through. We were too worried about his health, and everything. They sent it to the Mayo Clinic, but I don't really know much more..." she said.

"Well, he was really sick, and you were dealing with the surgery and everything," I said.

One could completely imagine how overwhelming it would be just to deal with a frightening health situation and surgery and looming prognosis. Typically you don't have to simultaneously gather evidence that your tumor may have been caused by an industry that surrounds your home with countless gas wells and pits emitting chemicals into the air that fill your house each night.

"I admit, I was pretty naïve about the way I dealt with this from the beginning," said Susan, as if she were blaming herself for not being more clever initially.

"I called the EPA and the State Health Department and they all acted like nothing was dangerous, that everything was fine," she said.

"Of course they did. That's what infuriates me so much, is that it's up to the person who's sick from this stuff to have to prove it. The

COLLATERAL DAMAGE

industry doesn't have to prove that it's safe to use these chemicals and do what they do in people's backyards; you get to try and prove it when it starts to kill you. That's why I am working on this, it infuriates me," I said.

I told Susan about our trip down to Denver this winter, and about meeting the new people whose symptoms and stories matched up with others so chillingly well.

"When I met Susan Haire and she started describing what was going on with her, the hairs stood up on the back of my neck. Chris Mobaldi was lying reclined and moaning in the very back of the SUV and I leaned over the backrest of the front seat, catching every word that Susan spoke about her illnesses. It was Saboteur symptoms all the way; it was the very same things that the people from Canada had up at the sour gas fields. I don't care if you call it sour gas here or not, that doesn't matter. When people have the same symptoms, and their animals die strange deaths too, that's enough evidence that the same kind of thing is wrong here," I said.

I told her about Dee Hoffmeister coming home one evening and walking into a visible cloud of fumes on her front deck, and dropping unconscious on the spot. And how her husband had carried her into the bedroom, and family from Glenwood came to take her away from the house and the fumes.

"She couldn't return for eight months, it made her too sick to come back. She had to stay in Glenwood. And she had the regular things, the headaches, the nausea, the exhaustion and forgetfulness. Pretty much the inability to do anything. That's the same way her neighbor Karen Trulove is, she didn't get knocked down but she got the horrible headaches when the fumes got bad, and she couldn't function normally anymore. She's probably in her early fifties or late forties, and really vibrant, but she had to quit working and can't do anything much. She's sick all the time. They're moving too, and although she went public and talked to the press and went to the Capitol, she doesn't want to go public anymore. They are trying to sell their place, and they've already been in the papers a few times," I said.

"Why isn't someone documenting this, and getting this information out? Why don't people know about this?" Susan asked in an angry voice.

"Well, that's what I'm trying to do. I'm trying to keep track of health impact stories, and to keep names and numbers, and to pass it on to the press. I'm working on a book, I don't know if it will get published, but I'm working on it. I can't stop. I just keep getting more and more people's names — that's why I called you. We had three documentary film crews here this summer, and the last one who came spent almost all day Sunday with your neighbor Rick. She interviewed a few other people, I gave her all the names I could, but some people don't want to talk anymore. They just want to sell and get out. Or they're sick of it, of fighting it and they just want to live what's left of their lives and give it up."

"We're moving, we already sold. We're moving north of the river where there is less drilling, at least now, anyways. I have some friends who live on Owen's Drive, and we talk about this stuff all the time. Some of them are sick, and I'll talk to them tomorrow and get back to you," said Susan. She had to leave and take her son to a football practice, and we said goodbye.

I called Theo Colborn, the scientist from Paonia Colorado, and asked her about that type of tumor. She said she would look into it, but was just then getting picked up by a friend for an outing, so we cut it short.

I am hoping that this new lead will turn into something more. As I said to Susan Wagstrom, I am putting my trust into the lawyers and the press. I feel that something is so very inherently wrong with this situation here: the unregulated chemicals, the entire hydraulic fracturing process, and the lack of regulation, I can't stop typing away on this computer every chance I get.

When I returned from the five-day raft trip last week, I didn't want to be here. Of course, I wanted to return home and see my animals and friends, but I didn't want to return to Garfield County and all its industry-related mess. Again, I felt jealous of friends who drove off to other nice places and homes not impacted by the gas industry. At least, not directly impacted, that is. The statewide extent of the air and water impacts remains to be seen. When we started our trip a week ago and drove west of Silt and Rifle along Interstate 70, our friends noticed the rigs and industrial equipment that loomed around the pad sites. They were shocked by what they saw, and listened intently to the answers to their questions. They empathized with us having to live in the midst of all this, but I knew that they had little-to-no true comprehension of the scope of what was going on and what was to come. No one ever can until they travel up into the ground zero areas where the entire horizon is filled from end to end with gas industry infrastructure. Why is everyone here so blind to what is going on? Why?

Mishaps And Problems

The Arbany And Schwartz Well Disasters

Saturday September 9, 2005

Today was a much-needed day to sleep in after a tiring workweek. I slept until an unheard-of 10:30, then got up and fed the animals. The weekend days are the only days I have to commit to getting any work done on the book, so I knew I needed to find a direction for the day's pursuits, but I didn't know what it would be. There were so many directions to go with this, so many known people and names and stories, yet one had to present itself at the right time to be pursued.

I ended up going to Nancy Jacobsen's up Dry Hollow to pick up posters for next weeks adopt-athon we were going to volunteer at. Assorted cats, kittens and puppies and dogs were available for adoption and the event would be held at the Divide Creek Animal Hospital. Nancy had the color posters all printed up, and as I had never been up to her ranch she invited me to come on up. I could meet her husband and dogs and cats, and get the posters and see the ranch.

The drive up Dry Hollow was now very familiar to me, and in fact I had been up Nancy's rocky and pitted country road years before to buy hay from ranchers who since have moved to Oklahoma. I steered the pickup carefully over the sharp rocks, and went at a snails pace once I left the paved road. A flat tire was not in my itinerary for the day, and this was the perfect road for a flat.

The Jacobsen's had a lovely home nestled in a horseshoe shaped rocky landscape, large junipers dotted the desert-like hillside, their trunks fat and gnarled. The entrance to their home had a lovely enclosed patio with cement flooring resembling dark natural rock, and a tasteful planting of purple fountain grass waving their plumes high above pots of colorful petunias and trailing ivy.

"This patio is beautiful! I love the tile!" I remarked.

Nancy described to me how it had been poured, one continuous cement pour that had then been scored, textured to look like rock, and then painted a blue tinged back.

"Our first patio got cracked by the explosion at the gas well — that's when our pond got breeched and our garage floor was wrecked, too. Encana sent some people out, but the inspector who came twice said they had nothing to do with it in the end. That was the same explosion that knocked Jim Ubank's house off its foundation. Encana settled with him, but they wouldn't pay for our damage," said Nancy.

I ended up visiting with Nancy and her husband Gary for a long afternoon, and inevitably the conversation strayed toward gas and oil. I had told Nancy over the phone that if she could give me more information about things we had talked about earlier I would be appreciative. Nancy had extensive files of newspaper articles regarding the problems up in the Dry Hollow and the Divide Creek seep moratorium area also. And much more than that, too…

We referenced a map of the area, and talked about the problems that had occurred up by the Amoses' place, the Arbany Well, and the Schneider Well which had been the source of the Divide Creek seep when due to a failed cementing job of the well, 115.5 million cubic feet of natural gas seeped into Divide Creek. Also, we discussed the Red Apple area, where the Mobaldis and their neighbor Ted had lived. (Ted hired a lawyer to file suit for health complaints that he, his family, and his animals suffered, and he and his immediate family have since moved out-of-state after settling with the gas company.) While Ted, as many others, now cannot make public comment on his case, it is known that he suffered the potpourri of symptoms typical near "bad wells:" headaches, dizziness, lightheadedness, nosebleeds, respiratory problems, and more….His animals suffered from symptoms similar to wasting disease, and their dog died. Ted went to a specialist in Denver to get tested for exposure to toxins, and his results were astounding. He said that he went to each of his neighbors with the test results, and they did not seem concerned.

What all these places have in common are water wells that have been blown out due to faulty gas drilling, and very sick people who live in the vicinity.

Nancy, Gary, and I enjoyed a rambling conversation on a variety of issues before I left about six-thirty. Somehow I managed to spend half a day up there.

As I drove home north of the Colorado River, down the valley away from the density of well pads and rigs, my mind was full of information. Hopefully I had jotted enough facts down in my spiral notebook for later.

It is too late now tonight to write much, but the one haunting image that will not leave my mind was the comment that Nancy made about the something that her neighbor had recently done.

At 2:30 a.m. he piled up about twenty dead goats by the well pad, about two weeks ago, and lit them on fire. Apparently Encana called the fire department to report it.

"I said to one of the workers, 'Don't you think he was sending you a message?'" said Nancy.

I cannot forget that image; a pile of twenty dead goats very likely killed from industry exposure, piled up and lit on fire beside the Arbany Well pad.

THE ARBANY WELL

On March 9, 2001 an open-air natural gas explosion occurred at the Arbany Well pad when workers for a gas well drilling subcontractor were drilling one of four boreholes on the Encana Gas and Oil-owned well south of Silt, Colorado.

COLLATERAL DAMAGE

The explosion was at the borehole known as Majic 10-2, according to a complaint that was filed on the incident with the Colorado Oil and Gas Conservation Commission. The gas well sits on the property of Rick Arbany.

According to the Mamm Creek Field Drilling Superintendent Richard Eberspecher, the drilling crew employed by Patriot Drilling from Casper Wyoming was using a technique called "managed pressure drilling." Eberspecher describes the technique that was being used on the Majic borehole as a faster, less conventional technique than typically used. In order to control the pressure of the gas during the drilling process, usually a large quantity of a mixture known as "drilling mud" is put down the drill hole. The pressure of the mud basically holds the gas down.

In managed pressure drilling, a choke device is in place at the top of the well to regulate the amount of gas being released from the drill hole. A gauge on the choke indicates the pressure of the gas in the drill hole, and the pressure is regulated automatically. Eberspecher reported that when the pressure gauge indicated an unusually abnormal pressure, the crew was unfamiliar with this choke system and managed pressure drilling, "so they shut her in and called Swaco." Swaco is the company that rents the managed-pressure choke systems. Half-an-hour later the Swaco company employees arrived and found the pressure at the Majic bore hole to be at 300 lbs. This is 200 lbs. over the usual pressure, so the employees shut down the well and released the gas. There were questions as to whether or not the choke system was functioning correctly.

The greatly pressurized gas shot out of the well and was then ignited in the flare pit. Per regulations gas that escapes at a wellhead is burned off in a pit surrounded by an earth berm. The employees continued to monitor the pressure in the borehole, and released the gas twice.

The nearest neighboring home belonged to the Dietrich family. Their house was less than 100 yards from the well pad. Stephanie Dietrich, who was inside her home, felt first one, then ten minutes later, a second rumbling under her feet. She believed that she was experiencing earthquake tremors.

Shortly thereafter she realized that the incident had something to do with the activity at the well. She walked up a rocky hillside with a neighbor to watch the ignited gas exploding into huge flames.

"The flares were just unbelievable," she said. "It looked like gas was just exploding into the air."

Neighbors and people driving by on the road were stopping to watch the 100-to-200-foot flames leaping from the flare pit. Even those used to the typical flares recognized that something was definitely not right, if not plain totally out of control.

Lisa Bracken felt the concussions from her home that is over a mile from the Arbany Well. After telephoning her neighbor, Bracken drove over with her video camera to the Dietrich's house and filmed the huge fireball that burned with roiling black smoke for over an hour.

The consensus of some of the neighbors who live in the area is that the gas industry is neglecting safety and cutting corners by utilizing the fastest and cheapest methods to get as many wells producing as quickly as possible. Concern over the lack of regulation and lack of enforcement grows with each incident that occurs to the landowners who live on the properties near the well pads. The fact that a crew inexperienced with the managed pressure drilling was employed to drill the bore hole at Majic 10-2 that resulted in two explosions and an hour-long out- of-control flaring infuriated the surrounding residents. Because Encana wanted to get to the gas more quickly than they could by using the traditional method, safety was likely compromised. And it's not the first time this has happened, nor will it likely be the last.

The deputy director of the COGCC, Brian Macke, reported that the Arbany Well explosion incidents didn't involve any violation of regulations or the law. He expressed the opinion that although the event was noisy, it most likely didn't have any effect underground. He also said that probably didn't have anything to do with the gas seep discovered in the nearby West Divide Creek on April 1st, 2004.

Not all are willing to agree with Macke's conclusions, and some speculate that the explosions did result in tremors that could have contributed to the underground seepage of 115.5-million cubic feet of natural gas from the nearby Schwartz's well pad into West Divide Creek. This would be difficult to determine, as the faulty cement job on the Schwartz's well was an obvious factor in the seep.

One thing that neighbors are convinced of is that the explosions at the Arbany Well pad were responsible for serious damages to their homes and properties. The Jacobsen's home suffered cracked cement flooring on their patio, cracks in the cement garage floor and also in their shop, and they later discovered a breeched irrigation pond that failed to hold water when filled. They lived quite close to the Arbany Well.

The Ubank's home was knocked off its foundation after the blasts, and Encana did pay for the repairs to the home. The Jacobsen's and the Ubank's are both about the same distance from the Arbany Well. Encana has never yet agreed to compensate the Jacobsens for their alleged damages, nor do they admit that the explosions at the nearby well were the cause. The Jacobsens continue to pursue payment for their damages, and are currently in negotiations with a lawyer. This is the second lawyer they have hired.

"The first lawyer did nothing for us; we spent a lot of money and we have nothing to show for it," said Nancy Jacobsen angrily.

The Jacobsens are not alone in their anger and frustration. Many of their neighbors and others who live near wells in Colorado are furious at what is being done to their lives, their properties, and their health. Some have lawyers, some have settled already (always out of court so the story and facts are never publicly brought forward), and yet many others lack the funds and the wherewithal to mount a fight against the almighty gas companies.

Sunday September 10, 2006

I had just turned the computer off after several hours of writing when the phone rang. It was Debra Anderson, the documentary film producer from New Mexico who had come up to Garfield County recently and filmed a number of people for a project. She had gotten in some good footage of people with health issues from the gas wells near their homes, and interviewed numerous other individuals also.

We talked about the latest events, and I told her about the Arbany Well explosions and the Divide Creek seep stories from some time ago that I had been working on. They were well documented in the press, and Nancy had given some copies of the articles to me yesterday, in addition to lists of impacted landowners with water well problems ranging from contaminated wells to dry wells.

Also, I told her about the compilation of landowner stories from the Colorado. In particular I read her details from the story of an instance on December 7th, 2004 when Devon Energy released hydrogen sulfide gas to the atmosphere on the Bedrock Unit 16-7 well while flowing back a frac. The hydrogen sulfide exposure on that occasion led to severe symptoms in several individuals, and a Notice of Alleged Violation was filed by the COGCC to Devon for the incident—specifically for failure to inform the agency of the incident, for failure to report the positive gas analysis for H2S, and for failure to provide an H2S safety plan. H2S was detected in one completion zone at a concentration of 30,000 ppm and in a second zone at a concentration of 80,000 ppm. Such concentrations will kill a person instantly.

(Details obtained from "Land Owner Stories"—Colorado Network for Land Owner Protection)

I have no doubt in my mind that hydrogen sulfide is here too—or a cocktail of industrial chemicals that induce the same symptoms; for the symptoms are the same. The dizziness, the nosebleeds, the massive and instantaneous headaches the landowners experience after breathing the fumes. That's what all these people here have had. Susan Haire, Chris Mobaldi, Laura Amos, Dee Hoffmeister, Rick Roles, and more. I'm not saying it's detectable everywhere in the air, but after events where things at a well go wrong, these people are suffering symptoms from it. Then they have the long-term toxic sensitivity to deal with, after such high level exposure. I'm realizing that the people who are coming up with the worst problems have been in proximity to a combination of that some of the bigger events, like blown-out water wells and seeps. It happened in three different areas here, at the water well at the Amoses, the Goad's water well, then the explosion at the Arbany Well and the failed cement job at the Schwartz well that led to the Divide Creek seep.

"When things are going wrong with the drilling or the fracking, then the property damage and the health issues are happening, too. It makes perfect sense," I said.

"So is it in the water or the air?" Debra asked

"It's both— it's a combination deal. The water can be contaminated if the gas or the fracking chemicals leak underground, and the air can be contaminated from the emissions, from the flaring, or from the evaporation process from the pits," I answered.

When I told Debra about Jim Ubank's house being knocked off its foundation after the Arbany Well explosions, and the Jacobson's house suffering cracks to the patio, garage, and shop, plus the ruptured irrigation pond, I could tell by her voice that she was shocked.

"Now I have something really awful to tell you: just two weeks ago Rick Arbany stacked up about twenty dead goats and lit them on fire next to the well pad at 2:30 a.m. Encana called the fire department. Nancy said she was talking to one of the workers and she said, 'Don't you think he's trying to tell you something?'" I said.

"Oh, my god... I wish I could have seen that!" said Debra.

"No one could see that, it was two in the morning," I responded.

But I knew what she meant. It was a horrific image, yet it symbolized perfectly the complete picture of what was going on here. People piling up the carcasses of their livestock they believed to have been killed by exposure to the gas extraction industry, then in fury burning the corpses by the pad for the workers to see. On that note, I am done for the day writing about all of this sad business.

TARA MEIXSELL

Sept 11, 2006

Today was a busy day, it was a typical Monday at the office and I did a fair amount of gas work in the evening. I managed to fax Debra twenty-four pages of documents containing landowner health and other impact stories, plus the list of about twenty trashed water wells near gas wells. Joe Brown e-mailed me a few times, and I was saddened and stunned by his news (after not hearing from him for a while because he was in Europe for some weeks recently), that he was re-thinking his whole time frame with the documentary, and exploring other avenues, which included the creation of a nonprofit entity with the help of others to help fund the film project.

I totally understood, however, that Joe was pursuing graduate school and had to deal with the realities of the time frame at hand.

Later I called Chris Mobaldi to let her know about the conversations with Debra, the other film editor who was also now quite interested in the issue after reading the *OnEarth* article featuring Laura Amos's situation and the whole scene going on down here in Colorado with gas development, including the controversy about the dangers of fracking chemicals. I told Chris that if she reached the point with her legal issues (now pending) where she could talk to the media in any venue, Debra was definitely anxious to interview her as was Joe Brown. Then I asked her about the call from the Montell show.

September 14, 2006

Several nights ago I got a call from Susan Haire from Texas. We will follow up soon by phone; we talked for a long time that night and I did jot down a few notes. She has no doubts, nor do some of her doctors, that she was exposed to hydrogen sulfide. In addition to having a lawyer now working on her case, she is also seeing toxicologist specialists.

I read her the story of the man from western Montrose County who documented his symptoms and exposures to fumes that turned out to be hydrogen sulfide. When I read it the first time, I knew immediately that it matched with hers, and many others' symptoms who lived near the wells.

COLLATERAL DAMAGE

Susan told me about a nurse named Deb Meader who knew some people who had recently died from cancer, which they and their doctor believed to be connected with their exposure to the gas wells. Their names were Bill and Lois Allen and their physician from Grand River Medical Hospital in Rifle had apparently written a letter to the Williams Gas Company stating that their exposure to the chemicals from the wells had caused their illnesses. Their home had been west of the Rulison exit of I-70; it was a ranch property. The Allens had had a vegetable stand where they sold produce, but they stopped selling it because they believed that the vegetables contained the toxins that permeated the area. In good conscience, they could not sell the produce to others any longer. Lois developed herpes in her eyes, and both Lois and Bill died last summer two weeks apart. In addition to the cancer that finally claimed both of their lives, asthma was also a serious issue.

Susan said that apparently they had children who were rightfully upset with the situation, and they had been bringing this information forward to the press. They also had pictures and film footage. She indicated that they would still be eager to talk about it, and that the nurse might be of help in locating them.

"I am so glad you called me," I said. "I needed a pick-up about now. One of the documentary film producers just told me today that he had to put the project on hold. After all the days I spent with him working it was a huge let down to think that all that might have been for nothing. Not to mention the summer weekend days spent driving around setting up and doing interviews from dawn till dusk. I swear I got heat exhaustion some of those days; in fact I know I did at least once. But it all seemed worth it, because I figured it was going into a film that would get distributed and seen. It kind of takes the steam out of your sails to think that it was a no-go, at least for now. I know that it is all worthwhile, nonetheless, but I have to say it was disappointing. So, I am very glad you called!"

We agreed to talk again later, and I said goodbye to Susan Haire.

The next night I looked up numbers in the phone book for Allens from the Rulison area, but the number that I dialed merely had a generic recording:

"The number you have reached has been disconnected."

That was disappointing; now it might prove hard to find and speak with the surviving children of Bill and Lois Allen. The first thing that crossed my mind was that perhaps Williams Gas Company had settled with them, too, for the death of their parents, and they no longer could speak about the issue. I hoped that was not the case, but certainly won't be surprised to find out that, yes indeed, that is what happened. Just another potential name to add to the list of people who can no longer talk....Stories that never happened...illnesses that never occurred, industry problems that don't exist...

Just in this small neck of the woods in western Garfield County a growing list of people who have settled exists. And perhaps many more unknown others, and more to come...

Tonight on my way home from work, I called Deb Meader, the nurse, and reached her by phone. I pulled into the parking lot of the local veterinarian to talk, and to take some notes. It turns out the wells had been in the Allen's area since ten years ago, and Bill and Lois had been active members of GVCA. As Susan Haire had mentioned, they also "went through all the right channels" to voice their complaints about their exposure to the fumes and toxins from the wells at their property west of the Rulison exit off I-70. The right channels would comprise the local Health Department, the COGCC, and the EPA among others.

Rebecca Clarren, a noted journalist who has followed these issues, apparently interviewed the children of the now deceased Bill and Lois Allen. Deb Meader mentioned that Rebecca had contacted her wondering where the Allens were now, as their e-mail and phone were disconnected.

When I told Deb that my friend Chris Mobaldi was currently unable to speak due to her legal situation, and that many others can no longer freely discuss their issues with regards to health and the gas industry exposure because they had taken legal action and received settlements that required signing of disclosures not to talk about the issues, she seemed surprised and shocked.

COLLATERAL DAMAGE

"That's interesting...but it's good these people are getting money," she said slowly.

"Yes it is interesting, because it works. You give people money to be quiet, so for all practical purposes these things never happened. No one else knows, it never happened," I said.

"So these companies know what they are doing to people then: they are killing people. They should have to pay for it..." she said.

"Yes, they are killing people, and they will pay for it by going to hell, if hell exists," I replied.

When I finally got home after a very long day at work, there was a message on the phone from Debra Anderson in New Mexico, the latest documentary film producer on the scene. She had gotten hold of Susan Haire in Texas, and the recording on the message machine tape said:

"You're right Tara — she's amazing. I am planning to get together with her in the next few weeks, and maybe go with her when she sees her doctor."

I can usually write about these things, people's symptoms, the issues, the explosions, the whole thing, but right now it is just too much.

The gravity of the reality has ceased to stun me, but two people who sat in the same meetings I attend, fighting for some hope of oversight on this issue, are now dead. And within two weeks of each other, just last summer. I don't know if I would recognize their faces—if I was at GVCA meetings with them — or if they were too sick to go to meetings by the time I started getting involved. Chances are very likely they were.

Two people are now dead, and their children who recently had been reportedly anxious to bring the facts forward about the gas industry's role in their deaths were suddenly unlisted and not communicating. What happened?

I remember telling Rick Roles recently that some of those meetings with the local officials, and representatives from the companies where everyone sits primly around the big lines of tables talking about these things in civil voices make me want to scream about what is really going on; they are killing people.

"I do scream at them, I stand up on my chair and scream at them," said Rick.

September 15, 2006

I am tired today, we are short-staffed at work and it has been a very long week. I left a message at Astrella and Rice, wanting to find out when Lance Astrella thought he would be up here so I could arrange my work schedule to be off to meet with him.

He returned my call later in the day, when I was at the feed store in Silt picking up things for the animals. He told me that he was going to Alabama on a case of airborne exposure to gas-related toxins, and he wouldn't be up here until sometime in October now.

"Has anything happened lately?" asked Lance.

"I talked to Susan Haire this week, she called me from Texas. She sold her place now. She told me about these people, a husband and wife, the Allens. Their doctor wrote a letter to Williams Gas Company and said that he thought that their illnesses and cancers were from the gas exploration. They're dead now; they died last summer, within two weeks of one another. They died of cancer and one of them had asthma too. They had a produce stand west off the Rulison exit of I-70, but they stopped selling when they realized their produce was most probably toxic. Their children were angry about it, and they have documentation and film footage or pictures…they talked to Rebecca Clarren and did interviews, but then Rebecca was trying to find them and couldn't. Their phone and e-mails are disconnected," I said.

I gave Lance all the information I had: the family names of the deceased and the children, and the doctor's name and the hospital where he worked. I had pulled of Highways 6 and 24 next to the horse farm where they train racehorses. For years I have watched the trainers galloping the colts around the dirt track in the field right by the highway. A few horses and a small pony grazed in the paddock to the west of where I was parked on the county road's shoulder, to talk to Lance and not be driving at the same time. Gas rigs were now just across the highway, they were cropping up there going east from Silt along the river like weeds now. Every time I went out there, there was another one looming. Gas workers' trucks roared past every few minutes, fuel tanks and pumps in the back off the white truck beds. The vacant lots were filling with huge long gray industry tube-trucks at a rapid pace, and pushing into the horse pastures and residential areas. Those big rigs hauled water and waste materials to and from the wells. They were everywhere, all the time. So were the fumes and contaminants, but no one seemed to be aware. No one talked about it, anyhow.

"Your group, the GVCA, should go down to the hearing about air quality in November in Denver. The EPA did studies and filmed gas wells at night with infrared cameras and they could see the amounts of VOCs being emitted. The Front Range is concerned about it now. They know it's a health problem, and they are worried it will drive business away. They will probably pass an ordinance that limits and controls emissions, but maybe only on the Front Range," said Lance.

"What is this hearing for, is it statewide?" I asked.

"Yes, it is held by the Colorado Department of Health, so it could be statewide. But, the industry won't want to do it statewide because of the cost," said Lance.

Lance said he would get in touch after he came back from Alabama, where they were beginning to get evidence that could be tied to airborne contaminants from the gas wells, and people's health issues. He planned to be up here in November, so we would meet up then.

Before I got home I called Rick. He had seen the neighbors recently — the ones with the salivary gland tumor. He had also been talking to a lot of press: Orion Magazine, and another, and Theo Colburn came up with several others this week.

"There was a guy from MIT; she always has someone from MIT with her. They got too sick and couldn't finish the tours. One of the women had been exposed to toxins before, and she had to leave right away, before they got to the first pits. The last group made it up to the big evaporation pit before they felt too ill to continue.

"There were red dragonflies floating in the pit, then live dragonflies flew and landed, and they turned red, and they were dead. The liquid in the pit was black last week, but this time it looked like run-off. Then when you looked at it, it was all different colors: red, and green. There is a pipe, where it comes in direct from the wells now into the pit. They have a rope around it, where the pipe feeds in, like the kind they have for oil spills, like the Valdez. Well, good luck to the Pacific if it is like the rope they have here. It is so saturated all the stuff just flows right through," said Rick.

We talked about the couple who died of cancer last summer in Rulison. Rick didn't know them, but he knew some folks who lived near the exit off I-70. Rick couldn't remember the guy's name.

"I just saw him today, but my memory is so bad from the gas, I can't remember his last name. I just have to go over to his house to talk to him," said Rick.

Sept 17, 2006

Today I did a few hours of editing on this computer, scrolling back through the text and adding pieces there had not been time to fill out during this busy week.

Yesterday was a busy day; I worked at an adoptathon for the Rifle Animal Shelter and helped unite new owners with numerous cats and dogs in need. We had tents set up in the gravel parking lot of the Divide Creek Animal Hospital, and the wind tried valiantly all day to rip the tents down. At the end of the day, one kind man who had merely come to adopt a kitten for his three children helped me hold onto the metal framing of the "EZ-Up" tent to keep it from sailing off like a kite. The other volunteers were breaking down the tents as fast as they could, while the two of us continued to keep the last tent in the cat adoption area from flying away.

Nancy Jacobsen had asked me to help at the event. After my long visit to her Dry Hollow ranch last weekend, I was now acquainted with her husband Gary, a friendly short wiry man with a black eye patch.

Of course, as we did our work through the day, Gary and I strayed into some discussion about the gas and oil effects occurring there in Dry Hollow and elsewhere.

Toward the end of the day all the volunteers were tired and quite cold from the chill autumn wind that blew all day. It was a rude ending to a hot summer. Almost all the animals had been connected with new homes, and the day's work had been successful.

The group said their goodbyes, and handshakes were passed around. Gary Gagne, Nancy's husband came over to me and said:

"Thank you for what you are doing."

"Well, thank you, and Nancy too. You are the ones doing all the work!" I said referring to the adoptathon. And they, and a few others, were certainly the ones doing all the work, hauling truckloads of kennels, table, crates, and more; and setting them all up and taking them down.

"No, I mean for the gas work..." said Gary.

"You're welcome, I hope it is doing some good," I said. To say I was deeply touched by his remark doesn't scratch the surface of how I felt. I understood what he was saying.

Last night before I went to bed I picked a bucket of tomatoes and cucumbers, cut clumps of chives, mint and oregano, and moved a few flowerpots inside. I wore a wool hat, a Carhartt coat, and shivered under the light rain as I picked produce. It would certainly frost tonight.

My husband went out later and gathered more, and covered some tomatoes and flowers with bed sheets. I was way too tired to even think about turning on the computer, and as I went to sleep under thick blankets surrounded by my own cats, I thought of Gary's comment.

I want to deserve the thanks; I want some of the truth and facts of what I have learned to come out publicly. The people being harmed by the gas industry are suffering and dying, and no one is stopping it. No one even knows about it, except for the handful of affected individuals. The truth has to get out, and get out big-time.

Monday September 19, 2006

Yesterday I placed a number of calls to people, but didn't get much call-back response, unfortunately. I wished that I had, because it was my one full day of the week to be able to do work on this. Oh well. Instead I edited and added in things to recent text, and I did talk to Nancy Jacobsen for a long time. Mostly we discussed the adoptathon results from the day before. She did tell me also that she just gotten back (or her lawyer did, most likely) a letter from Encana stating that they would not cover the damages that occurred to her home and property after the explosion at the Arbany well. Nancy did not seem unduly surprised by this, but she didn't sound too happy either.

"We spent $16,000 to have the repairs done, plus the lawyer fees. Could you handle a bill like that right now?" asked Nancy.

"No, I would have to declare bankruptcy," I replied.

Later I called Steve Mobaldi and asked him about getting a copy finally of the television taped interview with Channel 8 in Grand Junction.

"How is Chris doing?" I asked as usual.

"Not very well; her speech is really strange and she's been sick."

COLLATERAL DAMAGE

"She sounded really good the last two times I talked to her — her memory was better than it has been in a long time, and I could understand her really well," I said.

Steve paused for a moment.

"I don't think she is ever going to get better," he said, with more depression in his voice than I have ever discerned before. We talked some more, then said goodbye.

On my way home from work, when I was down by the police station in Glenwood I pulled into their parking lot to turn around and head back toward the interstate. I got inspired to go see if I could locate Neil, who was a Glenwood Springs police officer. There was a police officer just pulling into the parking lot, so I waited until he got out of the squad car and asked him if Neil might be in, and that I wanted to talk to him; not about police business but about natural gas issues, and that I spoken with his wife recently. The officer told me to park and go wait in the lobby, and he would send Neil out.

I reported to the clerk behind the glass dividing-wall that I was waiting to see Neil. A minute later a tall, handsome gray-haired man came out in full uniform, gun on his belt. I briefed him on my reason for coming to see him, and he was willing to talk.

"I don't know if the cancer I had was from gas exposure; the doctors didn't know. A doctor in Grand Junction at Saint Mary's did the operation, and it's cancer. I can tell you the name of the tumor; it's an acinic cell carcinoma. It is a very rare kind of tumor, a place up in Seattle has done a bunch of work on this kind of cancer and the say that there have been only two hundred cases of this type since 2002," he said.

"Your wife told me some things about it, and the reason why I am interested in this is because I have a friend who is really sick; she is very sick and had two brain tumors. Her name is Chris Mobaldi and she was from Rulison. They moved now, but her tumors were pituitary tumors and she is a mess now. She looks like she is in her 70s and she's in her 50s. She used to weigh 160 pounds and she barely weighs 100 pounds now. She cannot speak intelligibly most of the time, and she is sick all the time. She vomits and has diarrhea. After I knew what was going on with her, I started learning about some of the other people's health problems; the people who are getting air-borne gas exposure and have the headaches and nosebleeds and are sick all the time. Some of them have gotten knocked over by gas fumes or hydrogen sulfide, then their toxicity levels are so high they are a mess after that. Then there are the ones with contaminated water issues plus the airborne, like Laura Amos. I got to be real good friends with Laura. I am trying to keep track of as many health problems as I hear of, and there are more and more all the time."

"Well, we're moving away. We're moving north of the river away from where the gas development is now. And we hope it won't get there," said Neil.

"Yes, your wife said that when I talked to her. It's difficult to know where the development is going to go, because it depends upon what they hit. If they hit good production, they will stay there and go further. The last meeting I went to with Antero about proposed development plans they laid out a map plan that said they would go north of the river and I-70 up to about the middle of Silt Mesa, along the line where Silt Mesa Road runs. That's where they think it will taper off. As you get closer to the hogback, the geographical formation curls up and the methane has been dissipating. So they don't anticipate it going there. That's what they're saying now..." I said.

I drew a small diagram on the notebook I was carrying, and showed the proposed area where development would go. I hoped that the Wagstrom's new property would be in a safe zone; at the very least not have surface development.

We discussed the COGCC, the lack of enforcement of any regulations, the continued frustrations of landowners in affected areas, and other sick neighbors.

I told him what Lance Astrella had told me days ago about the statewide ordinances being proposed in November. He too had heard that on the Western Slope our emission level would likely be twice the level allowed in Denver.

He was as upset as I was about it.

"It's because we live in the country, so I guess our lives are worth half as much," I said.

I remember driving home absolutely furious and despondent when I had heard wind of that at a recent GVCA meeting.

I thanked Neil for spending the time to talk to me, and shook his hand and left the police department.

When I got home, before I could get the grain bucket filled for the horses, the phone rang. It was Rick Roles. Rick told me about the tour they did with Theo Colburn, the MIT student, the Washington DC diplomat, and the reporter. We talked for over an hour, and I took notes. Tomorrow is another day....my back aches, I cannot write any more now. I am glad for new leads, plus Rick's friend might know the children of the now deceased Allens. I hope we can find them and they can talk.

Pit Violations - The Runaround

Tuesday September 19, 2006

Patrick Barker of Western Colorado Congress called about the air quality hearing in Denver in November. We discussed plans for a group of landowners to go and testify, and then we started talking about news I had just heard that morning.

I told him about the waste pits near the Rifle airport that were being backfilled that still had a good amount of liquid in them, and he encouraged us to call it in to the state and gave me a number.

I called Jaime Atkins at the COGCC Parachute office about 2 p.m. and told him about the pit that was being backfilled on the bottom of C.R. 319 in the Mamm Creek area. I gave him all the well identification information, and said that I was told the pit was being filled in with a lot of production liquid still in it. He said they would send an inspector right out, and thanked me for the call.

At 5:30 I called Jaime Atkins back, and he said the inspector had just left to go out to the pit.

Later that night, there was a message on my machine at home from the GOGCC inspector Dave Graham. He left his number. After I did the ranch chores I called him back.

Dave, the inspector, said it was our civic duty to call these things in. He said he stopped them from filling in the pit, and told them to remove the liquid before they filled in any more. They took out three tanker loads that afternoon. He indicated that what they were doing had been improper, and it was good we called it in. He said he would get the County Environmental Health Manager Jim Rada up there tomorrow. He also said he would have to talk with the BLM to see about the regulations on their well sites; they might be different.

Dave wanted to know exactly what type of complaint was going to be indicated on the complaint form. Never having filed a complaint before, I was a little unsure as to what to say. He said one of the choices was an odor complaint. I told him we better have Rick speak to the issue as to what the complaint was.

"We called it in because the pit still had liquid in it when they were backfilling," I said.

Dave did not indicate to me that that was an acceptable choice available for a formal complaint.

I told him I wanted samples taken, water or pit-matter samples, and I would pay for the testing myself if need be. I told him that several times.

I am too tired to type any more tonight.

Wednesday September 20, 2006

Rick called me right when I was leaving for work, and said he drove by the pit and they were still filling it in, with over 300 barrels of liquid in it, at least. I told him I would make some calls and call him back; I had to leave for work.

I wasn't sure who to call, so I called Nancy and Gary. They had a NOAV (Notice of Alleged Violation) filed on a pit by their place when they caught the industry workers filling it in, so they should know the best thing to do. Nancy suggested calling Jesse Smith, the current acting liaison between the industry and landowners. I got hold of Jesse right away, and he said he would call Jaime Atkins. Not too long later, I got a call back from Jesse that COGCC had someone on it, and everything would be taken care of properly. Someone from COGCC was overseeing what was going on at the pit. I thanked him, and felt fairly confident that things were in good hands.

To my surprise, about half an hour later, I got a call back from David Graham the COGCC inspector. He gave me an update on the pit, and said that he hadn't taken a sample and that Jaime had not directed him to go back out there today.

I told him that it was called in that they were backfilling this a.m. again, with 300-plus barrels of liquid still in the pit. David said,

"They probably pumped at night."

Now I began to wonder who from the COGCC was currently on top of it, if David wasn't out there. I called Jeff Smith back and reported what I had just heard. He said he would go out there himself.

Then, the following night, Wednesday, after a number of other phone messages I left with Jaime Atkins, I called Jesse Smith. He said they (Encana) did everything according to regulation; they just pumped the stuff out of the pit to be responsive to the complaint.

"They pumped three tank loads out yesterday, and two more today," he said.

I found the tone of his comment to be very different from what David Graham told me the night before, when he said it had been a good thing we had called, and said that they shouldn't have been filling the pit that way.

Thursday Sept 21, 2006

This morning I called the COGCC offices in Denver and left messages at a number of extensions. Most people were unavailable or out of the office. It was early also, so that made sense. I had to try and make my calls early, due to the fact that I have nine-to-five work obligations. I also left a message with an employee at the local BLM office. Last night I had left messages on both of Jaime Atkins's phones asking for a call back to clarify some of the questions I had regarding the pit, the backfilling regulations, and the confusion over what had gone on and what I had been told by different people on different days.

COLLATERAL DAMAGE

On my way to Vail I got a call from Jaime Atkins from the COGCC. He was friendly and wanted to answer my questions. I explained, "When I first called the pit-backfilling in, I was told by Dave Graham, 'It was the right thing to do; we stopped them from filling the pit and told them to pump it all out, and it is your civic duty to call.'"

Thursday afternoon the Bureau of Land Management employee called me back and I spoke to him briefly. I told him what was up, and asked if he could be present when the core samples were done. I mentioned my concern that the west end of the pit is where all the sludge I saw on the top yesterday was, and that is where the core sample should be done.

I remember over a year ago talking to a friend whose son-in-law works for the industry. He had told them how it was done — the "reclamation" of the pits. They pull the plastic liners (or bury them) then fill in the pits and cover them with uncontaminated soil. And then tests are done on the uncontaminated soil. I can visualize it, like a big pan of liquid fudge, being squeezed up to the end of the pan, after some has been taken out. Then, the rest is mixed with non-contaminated dry soil, and bulldozed into the center area of the pit. Where, I wonder, do the tests — the auger samples—get drilled, and who monitors?

Today, Thursday, Rick said they were mixing the sludge at the end of the pit with the dirt. I wonder if it is because I mentioned it to Jaime Atkins in my two messages yesterday. Or, do they do it that way anyways?

Rick said when the auger samples come up, they bring up the regular red and brown soils native to the area, and then they bring up the black chunks of the diesel and chemical mixes, and the gray of the adhesives. I visualized something like Rocky Road ice cream. What part of those samples needs to past the test?

Last night, on my way home from work, I knew I had to go drive out and see the pad site. It is just past the Rifle Airport. It was dark and cold as the sun was setting behind the clouds, and no one was on the now completely smoothed-over pad. It was just a huge expanse of brown soil. Then, after I gingerly drove the truck down the gravel road closer to the pit's edge, I saw the large area of black and gray slimy and shiny sludge. I had asked repeatedly for a sample of the liquid to get taken, and none had. By chance I had a bowl in the truck, and I stepped out toward the goo and dipped out a bowl full. I put it in a sack and drove away. If no one else would sample this stuff, then I would satisfy my own curiosity. I made calls that night and left messages as to where to get it tested.

Before I went into the house for the evening I took a glass jar out to the shed where I had the sample.

In the dark, with a headlamp on, I dipped the jar into the slimy substance. It smelled of diesel type substances, and strongly. The two barn cats were excited to have evening company, and I was glad to have them there as I messed with the stinky goo, using rubber gloves and wondering really what did all this matter. Why was I in my shed at night in the rain, wearing white medical gloves for protection, transferring perhaps toxic samples from soup bowls dipped from gas pits on Mamm Creek into glass jars?

As I told Dave Graham from the COGCC the other night, I don't want to be doing this; spending large amounts of time concerned with the impact and pollution from natural gas production in our valley.

"I want to be enjoying my life..." I had said.

Fortunately, (or unfortunately depending upon the perspective one might have), I currently find it impossible not to address the concerns I have regarding the negative health impacts of the natural gas industry.

A woman named Charlene whom I didn't know from Encana Gas Company called this afternoon. I was surprise. She wanted to talk about things with me.

"I didn't call Encana, I called the COGCC," I said, somewhat confused.

COLLATERAL DAMAGE

She said they were calling about the issues, and I got a few phone numbers to call back. I couldn't do it at that time. They wanted to hear what I had to say about the complaints at the pit. I did not call Encana back—they were not the regulatory entity in charge of overseeing—they were the company responsible for what had gone on up at this pit...

Friday September 22, 2006

I reached the employee at the BLM about 2:30. We discussed the issues about the Mamm Creek pit backfill. I had briefed him earlier the previous day. I told him about the conflicting regulatory information I got from various people at the COGCC. He wanted to know if we had pictures, and said that another individual had also called in about the filling of the pit.

"I wish I had been called earlier, so I could have gone out and seen it..." He said.

"Well David Graham from the COGCC said he would talk to you, and I didn't know that we should call you. Next time I will. Live and learn." I said. I was disappointed that in spite of the time and effort that was spent, the employee had not been able to get more information. He said he would update his boss on Monday and get back to me, and that the COGCC required a soil sample from the pit.

Later that afternoon, Bob Chessen from the COGCC office in Denver called. He was very polite, and I was also. I asked a number of questions, and he said they did need to drain the pit of liquids before filling it in. I told him the differing things I had been told by COGCC employees regarding the regulations, etc. I also told him what Jeff Smith, the standing Garfield County gas and oil liaison told me on Wednesday night. He indicated there was no violation, and that they basically drained the pit to react to the complaint.

Saturday morning I talked with Theo Colborn of the TDEX—the Endocrine Disruption Exchange think tank in Paonia. Theo is a world-recognized scientist on the effects of human exposure to toxins affiliated with gas and oil development.

"I meant to call you back earlier, but I have been traveling," she said.

"That's fine, I just wondered if you had any ideas about what I could do with this sample..."

We talked at length, and basically there was no recourse to research the sample—as it is a federal crime to mail potentially chemically hazardous waste to a lab. I found it ironic—we are told all is OK, yet in order to check the sample and mail it to lab for scientific analysis it would be a crime? Go figure...

I stopped in at the BLM office this afternoon and talked to my contact. He said his boss felt the issue was sufficiently taken care of, and basically it was over. I shared my questions with Jim, and indicated that I want to know more about the status of what happened and when and why.

Jaime Atkins from the COGCC called not long after, and said he heard from the BLM that I still had issues from the complaint that was filed on the Mamm Creek well. I told him I would call him back later as I was at work, and said yes, I did still have some questions about the situation.

I talked to him that evening for about a half an hour. He wanted to know what my continued complaints were. I said that I had talked to Bob Chessen in Denver from the COGCC and had been told that all liquids were to be removed from the pit. Jaime said that was not the regulation.

He reviewed with me the substances that were in the pit, and explained everything to be naturally occurring for the most part. Peat moss, coal filings that would explain the black color of the goo I saw, and some other materials. He also explained that the liner was supposed to stay in the pit, they wanted it in there. And, he said that a lot of the material should stay there, because

"What would they do with it?" he said.

"Why did it smell like diesel, then?" I asked.

203

"David Graham didn't smell it, so I guess it's just a question of a person's sense of smell…" said Jaime.

He continued to ask if the basis of my complaint had been resolved.

I said I still had questions: why was I told different regulations at different times? He explained that David was new on the job, and not real clear on the regulations. He said that the filling only occurred after David gave them the go ahead to do it. I pointed out that David had told me on Tuesday night they would continue the rest of the pumping the next day, after the first three tankloads he ordered on Tuesday occurred. Then, Wednesday morning a citizen's call came in that they were filling in the pit with 300-plus barrels of liquid still in it.

"Jesse Smith said someone was on it when I called him Wednesday morning, then shortly after that David called back and said you hadn't sent him out."

Jaime seemed angry and asked if I wanted to argue about who called who when…

"No, I don't. I want to know if it is violation if a COGCC official comes out and tells an operator to stop filling until they pump, then they continue filling the next day."

Jaime told me the number of the complaint form that had been filed.

"This complaint is not closed yet, they still have to have an auger sample done."

He invited me to be present when it was done. I said I would be interested in being there for educational purposes if I could. He said there were no instances of reserve pits being the sources of pollution. He did say that other things, like the bad cement job at Divide Creek seep site were problems. I indicated that I was aware of these issues.

The tone of our conversation was now rather politely testy.

I asked him if the pit had had fracking chemicals in it. He said yes, it had.

Then I mentioned that the BLM had a surface agreement that stipulated removal of all liquids from that pit. He seemed to be getting angry, and said,

"Well the BLM will have to talk to us about that. I guess it's an issue of what's considered a liquid."

By this point I was trying to contain my anger.

"The job I work at has a lot of regulation in it — my question is what the bottom line regulation-wise with the pits is. Just doing best practices always when they're not required is different than meeting the basic regulations. That's what I want to know. I'm getting a lot of different answers from different people," I said.

And a lot of those people were from the COGCC.

Again, I thanked him for his time.

It is interesting that in his detailed conversations of all the harmless things in the pit, all naturally occurring substances, he never mentioned fracking fluids being in there. I was the one at the very end of our conversation who brought that subject up. Plus, the fact that the BLM surface agreement stipulated the removal of all liquids from the pit. The tone of our conversation was not so conciliatory when it ended as it had started, but I made a point not to be uncivil.

I got home with a roaring headache and fed the animals. I called Laura Amos to talk; hadn't talked in some while. I vented with her about my frustration over the COGCC conversation, etc. She found it interesting. One of the things she told me tonight was about a guy she met who had worked for the industry. He was really clean-cut, and religious. She asked him what it was like to work the gas fields. He said it was a bunch of druggies, methamphetamine abusers ("meth-heads") who also got drunk a lot.

COLLATERAL DAMAGE

He said he got more money selling them clean urine for their samples than he did for pushing tools. I was stunned. I remember getting drug-tested to work for the Breckenridge Ski Corporation when I taught skiing in Summit County. They give you a vial but let you go urinate alone…

Then I knew, the gas field workers must have had the clean urine sample on them somehow, carrying it around.

"He said they carried the urine in little baggies, would put it in their lunch boxes or hide it in their clothes, so they had it with them when they got called for a test," said Laura.

Great. Yet another little bit of uplifting news. How some workers on the crews managed to be drugged and or inebriated and duck the radar. Just like the gas companies.

I told Laura about the air quality hearings, she said she would try to go. I told her about what I was trying to do writing about all this, and my questions about it.

"I know they will deny the validity of everything, but that is what I want to point out. Why does the injured landowner have to prove what they are doing is damaging when they don't have to prove it is safe? And the other thing is all the lawsuits that get settled and the money gets paid and it basically never happened as far as anyone is concerned. That part really bothers me. I know it is crazy to do this, and I don't know if what I am working on is publishable, but I have to do it."

I would like you to read some of it, and tell me what you think…" I said.

"Why, so I can tell you you are crazy?" she asked.

"No, because I know that what I am doing is about the truth, but it's crazy that no one really sees what is happening, and that people are dying from it. And no one cares," I said.

September 26, 2006

All quiet on the gas front. That doesn't mean that nothing is happening…

I had hoped that Judith Kohler's Associated Press article was coming out this week, so I am again dealing with failed expectations. Similar to the one film crew director's delayed agenda; it feels like losing time, valuable personal time so spare to have, set aside for these projects. Like I said to Laura, you have to assume that some of the things you do will be successful, and some will not. It is still painful to think that some of that precious time and energy will go nowhere. I will do what I can to see what I have been involved in gets done. At least most of it, hopefully…

Laura said the right things to me last night. I needed to hear it. I recalled our conversation:

"I would go to the meetings, and Larry would say when I got home, 'What did that accomplish? What got done?'....It's all a million little pieces, all put together," said Laura.

She is right. And it involves herding the press, herding the sick and affected and impacted people, and herding your own commitment to continue on. Herding cats…

"Sometimes I want to quit. I am sick of it; but I can't. I can't leave here," I said.

"You want to leave?" asked Laura.

"Yes, because of the gas. Otherwise, I love it here. But my husband won't leave, and so I won't either. But sometimes I just want to quit this all—to go away and forget about it like everyone else does…" I said.

"You can't, you're too deep in," said Laura.

COLLATERAL DAMAGE

Thursday September 29th, 2006

Yesterday and today have been spectacular and flawlessly beautiful Colorado autumn days. After the sudden rainy and freezing period we had last week, it is wonderful to see some of the beautiful fall weather come around before winter sets in.

The sun has long since set behind the mountain, but the bronze and orange red of the scrub oak and shrubs still shows against the greenish-gold field and hillsides. The weather is so pleasantly mild; it hit around eighty degrees again this afternoon, as it did yesterday.

Last night it was so beautiful when I got home, I did no gas work. I lingered over feeding the animals, and enjoyed a prolonged sunset over the valley and the mountains from the now mostly frozen vegetable garden.

At a meeting yesterday, it was mentioned that Bob Beauprez, running against Democratic Bill Ritter for governor, said that gas and oil development leaves the areas in Colorado better off than they were before. I had to interject, I was too angry to not say something.

"How could he possibly say such thing? I spent a good part of last week arguing with the COGCC people about a pit they were filling in before they drained it. They told me, Jaime Atkins did, that they need to leave the liquids in the pit because 'Where else could they take it?' And that's why they bury the liners, to hold the stuff in. Of course, they were ripping up these liners as they mixed and buried because at least one of the two liners was floating and totally failed. Jaime told me about all these naturally occurring substances. When I asked if the pit had been used for fracking , he said yes. He did not seem to happy to discuss it, nor was he very pleasant when I kept asking him detailed questions, including bringing up the subject of the BLM's surface agreement that stipulated removal of liquids from the pit before filling. I hope Bill Ritter knows the details of what really goes on out here, and what gets left behind; the toxins in the pits and the air so when he debates Beauprez he can take him on," I said.

TARA MEIXSELL

I still haven't quite decided what course to take with my issues over the COGCC response to my complaints. I have several thoughts in mind, but don't want to waste too much time on pursuing complaining to deaf ears. It's better to keep on logging all the violations, illnesses, and deadly toxic exposures.

Saturday September 30, 2006

This morning I was torn between wanting to drive to Aspen to the 1:30 showing of *Land Out of Time*, the documentary on oil and gas development impacts, or staying home. On the radio yesterday morning I heard a piece about the film, and heard tapes of Randy Udall's comments on the whole issue. It seems that the movie will focus a lot on the fight going on up in Wyoming, and is coming from the perspective of one of the local landowners. Udall commented on Gloria Flora's successful efforts to close off the spectacular Wyoming National forest area below Yellowstone. Gloria was instrumental in stopping development of the area, thus saving it from devastation. I heard Gloria speak as the featured guest last summer at the GVCA annual picnic up at the Bell's ranch in Dry Hollow. She is a determined, savvy, and an extremely dedicated environmentalist.

Udall went on to say that up in Wyoming a large group of citizens had been successful in their efforts to block gas development in pristine areas, and he likened it to the efforts here in western Colorado with the Roan Plateau issue. He commented that here there are a number of very sharp and effective people working on the issues surrounding gas development impacts, but there aren't nearly enough people. He also said that the people fighting to see that the land is not devastated by development are from very differing segments of society; groups of people you don't usually find working together. Hunters and hikers are standing side by side at the meetings, asking for the preservation of the land. He indicated that "these are not Haight-Ashbury-type people" (meaning stereotypical drug-crazed extremists or eco-terrorist types) but they are solid citizens from many walks of life.

The film would be shown first in the Aspen Theater on Saturday as part of the festival premiers, then in Glenwood Springs on Sunday at 5:30. I wondered who from this area, besides Udall, would be featured. I had not heard of this film producer myself, but I'm sure he filmed some people I know. Although I hate disrupting the all-too-short weekends, especially in the evening to drive to the city one extra time, I knew I should make an effort to see the film. Laura Amos talked about coming down for it, but she wasn't home last night. She's probably back up at outfitters camp.

Patrick Barker from Western Colorado Congress called yesterday midday and gave me an update on the meeting in Denver for the Air Quality Initiative. He said it went very well, and that even the COGA (Colorado Oil and Gas Association) people seemed to support the decreased level of allowable VOC emissions from the gas wells. That was good to hear. I told him that Dee Hoffmeister and Rick were both willing to go, and that I had done a very good hour-and-a-half tape with Dee last weekend.

"Hey Tara, I have something else to ask you. Do you mind being quoted in this article that is going to go out with the air quality press release? We expect it to be released next week, at the same time that Judith Kohler's Associated Press article goes out. We hope that national press picks hers up, then. They didn't release them this week because of the school shootings in Colorado, that's all the news was covering…" said Patrick.

"I know; it was all over the national news too," I said.

How sad it had been to see the pictures of the teenage girl from Bailey, now dead after the gunman entered that rural high school. Many times I had driven past that building on highway 285 going to and from Denver when I lived in Park County.

"I talked to Judith last week, and she wasn't done with the article then…she said she had so much she didn't know where to go with it. Then I gave her Susan Haire's number in Texas, which she had been looking for. I hope she talked to her; Susan's story is powerful, and bad. And Susan is very, very angry and convinced she has suffered irreversible damage from chemical exposure. Also, I gave her Rick's number. I know she talked to him that day, he told me that night she called."

I was happy that after all, the article was finally going to go out and that it was in fact as Patrick had first indicated; going to go out simultaneously with the air initiative press release.

I know I have incredibly high expectations for this article, but I can't help it. The Mobaldis spent a whole afternoon with Judith doing the interview, and Chris is the most seriously impacted (still living) person that I am aware of with a gas-related illness in western Colorado. Not to mention Dee Hoffmeister, Karen Trulove, Rick Roles, and the now deceased Allens. If an associated press article goes national with Chris Mobaldi's horrifying story, that will be about the biggest thing that could happen. The chances of her legal case going to court and going public seems remote, as all similar cases seem to be settled quietly, out-of-court, and with non-disclosure clauses attached so the injured parties need to keep quiet about the issues in order to get paid off the agreed-upon amount.

I saw Carol Bell, the landowner with several wells on her property in Silt the other day. We talked about the release of this documentary on gas impacts. Carol said Peggy Utesch had talked to her about it, and indicated that it was about time one was released after all the film people who have come here to do interviews. One gets that impatience over time. I certainly have, especially after the disappointments to find out that hours and days you have spent on one project may be for nothing, at least for now. That is hard news to swallow after you have been told timelines and venues of release. I guess disappointment is what one would accurately call it. Just don't put all your eggs in one basket, or as Laura Amos said this Monday night:

"It's a million little pieces."

Patrick Barker from Western Colorado Congress encouraged me to at least write a letter to the editor about what happened with the reports to COGCC about the filling of the pit. I told him how frustrating it was to spend so long dealing with the different entities and never get a clear answer. I hadn't yet decided what I want to do, but I documented everything pretty well.

COLLATERAL DAMAGE

So this morning I turned on my slow home computer and fought with it for awhile, then sent off a letter/article to the local Glenwood Springs Post Independent. I also forwarded a draft to Patrick and a few other places. One to Tresi Houpt, the one out of three Garfield county commissioners who has taken a stance in terms of gas-related health issues and the like. Maybe I will send a draft to the other two commissioners, but I didn't today. I really don't think it would do much good. But I know I should, anyways. Here's what I wrote:

I am sitting here at the picnic table typing outside. It is a flawless, perfect fall afternoon in Colorado. The film festival screening of the gas documentary is going on now in Aspen. I hope it is going well. There will not be many more days like this left this fall, with brilliant foliage displays and a few splashes of yellow and purple and orange from the flowers that survived the last frost. It is warm enough that the flies have come back (in small force) to annoy the horses. After four or five hours typing about gas issues, it is time to log off and go enjoy life.

October 1, 2006

A change of weather is blowing in, and the sky was clouded over today. This may be one of the last brilliant days of fall before the leaves are all ripped down. I am antsy to go into Glenwood for the 5:30 showing of the gas documentary. I hate driving in on the weekend, and usually balk at any weekend activities, especially in the evening. With the intellectual capacity and attention span of a gnat, I forced myself to edit for a little while mainly to kill time and at least get something done on the book on this weekend day before Monday comes again oh so soon. Now the time has come to do the evening feeding routine and go to the film festival. I am curious how many GVCA people will be there, and who else will be drawn to this unusual film. I hope to find it motivating. As this week begins with yet more anticipation over press releases, I brace against the disappointment of dashed hopes. Or perhaps better said, delayed hopes.

Just got back from Glenwood from the film, and just like I had hoped, it has inspired me. There was a long line outside the theater a good twenty minutes before the film began, and most people had purchased tickets in advance. Listening to the talk in the line, I began to wonder if I could even get in at all. Luck would have it, I did. I was so happy to see that the theater was packed.

Parts of the film made me angry, not at the filmmaker's intent but at the futility and reality of the whole situation. Arial footage of the maze of roads and well pads drew gasps of shock from the audience, in chorus. I have seen those before, oh so many times now I don't gasp anymore.

Seeing Dick Cheney drone on with his agenda especially makes me angry, as does George Bush's poker face talking about energy. They, and their team of industry cronies are the masterminds of this uncontrolled and unmonitored plundering and ruination of the West. That's what makes me furious.

My champions in this film were Gloria Flora, the valiant and brave Forest Service Director from Wyoming, and Randy Udall. Their voices rang strong, loud and true. I could not have agreed more with their every word.

The hunting guide from Garfield County made strong and believable points, that a few years back he never would have gone into a room with environmentalist…but now we were all getting together on the same side of the issue and we had to if we don't want to lose it all. He talked about losing his long-time clients who came to Colorado year after year to go on hunting expeditions, mostly to camp and be out of doors in beautiful wilderness areas. They have told him they won't come back here when the place gets too overrun with wells.

I finally got to see on film Tweeti Blancett and her husband, the hard fighting New Mexico cattle ranchers I had heard so much about. Alan Rolston, the previous Western Colorado Congress appointee to the GVCA had said,

"Tweeti Blancett is my favorite person in the whole world…"

After seeing her in this film I have an idea why Alan thought so highly of her.

COLLATERAL DAMAGE

In the movie she and her husband discussed their issues regarding gas development on their New Mexico ranch as the camera panned along. The two rode horses and drove cattle over well pads and beside waste pits filled with toxic substances and lined with ripped black plastic. They discussed their lives, their disappointment with the government's lack of concern over what was happening to their land and their grazing lease. Footage of dead cows stretched across the ground after drinking contaminated production water chilled me.

Barely containing the tears that welled in his eyes, the tough rancher stated how hard it had been to sell off his herd. They had to get out of there. He said his family had come there to get away from a bad place, and he had to do the same for himself and his family.

After the film there was a question and answer time and some special guests were introduced.

The pilot who flew the plane during the aerial filming (the World War hero) was there: a scrappy, determined man who loved his home country fiercely. It was what he had fought for; it is what gave him hope. He was not going to give up easily.

Randy Udall was also introduced.

Questions were asked, and answers were given. Many in the crowd were eager to figure out how to get involved, how to make a difference. That was heartening, hopefully they would stick with it.

I think about what the film producer Mark Harvey said. He wanted to make the film because he, as a native Coloradoan, knew that something had to be done to bring what was happening to light. That people needed to set aside time to pay attention to the issues that effect out public lands. He gave the example of one day a month. I think the average citizen will be hard pressed to do that, but at least if people would pay more attention than they are now, it would be an improvement. His film will be just the mechanism to bring more of that about.

The average person, even in western Colorado, knows or cares so little about these issues it is pathetic. People are under the illusion that nothing bad would ever be allowed to happen in the United States, and things are all being watched over carefully and managed properly. This must, must change and fast.

And, as Randy Udall said tonight (I cringed as he said it) it is just beginning here in western Colorado. Ten thousand more wells are going to be drilled...

I wonder where I will be when that 10,000th well gets completed. Will I be in Garfield County? If I am, will I want to be?

Monday October 2, 2006

After a few calls to the Glenwood Springs Post Independent voicemails today, I finally reached an operator and left a message for the Community Voices Department. I got a call back almost immediately, which was left on a message on my phone as I had another call at that time.

The employee from the Post said that they thanked me for my contributions that they had received, but the limit to letters to the editor was 350 words and my contribution was over 900 words. She requested that I send an edited version in, under the word limit. Actually, I had done a word count while writing what I knew was gong to be a longer-than-acceptable letter to the editor, and sent it as a community section. That is neither here nor there; the paper got it, plus my update with the correct spelling of a few people's names, and my contact information.

Not a minute after I got the call from the paper, I got another call from Jaime Atkins from the COGCC. Although he didn't know my name correctly, he was very, very polite. Different from the last tone of voice I had heard from him...

He wanted to inform me that a sampling of the soil from the evaporation pit at the Encana well GMU 23-8C (H23NW) would be taken next Tuesday morning at 8 a.m., and they would be doing it in the "hot spot" areas. I was invited to be there. I thanked him for that, and let him know that at the moment I was extremely busy at work (which I was), and that I was unsure that I could make the appointed time. I had other pressing things going on.

Then I told him I wanted to know how to most easily get a copy of the report of what went on with the complaint at the pit.

"You mean your complaint?" he asked.

"Yes," I said.

Then it made me wonder about the other person who called the complaint in, how was that one going?

"I would like to be there if I can, it would be informative," I said. Again I mentioned something about the substance I saw at the west end of the pit that smelled of chemicals, of diesel.

I called Patrick Barker and told him that if he could help chop my 900-plus-word writing to 350 words, I would be able to resubmit the letter to the editor. I described how conciliatory Jaime had been in the phone call, and how interesting it was after our last rather heated conversation. I still have had no call back from Bob Chessen at the COGCC. That too is interesting…

When I complained to the BLM, a concerned high-ranking employee informed his boss. Then, he told his boss to contact me. Next, I get a rather heated call from Jaime Atkins at COGCC. No call from the BLM boss, who told his employee and Jaime everything was taken care of. That's the message I basically got.

Why, I wonder, why, was Jaime who seemed quite angry about my questions last week, now so friendly?

I asked for a copy of the official report on the complaint to be sent to me, in case I could not be out at the pit for the core samples at 8 a.m. next Tuesday.

Forced pooling

October 5, 2006

This morning I had to stop at City Market to get things for work. It was early, and not many people were in the store. In the bread isle, Linda Murr saw me. She works in the bakery, and I know her and her ranching husband Richard. I sometimes buy hay from them, and we know each other from gas meetings. When I go load-up on hay, Richard and I discuss the latest on development plans for Silt Mesa. He owns two big producing hay properties at each end of the Silt Mesa area.

Linda had a stressed, pinched look on her face.

"I was hoping I would see you; I've been keeping my eye out for you. The gas company is going to put us in a unit, we got a letter that said it. Richard talked to them, and they said that since he didn't sign with them before, he won't get anything and they are going to go ahead with the unit," she said.

"So he didn't sign anything, good. As long as he doesn't sign anything then you still have options. You need to talk to someone about this, you should call Peggy Utesch, she does surface agreements for owners and she's one of the main people with the Community Development Plan. Do you have a pen?" I said, as I was rummaging in my bag for something to write on.

We talked a bit more, and then after I paid for my order I went outside to the phone booth and wrote down Peggy's number and brought it back into City Market for Linda. I said I would call Peggy too. I also gave her Liz Lippitt's name; she had a horse place right near the Murr's west hay field. She, along with Peggy and a handful of others had developed the Silt Mesa Community Development Plan.

COLLATERAL DAMAGE

I got hold of Peggy in the late afternoon, and told her about the Murrs. I was unsure sure of what kind of unit it was, a Federal Unit is one situation where private land can be grouped together with 10% federal land and regardless of a landowner's 100% mineral rights ownership, companies can drill on your property regardless of your agreement. The other situation is a grouping of mineral owners together all agreeing to be in a unit, and they all get compensated and can set the guidelines with the company about details. She was concerned if it was Antero doing the unit, because they didn't disclose it to the Community Development Plan people. She recommended that they call an attorney, and Ralph Cantafio was the one she recommended. Peggy said she would call the Murrs tomorrow; she just got back from an out-of-town trip today.

On my drive home from work I heard an Aspen Public Radio newscast. They covered the governor candidate debates held in Rifle last night at the Rifle High School before a packed crowd. I grew angry as I listened to Bob Beauprez's stance on energy development here. He discussed the Roan Plateau, and talked about "compromise: and similar language. He chastised the viewpoint of the vast majority of those who wrote comments in to the BLM saying they wanted the top of the Roan Plateau to be off limits to any gas drilling. It makes me think of the movie from last Sunday night, *A Land Out of Time*, and the need to save the last few special places. Has Bob Beauprez ever gone up to Hunter Mesa? How about Mamm Creek? Has he ever been hit by a cloud of noxious gases off a well site that will severely impair his health for the rest of his life? Would he be willing to have a chat with Chris Mobaldi, Dee Hoffmeiester, or Susan Haire about sensible gas development? I doubt it.

After the debates, the reporter interviewed Nancy Jacobson who was standing near a group of people holding a sign that said "Republicans for Ritter."

"I am not voting for Ritter, I am voting against Beauprez. They both need to do their homework, the western part of Colorado had become a sacrifice zone, and the people in Denver are letting it happen," she said.

The reporter identified Nancy as a member of a group of landowners who were fighting the gas development. Yes, Nancy is part of a group, her own group. She is somewhat splintered from GVCA, she doesn't go to every meeting anymore, but she comes out swinging and writing when it is time. You go, Nancy! After finding a pit filled with contaminants being quickly backfilled by the gas company without removing the noxious liquids by their property, Nancy and her husband Gary know all too well the way the gas industry operates. They have lived with the fumes, the explosions, house damages, and headaches. And Nancy is one sharp lady.

On Friday morning I called the COGCC number and left a message for Trisha Beaver. She called back that afternoon. I asked her quite a few questions about the unit being proposed in Silt. I turned out I had been told a few incorrect things. The hearing was not until the end of October, and next Tuesday October 10th was the deadline for written protests to be received at the COGCC Office and at the gas company legal office, which in this case would be Antero.

I inquired as to the implication of the creation of a spacing unit for a landowner who owned 100% of the minerals and who had not signed them away. She explained that they could not be force-drilled, but if they leased the minerals down the road then it could affect their royalties. She also wanted to know if they were actually in the unit, or just on the perimeter.

"Why would they be involved if they were on the perimeter?" I asked.

"Well, issues such as setbacks would come into play," she said.

I asked again if the bottom line was that if they owned 100% of their minerals and hadn't leased, then they could be forced-drilled on as was the case in the creation of a federal unit. Yes, she said, that was correct. I asked if they could be directionally drilled under. No, they couldn't, she said.

Trisha confirmed that the spacing was ten wells per fifty-two acres in the proposed unit, and when I asked if it was surface she said no, one well pad on the surface of that fifty-two acres and four directionally drilled from that pad. That, I believe is good news—it is certainly better news than hearing that it would be five surface pads on fifty-two acres.

"So is the reason why someone holding all their unleased minerals gets notification of the proposed formation of a spacing unit because of what will happen around their property?" I asked.

Trisha said yes, plus the leasing options down the road for that landowner might be affected.

I thanked Trisha for her call back; she had been very informative and very helpful. I had jotted many notes down on a large sheet of paper as to how to negotiate the COGCC website to get the requirements for filing a formal protest to the unit formation, plus dates and other information.

Shortly afterwards on my way to Rifle I stopped briefly at City Market to see Linda. I had told her the night before I would, and also that I would be happy to help send information to the attorney if they so chose. That had been suggested by both Peggy and Bob Elderkin…

I found Linda eating a late lunch in the break room.

She saw me at the door, and came out immediately to talk to me.

"I stepped in where I shouldn't have. I'm going to have to let Richard deal with this," she said.

I told her what the COGCC office had told me, in detail. I read off the notes I had taken. I offered my help if Richard wanted, but said to call Saturday. I would need the time to do it over the weekend with work again on Monday and the protest deadline on Tuesday. I felt a bit deflated over suddenly hearing the help she had asked for was no long wanted. More phone calls made, more time spent, and for what? The same feelings I had about the calls to the COGCC and the county over the filling of the pit…

I held out the sheet with all the notes from the call to Trisha Beaver at COGCC.

"So do you want this?" I asked uncertainly.

"Yes!" she said quickly and took it from my hands.

"Thanks for your help," she said.

"You're welcome. Tell Richard to call me if he wants to," I said, and then I left.

Friday night, October 6, 2006

Al was leaving on a big raft trip down Cataract Canyon despite the cold and wet weather. It had been raining hard for the last few days, and the group was beginning to get a bit concerned about the high water flows projected. He stayed at work late that night, going online to get the last-minute updates about the flows. Rafting friends were coming from Summit County and from Steamboat Springs and other locations around Colorado. A group was going to spend the night at our house tonight, then take off tomorrow. Before they arrived I called Nancy Jacobsen.

"Well, I heard you on NPR last night, on my way home from work. You were great!" I said. I could hear her husband Gary in the background asking questions.

"It's Tara; she heard me on NPR, on Aspen Public Radio last night," she said to him.

We had a good long talk, and I told Nancy about the latest dealings with the proposed creation of a unit. She pulled out her big resource book, The Landowner's Guide to Oil and Gas Development.

As we rambled on about things, one comment she made was that the crowd at the debate in Rifle between Beauprez and Ritter did not want to hear that Ritter spent half the day doing an Encana tour.

COLLATERAL DAMAGE

"He should have been out talking to the people, not touring with the gas company. It wasn't what they wanted to hear…" she said.

October 7, 2006

I went out to feed the animals tonight, and got the mail from the box by the road. A silver PT cruiser was sitting parked on the road. Another person lost on these confusing country roads. As I got my mail out of the box, the car began backing up toward me. There were two middle-aged women in the car. They rolled down the window and asked where a numbered county road was that I was not familiar with. Then they asked how to get to Peach Valley Road.

"Where are you trying to go? It's pretty confusing how to get there from here; you'll have to go back to New Castle and go out on Highway 6 or you'll get lost," I said.

The lady closest to the window read from directions scribbled on a large sheet of paper.

"We're trying to find some property on Peach Valley Road," she said.

"Here, let me see those directions," I said, and they handed me the paper.

"Yes, go back to New Castle on this road, then get on Highway 6 and head west, then follow these directions here and turn on that county road four miles out. That should get you there," I said.

"Do you like living here? We're thinking of moving here," The one lady with large earrings said.

"I would like it, except for the gas development. When we moved here four years ago, I didn't know anything about it…if I had I wouldn't have wanted to come. The development is moving into the Silt area north of the river now, and it's creeping up from the west. Have you driven down Highway 6 toward Rifle yet?" I asked.

"No, not lately," the woman answered.

"Well, you need to. And you should drive south of Silt and go up Dry Hollow. Take a look at what it looks like when they are developing for natural gas. They can put the rigs within 150 feet of your house; that way if they fall over they won't land on your house. That's how tall the rigs are," I said in a flat voice. Unfortunately, I was not disguising the truth. If they really wanted to know what they could end up dealing with, I might as well tell them.

"We went and saw that movie about gas last weekend, *Land Out of Time*. We're worried about what is in the water, or what can be in the water," the second lady said.

"You should be worried about the water, and about the air too. I have friends who are very sick very likely from having been exposed to contaminants from the wells near their homes. They can use carcinogenic chemicals when they drill the wells, and it is a bad thing. I'm sorry to say, I'm not making this up…" I said, realizing that yet again, most people had no idea that anything could possibly be harmful if it were allowed to take place in people's yard and right by their homes. I thought of Chris Mobaldi for a moment, and Susan Haire and Dee Hoffmeister. It was too depressing to even talk about to these two women I didn't even know.

"I'm pretty involved in these issues, I'm in a few of the groups that do work on landowner concerns," I said as way of an explanation for my blunt and discouraging comments about their consideration of moving here.

"What groups are those?" the lady in the passenger seat asked.

"Grand Valley Citizens Alliance, and Western Colorado Congress. They're good groups, they're grass roots groups of landowners working on these issues," I answered.

"Where do you live now?" I asked.

"Carbondale," They replied.

COLLATERAL DAMAGE

"Well, stay there, or stay up near there. Whatever you do, don't buy anywhere on Silt Mesa between the river and half way north up toward Silt Mesa Road. They predict that the gas will be viable and recoverable in that area. Of course, they never know where they are going to hit, and when they hit good then they move out from there. Sometimes there wells come up poor, so they don't develop more there. You just never know," I said

They thanked me for the directions, and began to drive off, then they came back to retrieve their pen which I still had from drawing directions for them.

"We need our pen!" they said.

I laughed, and said, "That's what I do, steal pens from people asking for directions out in the country."

Then I walked slowly up the gravel drive to the house with the mail in my hands. I was angry and bitter as I glanced at the fliers and bills. How come I couldn't even go out to the shed to feed the cat and get the mail from the box without getting into discussions about gas? Good grief. Instead of enjoying a quiet, overcast Saturday fall evening in the tranquil narrow valley I was thinking and talking about all of this…

I felt like it was the right thing to do, to answer their questions honestly. I would not wish it on a friend or stranger to move out into this area uneducated about what could happen to your property, your health, and your life. Fine, if you know all about it, you can take the plunge well- informed. But I will not hesitate to be honest about what can and does happen if someone out-and-out asks me.

There are other reasons for my anger and frustration, I know. Having been approached in the grocery store by Linda Murr a few days ago when she asked me for help, then after spending a concerted amount of effort and time finding out information that they may or may not follow up on, is discouraging. Offering to help, full well knowing that people who are not computer and internet savvy would have a very difficult time moving forward with a protest to the COGCC and negotiating a website, not to mention drafting and sending off the proper format of a protest to both the COGCC and the attorney for Antero Gas Company by the Tuesday deadline.

Although I was a bit under the weather from the flu bug that had raced through the agency offices this week, I would have been able to do work on their behalf on my two days off. And, after the call I had with Trisha Beaver from COGCC, I had plenty of information to steer me where one needed to go. We also had resources and assistance available from some GVCA and Western Colorado Congress people most likely.

Providing information and easily understandable resources to landowners was the whole reason I did work last year on the landowner informational meeting we held in Silt. I gave a short talk, and passed out Xerox copies of basic information on how to proceed if the gas companies came knocking on your door wanting to lease your minerals.

There was information on surface agreements, water-well testing contacts, and the names and numbers of the most knowledgeable lawyers who were familiar with drawing up surface agreements for landowners. Ralph Cantafio's firm from Steamboat offered a boilerplate-type agreement for only two hundred dollars. Also listed was the number of a water testing company from Grand Junction that offered reduced rates for groups — typically neighbors who pooled together.

It is the baffled and overwhelmed landowners—whether suburbanites, retirees, or farmers who need this information. That's why we compiled it and held the meetings last year.

I voiced my frustrations to Patrick Barker of Western Colorado Congress on Friday afternoon.

"You can't help people who don't want to be helped," he said.

"I know, but she came to me for help. That's what is frustrating," I said.

"Well, maybe getting the news to the Community Development Plan Committee will do some good; Peggy Utesch was pretty angry to hear about the creation of spacing units. She said Antero was supposed to be informing the Community Development people about these kinds of things, so people could be informed as to what was in the works."

"And I still don't understand why Peggy implied that creating these spacing units means people could be force-drilled, even if they own 100% of their mineral rights, while Trisha Beaver from the COGCC said that was not the case," I said.

"Well, I again commend you on your efforts. There need to be more people doing this," said Patrick.

"Thank you Patrick, I know. But there aren't. I just want something good to happen, and I do know that we have helped some people. I really hope that Dee Hoffmeister can get some help," I said.

I truly worry about Dee, and her family, still sitting up there in Dry Hollow experiencing bad fume events from the wells right by their homes. They do not have the wherewithal to relocate, and she has been severely sickened from it. Dee continues to complain to Barrett Corporation, but has not gone to a lawyer nor gotten medical evidence of her condition. She has also called senators, and they have told her because of the way the laws currently are dictating conditions under which gas development occurs; there was nothing they could do for her.

I have written for a few good hours now, and it is Sunday afternoon. I feel better than I did yesterday now that I have gotten some work in on the computer. I feel like I can relate to the title of the documentary from last weekend, *Land Out of Time*. The two-day weekends are so short, and the hours I take out of my home time to write feel like hours I am missing from my quality time. On the other hand, I enjoy writing and miss it when I don't do it. The difference with this project is it is not an historical fiction piece of work as my other projects have been; it is something that has the feel of pressure for completion behind it. It is a totally different endeavor than letting your mind loose on something that is to be pleasurable; this book is supposed to informative, timely, and factual to the best of my abilities.

I wish I had more time to work on this story of landowner impacts from the gas development here. But, with full-time work obligations I have what I have, and that's the best I can do.

Sometimes the pressure feels great, and the expectations one puts on oneself are large.

Now however, the sun is shining after a number of days of steady rain. There may be an hour or two or three before the next roiling white and gray cloud banks visible to the west blows in. It is time to enjoy life, to do chores and clean barns and pet animals. Time to admire the covering of brown and yellow almond-shaped elm leaves littering the damp ground, and the green and yellow leaves still clinging to the half-bare branches of the huge trees around the house and along the gushing creek.

The herd of over 1,200 ewes just came off the mountain from summer pasture. They move up and down the field like a slow and steady tide, and I hear the sound of the bells hanging from the necks of several clanging over the roar of the creek, which is fat with recent rains.

Thursday October 12, 2006

After many days of rain and overcast sky, today is wonderfully sunny. The muddy corrals are beginning to dry up somewhat, and the horses and llamas are basking their dirt-caked bodies in warming sun. The foliage that clings to the large elms is still pretty enough to be beautiful; the backlit brilliant-yellow leaves slowly twirl to the ground.

I took today off from work, and now that jury duty has been cancelled, I have a rare break in the busy week to write and work on gas issues. I called Nancy Jacobsen to ask if her husband could talk to me about translating audiotapes directly into the computer. Gary is a computer specialist. As usual, Nancy and I got into a lengthy discussion, most of which focused on oil and gas talk.

"Did you see yesterday's Sentinel?" she asked.

"No I didn't, where is that paper from?" I asked.

COLLATERAL DAMAGE

"It's out of Grand Junction. They had an article about a family who is suing Williams Company in a class action law suit for failure to pay royalties. I about did a dance when I read that! I guess I can still get excited about these things, which is good," said Nancy.

I asked her for more details about it, and then we talked about the recent articles discussing Mary Ellen Denomy's abrupt leaving from the job of county oil and gas auditor.

Nancy said she would have Gary get back to me about the audio transfer for the computer. Then we got onto our other pet issue, The Rifle Animal Shelter's goings-on. There was going to be a silent auction and wine-tasting up at the Rifle Golf Course in October. Nancy asked if I wanted to go, either as a volunteer or as participant. I said I would if I could; it depended upon how work was going at that time, as always. The Rifle Golf Course was in an absolutely gorgeous area, just below Rifle Gap Reservoir and State Park. Sadly, however, the last time I drove up there I saw the new gas drill rigs encroaching.

Nancy and I swapped notes on the foster cats, and Black Max, the luxurious longhaired black Persian cat that was beating up her older females. He could no longer be fostered at her house. Nancy and I always have two subjects to discuss, one is gas and oil, the other is the shelter animals. Finally, we said goodbye. Nancy said she would e-mail the Sentinel articles.

This morning I was awakened by a phone call a woman from New Castle who read my letter to the editor and wanted to tell me about a committee they were starting in New Castle. Also, Peggy Utesch suggested that she call me. Bleary with sleep, I jumped out of bed to get to a pen a paper and jot notes. It turns out the committee involved solar homes, weed control, xeriscape gardens, and recycling and other important and worthy causes all of which greatly interest me...

Once I realized this committee was not concerning oil and gas issues, I confessed that I was interested, but already overburdened time-wise to join the committee as an active member. Already between full-time work, GVCA, and the book project, I had to pick and choose what activities to attend and what to skip. Like last night's award ceremony in Rifle, where Antero was receiving an award for their participation in the Community Development Plan. It would have been nice to go, but this week had been overbooked already and sleep was needed.

Tuesday morning I got up early and quickly dressed, then slammed down a few cups of coffee, hayed and grained the animals in the rainy, muddy corral, and then raced off to Mamm Creek to the pit where the COGCC was doing the auger samples. I had been graciously invited to attend, and I wanted to go, mainly to see what the whole process looked like and to get a face-to-face chance to ask a few more questions about what really happened at that pit, especially on the second day when they were supposed to be draining more out but Rick witnessed them backfilling with over three-hundred barrels of liquid still in the pit.... And this was early in the morning.

Dave Graham, the COGCC inspector told me himself later that morning that he had not been sent out to check anything again. That is when I began a string of phone calls which continued throughout the day, only to be told abruptly by Jesse Smith, the acting Garfield County gas and oil liaison that no violations had occurred, and that three tank loads the first day were taken out, (and then after my pestering calls) two more loads were drained on Wednesday.

Driving west through the rain toward the pit near the Rifle Airport, I fought the urge to stop the truck and retrace my tracks back toward Glenwood. I got onto I-70 and took the airport exit, then drove up and over the steep hill where a few houses and small barns huddled near the massive new drill rig just across the road. It was bustling with activity; pipes were being loaded and the well pad was thick with equipment and people.

After one false turn down a wide muddy road prior to County Rd. 319, I inched my way up the pitted mud-soaked road until I dared do a five-point turn on a wider spot. The last thing I needed was to get stuck in a ditch out of cell phone range up Mamm Creek early on a workday. God forbid. I returned to the Airport Road, then made my way successfully to the well pad on County Rd. 319.

COLLATERAL DAMAGE

There was no activity at the smoothed-over pit, no trucks, no auger-testing people. The rain fell down on the dismal landscape under the gray sky. Oh well, an hour's driving in the early a.m. for nothing. Maybe I had been too late?

That night when I returned home the message light was blinking on the machine. It was Dave Graham, the COGCC inspector letting me know they had postponed the auger test until next Monday morning at 8. They couldn't do the test in the rain. Well, I had to trailer the horse into the Glenwood Vet Clinic for x-rays Monday a.m. I would miss the auger sample. Too bad I didn't get the call earlier; I could have saved some gas and time.

I do not begrudge the trip, however. There are times when I have to listen to my instinct of what to do, and this was one of them. That morning when I drove back down the road toward the interstate after leaving the pit, I was overwhelmed by the sight I saw at the rig by the small homes on the crest of the steep hill. Perhaps one day this had been a charming spot to live, a beautiful vista to look down upon. That was not the case today, nor would it be for many years, I am sure. It was now a full-blown industrial sight, right next to the country homes.

As I got to the intersection where the on-ramp leads to the east on I-70, I again drove by the group of water trucks parked with their back ends dipping into the ditch with large hoses. I paused and looked over. How I wanted to park and just walk over there, and get a bucket of what they might have in those gray tubular trucks. How I wanted to know where they had been, and what they were picking up or dropping off out of the Mamm Creek Ditch. Instead, I turned onto the on- ramp and headed off through persistent rain to Glenwood and to work.

Tuesday night the regular group assembled for the monthly GVCA meeting. Peggy immediately talked about the forced pooling that has been discovered on Silt Mesa, and how disappointed the Community Development Plan people had been that Antero was not forthcoming with the creation of spacing units. Antero seemed surprised that the public would want to know about the units, was what the spokesperson from Antero told Peggy. It was obvious that Peggy did not share the sentiment.

"We had a lot of phone calls over the weekend, and we have decided to still give Antero the award tomorrow night, but we have realized that the honeymoon period is over..." she said.

After some discussion back and forth about the specifics of what a spacing unit is as apposed to a federal unit, there was a general agreement that GVCA would continue to support educating the community as to upcoming disclosers of such things as spacing units and their implications for landowners.

Then Alan V. talked about Presco's proposal to drill within nine-hundred feet of the nuclear blast sight on Battlement Mesa. The groans of disbelief were audible in the crowd. Some people put their faces in their hands as the news continued.

After Alan was done and petitions were passed around and signed to object to the drilling by the blast sight, I walked across the room to hand two petitions and a note to Alan. On it I had written Judith Kohler's number with the Associated Press and Deb Anderson's number.

Patrick Barker of the Western Colorado Congress addressed the air initiative, and talked about the positive outcomes that were forecasted. A lot of discussion was had about condensate tanks being left open, and the emissions that were not being included in the industries projections of emissions levels. Duke Cox talked about the pollution permits that were given for various condensate waste pits in the county; the unfortunately allowed sites where some of the condensate was shipped and dumped into huge pits to evaporate. One of the landowners from Parachute who had gone to Denver for the air quality hearings lived a few miles from one of the huge licensed pits. I felt so badly for him, and his family and neighbors.

Other business was covered in the fast-paced meeting. Everyone got to make a remark if they wished. I raised my hand, and made a public thanks to Patrick Barker. I re-told the story of the pit filling, and said how Patrick urged me to call it in to the COGCC. Then, I said that as a result of the network of people from GVCA, I had been able to call Peggy when I was asked to help with information on the forced unitization on Silt Mesa. I was so glad that the Community Development Plan people were there to pursue the information, and come up with a plan of action to help other landowners in the future. Duke Cox, the President of Grand Valley Citizens Alliance, stood up and thanked everyone for the hard work they were doing, and said that Senator Ken Salazar had recently called and asked advice on what were the important tasks at hand to focus on. He talked about what an important model entity GVCA had become. Then he said,

"The work that each and every one of you is doing today is going to shape the future of what this community and valley look like a few years down the road. I know everyone is working hard, but don't give up; it is very, very, important."

Soon it was after nine and the meeting was closed.

Last night I called Deb Anderson in New Mexico. We talked for over an hour, and I was overjoyed to hear that she was seriously working on the film. She already submitted for a grant with the State of New Mexico. This weekend she was going to film Tweeti Blancett. I told her about the film screening of *A Land Out of Time* recently here in Glenwood Springs with the Aspen Film Festival. The Blancett's ranch in New Mexico had been filmed, showing grim images of dead cattle near production pits filled with drilling liquids on the ranch lands. Tweeti had spoken eloquently and sadly about their struggles and their hard choice to soon leave the family ranch.

Today we talked again, and discussed the various entities that could be approached for funding for the film. After our call, I called the San Francisco Office of the Natural Resources Defense Council, and then was directed to the New York Office. I had a call with that office, and was told to submit an e-mail with description of the project, and to mention that the *OnEarth* article "Wrecking the Rockies" by Michelle Nijhuis had spurred Debra's interest in this project.

I called Deb back, and passed the information on.

We will see….My fingers are cold now as I have been typing outside, been working on gas things all day, time to stop now.

3:00 p m

On impulse, I called John Martin, one of the Garfield County commissioners just now. Usually, he is pro-gas and I do not see eye-to-eye on his stance on the issues. I called to tell him that I appreciated and agreed with the stance he took in the paper yesterday opposing the leasing of minerals in a forced-pool unitizing situation on Garfield County property by the Rifle Airport.

I told him that friends and hay ranching neighbors of mine had recently gotten a letter from COGCC being told they were being force-pooled and put into a unit.

John said he was trying to work on changing legislation on it. We discussed the ways in which it was not fair to the landowners.

I spoke honestly.

"These ranching neighbors and landowner friends of mine, the people who don't want to lease out, they need lawyers to be able to deal with it. Not all of them will go to lawyers. All the law and rules and regulations are set up for the industry to be able to do what they want anyways….It is not fair, and I will fight against it as hard as can. That is why I called you, because of what was in the paper," I said.

"This place is getting wrecked, and the laws are set up for them to steamroller over everyone, even those who have 100% mineral rights. It's not fair," I said.

"And I don't know why Trisha Beaver from the COGCC told me in two different phone calls that in these forced-pooled spacing units they can't drill under or on your property if you own 100% mineral rights. Why don't these state regulatory agencies know the rules?" I was angry.

"She's an attorney for the COGCC. You have my number, call me again," said Commissioner Martin cordially.

COLLATERAL DAMAGE

If he continues to work on legislation for this issue, I certainly will. For some strange reason, I felt good after I called him.

It was like a feeling of crossing party lines for an issue.

I feel the same way about the steady parade of hunters and their big rigs and trailers and ATVs rolling all day in front of my house. Today must be the day before an opening season; it's been constant traffic all day.

I am not a hunter; I have problems with poachers every year. This year however, I feel less hostility boiling over.

They are partners in having a stake in preserving wild places from gas development.

Sunday October 15, 2006

Yesterday Rick Roles came over and we did a three-tape audio interview here for a number of hours. The weather was indecisive, but warm enough to sit outside on the small deck at the picnic table. I had the tape and the mike all set, and had extension cords run from the outlet by the door. I have met with Rick enough times now that we are comfortable as friends, and as he got out of the big white pickup he was the picture of a red-neck mountain-man type of cowboy. He has sun- weathered skin, and lines around his eyes, and beneath his dirt-encrusted cowboy hat his long brown hair was braided tightly into a tail that hit his shoulder blades. Although Rick on first glance has a somewhat unpolished demeanor, he also has impeccable manners, and his conversation is punctuated with

"Yes Ma'ams."

As he was talking into the recorder, I regretted that I did not have the video camera also rolling, but for now I was just doing the audio. As he relayed information to me as we sat in my kitchen (the cold and light rain had forced us in from the deck), there were moments when I was struck by a very strong sensation that the things I was hearing were potentially important pieces of information. That is, if they get into the right hands...

Rick left in the early afternoon in his big white pickup to meet family in town. His boisterous black dog Rowdy was ready to get out of the truck after waiting patiently through the long taping. There were too many animals at the ranch to invite Rowdy to come out of the truck. Between my numerous cats, the llamas, and the neighbor's free-ranging-by-day chickens, Rowdy could have easily lived up to his name and gotten into a bit of trouble.

The weather never cleared that afternoon and I stuck close to the house and barns. Al called from Moab, they had gotten off the river at Lake Powell a day earlier than previously expected, and he would be home that night. I tidied up the house a bit, and began to listen to portions of the tape from earlier that day. Although we had rambled over a variety of topics freely in our discussions, the tapes sounded quite good. I was pleased.

As I lay in bed that night, some of the things Rick had said began to sink in. I thought of him trying to sleep in a chair, with great difficulty. I could not imagine having to sleep in a chair, as I lay in my own comfortable bed under thick comforters. Rick could not bear any weight on his shoulder blades due to the pain.

At the end of the taping that afternoon, I asked Rick what had inspired him to get involved with the press, the documentary film makers, and others interested in hearing his story.

He said he figured his chances were over, but maybe he could help other people in the future avoid having similar health problems. Although Rick has lived amongst the wells and evaporation pits for over ten years, it's only been recently that he has gotten involved with any other land owners. And that decision was spurred by his discovery of a gas well pad being put in by his grandchild's tee-ball field in Rifle; right by it at the elementary school. With those thoughts rolling over in my mind, I faded off to sleep.

COLLATERAL DAMAGE

Sunday

Today Al is home from the rafting trip, unpacking his rain-soaked gear and hanging it over fences to dry. They had a very rainy trip through Cataract Canyon this week, with only a few days of sunshine. It is good to have him home, and I am glad I made good use of my solitary time doing phone, computer and taping work when the small house was all mine.

I drove into town to pick up the local papers. Page two of the Glenwood Spring Post Independent has a headline reading:

"Parachute rancher files class action suit against Williams Production."

The story proceeds to cover the reports that Sid Lindauer has filed suit against Williams for underpayment of royalties. Lindauer's wife and brother are also plaintiffs in the lawsuit. Also included are all the mineral rights owners in the county. The court filing reads that the underpayment occurred "because it has failed to fully account for the proceeds that it receives from the sale of the gas products," and additionally, Williams charged expenses to the mineral owners for placing "gas in a marketable condition to deliver to the commercial marketplace".

The spokesperson for Williams, Susan Alvillar said, "We're evaluating the claim by the plaintiffs, but because it's in court we are not able to comment any further."

The October 15, 2006 article by Donna Gray continues, saying

"With the Lindauer suit, [Attorney Nate] Keever said a class-action approach seemed the best avenue to take.

'You can't go against Williams if you are small financially', he said. 'the idea was how to put together [a suit] that would help everyone… so we don't have to do this over and over again'.

"He said he has 'a long string of people who are interested in suing Williams over their royalties'.

"Williams is not the only target of lawsuits. Encana Oil and Gas (USA) has also been named in a class action suit filed by mineral owners across the state. That suit accuses Encana of 'general skimming', Keever said, 'Taking improper deductions and pricing incorrectly'.

"Keever said that he hopes the Lindauer case will put the matter of underpaid royalties to rest. 'I hope this is the last royalty suit against Williams.'" (End of quote from Post Independent Article.)

On page three, the top headline reads:

"Fire erupts at well pad south of Silt."

The article goes on to describe an early morning fire that broke out on a Barrett well pad. Two workers were sent to the hospital for smoke inhalation. Brit McLin, the fire chief for Burning Mountain Fire Protection reports that it was 5:30 a.m. when they arrived at the well pad.

"'There was one tank opened up and a 500-barrel capacity [tank] fully involved in flames. There was also half a dozen ground flares,'" said McLin.

Barrett Corp. shut down the wells on the pad, and the flow-back procedures at wells elsewhere in the area pending completion of the investigation.

I am curious if this well pad fire might have been the plumes of black smoke that Rick Roles saw yesterday, or if there were also pits being lit-off simultaneously. The times seem to match up, but I don't know if the location does. I am also curious what kind of smoke/fumes the gas pad workers were breathing, and how they have fared. Did anyone take an air-grab sample this time, unlike other airborne events (when chemicals in evaporation pits were being burned off as opposed to removed from the site for disposal) when Jim Rada from the Department of Health took no samples?

October 19th, 2006

This week has been too busy at work to write any evening, but I'll take a few minutes now.

COLLATERAL DAMAGE

The papers have been full every day of letters to the editor about the candidates coming up for election shortly. Oil and gas issues are a common theme; and the county assessor's race has been a bit ugly lately, as have some of the recent comments made about the commissioner now running for re-election.

I had a long call last Friday night with Susan and John Gorman. John is running for assessor and he is basing his campaign strategy on conducting and completing audits of the gas companies. These were ordered but as of yet, not one has been completed by the current assessor Sharon Hurst.

I called the local television channel from Rifle, and got information about obtaining copies of Chris Mobaldi's and Laura Amos's testimonies from last years COGCC hearing on the Divide Creek seep. Ironically, or better, sadly—that was the same day the tanker truck rolled on the county road up Dry Hollow and dumped a load of condensate.

In between the increasingly fiery and downright nasty political letters to the editor endorsing one or another of the candidates was this letter from a landowner in the gas fields. It caught my eye.

Glenwood Springs Post Independent-October 16, 2006

"THE GAS INDUSTRY HAS RUINED ALL THAT WE CAME HERE FOR

Dear Editor,

I used to be able to count seven oil/gas derricks from my back deck—now they have erected one more. The pall of noxious chemical and diesel fumes hanging over our part of the county cause the gag reflex to work in hyper drive.

One is concerned with being able to send our children out into the yard to play, for fear of impending benzene-induced leukemia and lung damage. There is fear of even bathing in the water, and we can never drink our water anymore (good for the bottled water industry). Several homes and ranches have been abandoned and are for sale.

People have given up or have been forced to leave because of medical problems associated with the negative effect of the gas/oil industry. Those of us, who because of economic concerns are unable to leave, are slaves to the oil and gas industry, and are slowly dying from the respiratory effects of this industry. It is lucky for the industry that the deadly ramifications manifest themselves so slowly; they will be able to deny any responsibility no matter what the empirical evidence shows.

No one is able to sell their homes for anywhere close to the assessed valuation because no one with any good sense or sanity, between the noise, the smell, the contamination, the horrendous amount of truck traffic, the dust, the smoke from burning ponds, the venting of the wells, would spend money to buy one of these contaminated properties.

All of our reasons for living in the pristine quiet country disappeared in the wake of governmental policies, along with unbridled greed and avarice of the politicians.

For anyone, including the politicians, who think the gas industry does not negatively impact their country, their county, their wilderness, the biota, the humans and wildlife, come and sit on my back deck and admire the scenery, smell the fresh air, and bring your respirator.

McKenzie Richards

Silt"

I e-mailed Debra Anderson and left a few messages.

She was glad to hear that access would be available for that footage, and wondered if there would be an issue with putting it into her film. I replied that I had been told, no, there should be no problem.

Debra said in her e-mail today that her trip to Farmington, New Mexico last weekend had been productive, but it had been depressing to see how gas and oil development had ravaged the land.

She continued,

COLLATERAL DAMAGE

"I wonder if this is what Garfield County will look like in ten years."

It made me shudder inwardly to hear her words.

October 21, 2006

Today is the first day of hunting season near our property. I was drinking my coffee this morning, and looked out to see two ATVs parked just past our mailbox and near my multiple No Trespassing signs. I went out and yelled at them,

"This is private property, for quite a ways…"

"We were just looking," they yelled back, and then they took off.

One of the passengers in the second ATV made some kind of gesture at me with his arm, then kept looking back at me as they drove off and as I walked down our driveway toward the county road.

Shortly after, I went to the hardware store and bought a large amount of signs and flagging tape, and spent the rest of the afternoon posting the already posted property. I walked way up the hill, and put orange tape around stumps and trees after I flagged the pasture fences where the horse and llamas were. Earlier, I had put up signs and flagging along the county road at the spot the hunters seem to consider a good place to stop and survey, if not park and hunt.

This year I had fleetingly thought that perhaps this hunting season I would have no trouble, encounter no sloppy hunters on my own property, or see no illegal hunters poaching from friends' pastures as I did last year up Main Elk Road. The hunter knew he had been seen as two trucks drew up across the creek from where he had shot the large elk on private land. He attempted to hide behind a large juniper tree in his bright orange suit as we looked at him. That had turned into a multi-day event as many calls between police dispatch and the DOW occurred over the poacher and the group of Louisiana hunters who set up shop in my neck of the woods.

The day after the poaching incident, the Louisiana guys drove by our place, and I went out to the road to talk to them, I flagged them down. I explained that we lived on private property and had problems with trespassing hunters, and that someone had been poaching nearby. Oh no, it wasn't them, they assured me....They described where they had been hunting the day before, behind the steep mountain of which we own twenty-eight acres on this side.

"Where's your property end here?" one of them asked as he looked up the mountain.

At that point, all my politeness skills vanished.

"If you don't know where private property boundaries are; then you should not be hunting in the area! I come home from work and I have people in my driveway ready to go hunting there! Right by my house and my pasture!" I said angrily.

The Louisiana hunters politely said goodbye, and drove off in their big muddy truck with the trailer of even muddier ATVs.

This year is pretty much the same...

Friday afternoon while at work I got a call from Jamie Atkins from the COGCC, informing me of the findings of the soil sampling done at the pit-fill I called in on Mamm Creek recently. He was exceedingly polite; I wonder if he read the letter to the editor I wrote about the pit-filling incident...

I gave him my e-mail and he said he would send the scientific analysis of the sample. Yes, there were contaminants found in the sample, but due to the location of the pit they were under the allowable level of contaminants for that area.

I then asked him about the realities of the spacing units, and what the true answer was as to a 100% mineral owners right's to be exempt (or not) from being drilled on or under if they were force-pooled into a spacing unit. He said they could not be drilled on or under in that case. I explained that I had heard differing things from various sources, including Trisha Beaver of the COGCC.

Again, I am unsure what the answer really is in this case. I am still under the impression that the county land was going to be in this category, and they owned 100% of the minerals and chose not to lease but were forced into doing it or taking a financial beating. Another project to pursue....That one should be easy though, because of the county information. I can call John Martin back, or just read the old papers if I can find them on-line or elsewhere.

Nancy Jacobsen told me yesterday when I called her that she was told by Jamie A. that the reason for the recent fire/explosion was "land-owner caused."

"What, did they light the tank?" I asked, totally confused. Was it perhaps sabotage? Here in Garfield County?

"No, Jamie said it's because they wouldn't let them flare, they didn't what them to flare, so the sand somehow cause a fire to start....A huge black cloud of burned off chemicals blew over our valley and hung; it was early, but boy did we see it..." said Nancy.

I had actually called to talk to Nancy about the hunters, as she had been prepping her place for their arrival also. Gary and Nancy had a bigger spread in an area where neighbors leased property for hunts right by their land, and they had many problems, including bullets whizzing onto their land from over two miles away.

"Don't feel bad about protecting your property; you have to or they will walk all over you..." she said.

"I know, but this place looks like some kind of Halloween festival I put up so much orange and black and NO HUNTING signs every where; I even nailed a bunch of signs to a pallet and dragged it up on the hill, and I duct-taped a sign on top of a spare election sign I had; those work really well!" I said.

"I know, I did the same thing!" said Nancy.

As we rambled on about hunters, Nancy told me about a fiery 70-plus-year-old neighbor rancher of hers who sat in her truck with her rifle leveled off on the door at hunters who came near her property. She was a fierce old-time defender of her property rights.

I told Nancy about my call from Jaime Atkins, and she told me about another COGCC guy she talked to recently about last weekend's fire.

"He's a good guy, and he told me not to forward his e-mails to Jamie A. I told him to tell Jamie I was writing a book about all this…" said Nancy.

"Are you?" I asked curiously.

"Well I certainly could, I have enough information compiled. I probably should!" said Nancy.

I agreed, she should. Nancy was a wealth of information and lives right at "ground zero" amongst the high-producing gas and oil wells on Dry Hollow. And she was sharp, and seemed to have adequate time to write. The question is however, would she, and if she did, when would it be?

Sunday afternoon is proceeding, and it is cold in the little greenhouse now where I have been typing for a few hours. It is cheerful in here, with the plants and the red tomatoes and herbs and mint. It is also quiet and a sanctuary away from the noise of the small farmhouse kitchen and television or music.

Mustard-colored leaves stick to the translucent roof of the greenhouse, and the sun shines brightly through. A few of the cats slide slowly down the slick clear plastic roof before they jump onto the grass after their naps atop the skylights. I only heard two rifle shots, and from a fair distance, as I wrote today.

Hopefully my yard and pasture decorations will keep the hunters away from our place this week.

Time to go out and enjoy the day, and be with the animals and do chores.

COLLATERAL DAMAGE

Tuesday October 24, 2006

McKenzie Richards had another letter to the editor in yesterday's Post Independent.

I called a few people and left messages, asking if anyone knew who McKenzie was, as the number was unlisted.

Oni Butterfly called me this a.m. at eight o'clock, just as I was getting ready to leave. I was surprised to hear—and happy for her—that she had sold her home and was moving to Carbondale.

We talked for a good while, and she told me that while she had recently been in Oregon, apparently the fire/explosion that occurred had been at her property. A neighbor had told her that fire trucks had surrounded her home, and the explosion occurred at the frac tank.

Oni said that in her discussion with a Barrett employee, Jim Felton, he had said that it was not an explosion; it was a fire. Oni said,

"I said 'Jim! Carol and Orlyn Bell heard it at 4:30 a.m. from a mile away, and Pat and Mike Smith heard it! It was an explosion!'"

Then Oni went on to say that in an e-mail she got from Felton, he inadvertently let the previous e-mails go along with it, and they read as an attempt at a cover-up.

I asked Oni to forward the e-mails but she was hesitant, due to some of the neighbors who were stilling trying to "deal" with Barrett.

His e-mails to co-workers indicated that they needed to downplay what happened, and that 'if' the injured gas workers got out of the hospital OK, then they could say it was just smoke inhalation. Something to that effect.

Oni was coughing during our conversation, and mentioned that her doctor at Silt Medical Clinic said that she had seen a large rise in the numbers of respiratory-related diseases over the past years.

I told Oni I was glad for her she was leaving, and then I went off to work.

Today's papers were full of front-page articles about the five mayors' endorsements of John Gorman for county assessor. Yesterday's paper had an ad running the five mayors' support of Gorman over his opponent Hurst.

In addition, there were more letters to the editor on both sides of the issue, and one from John Gorman himself discussing the reasons why Mary Ellen Denomy, the certified gas and oil accountant, had resigned her appointment as county gas and oil auditor. Much of the blame was said to lay at the feet of Hurst, for not following through with the audit. That is what the letter stated. Earlier, I received a copy of Mary Ellen's own resignation letter to the county commissioners.

October 29, 2006

Talked to Susan and John Gorman Friday night. John was exhausted and lying on the floor after a day of campaigning. We talked at length.

I asked him,

"John you don't live in the gas fields; why do you care?"

"I am a realtor and I have sold people properties out there; I have an obligation to care," said John.

I wrote a letter to the editor last night on the computer and saved it, then today revised it and ran into computer problems. I must have rewritten it five times now...

Debra Anderson called me last night and I told her I had the three DVD discs to send her. They included testimony from Laura Amos, Chris Mobaldi, Susan Haire, and Pepi Langegger at last summer's COGCC hearings on the Divide Creek seep moratorium.

Reality Check

Halloween October 31, 2006

Today I took my horse Buddy back into the Glenwood Veterinarian Clinic. He would be there for two nights, and his nervectomy surgery was tomorrow at 9 a.m. This morning Buddy loaded successfully into Kathy Miller's trailer, a much easier scenario than last week when we took him in for the nerve block.

I left him at the clinic, nicely situated in a roomy stall in the large surgery barn. Kathy and I stopped for lunch at the restaurant in New Castle in the old brick building from the mining days, The Elk Creek Mining Company.

We talked about the horse stuff, and the recent news in the papers. Invariably and somehow I got to talking about gas-related issues. Earlier that day I had asked Kathy if she had voted yet and she said no. So I began a pitch for her to vote for John Gorman for county assessor. Kathy, a sharp and shrewd woman immediately asked,

"Why?"

I launched into numerous reasons why John Gorman was the right man for the job.

As we waited for our food to come to the table, I went to get the local papers. I immediately flipped open to the letters to the editor. I was pleasantly surprised to see the letter I had written in support of Gorman already in; I had just finally successfully e-mailed it yesterday after fighting with my dial-up server all weekend. The letter came out better as a result of over five revisions. I had the sense to hold the first draft overnight, then after revising and running into dial-up failures with the computer I got to written copy by Sunday night. I believe it helped me boil my points down well.

(Inset 10-31-06 letter by Tara in GWS post)

"Gorman will assure fair royalties from gas companies"

I handed Kathy a copy of the Post.

"Here Kathy, you asked me why I was voting for John Gorman. You can read why..." I said.

She took the paper that was folded open to the page containing my letter to the editor. I could see by the puzzled look on her face as she began reading that she didn't yet know that I wrote it. Then after a moment or two, she looked up and said

"I just figured out that you wrote it..."

For the next hour-and-a-half we launched into a gas-related dialogue. I told Kathy about the various things I was involved in, and talked about industry practices including hydraulic fracturing, the exemption by the EPA from the Safe Drinking Water Act of the carcinogenic chemicals used, and more. Kathy had a stunned if not horrified look on her face as I spoke. She said the same kind of things that others say in dumbfounded voices when initially reacting to the information,

"That can't be, they can't be doing that, there must be some agencies overseeing..."

In a flat voice I told her what I knew, and when she was questioning me about some local group, the "Alliance" group, not affiliated with the state or the industry, I said,

"That's us, Kathy, the grass roots. It's the Grand Valley Citizens Alliance, a grass-roots citizen's group with a committed core group here and a few hundred members. We are not a regulatory entity, we a real grass-roots group of landowners," I said.

Because I was so relieved to have had the horse successfully delivered for a big surgery that would hopefully save his life, and because I was in the company of one of my most respected friends from Colorado whom I met during Hurricane Katrina animal rescue efforts, I could not seem to stop discussing issues about gas development. Kathy is a vet tech and a medical-lab tech and a very smart lady. She lives on the edge of natural gas development land and knows so little. When I began to tell her about the illnesses that people were coming up with, she was astounded.

"Why don't we hear about it? Why isn't it in the papers?" she asked.

"Well, I've been told the papers don't want to get into it, except for the Aspen Daily News locally. I've called all the papers about the big stuff, and the only ones to pick up it so far are the Aspen Daily News and the Associated Press. I worked for a long time with the Associated Press reporter Judith Kohler on an article about a lot of the people here with severe health impacts from living near wells. I don't know if the lawyers are telling them to wait....She told me about two weeks ago that she submitted it to her editor," I said.

We also talked about the film people and their interest, and the work I was doing now with Debra Anderson.

As I sit here at home later, tired from the day, I remember the look of bafflement on Kathy's face as she said,

"Why isn't it in the papers, why don't people know?" as I told her of the conditions and symptoms of those I knew of with problems who lived by some of the bad gas wells.

I looked her straight in the eye.

"I am doing all I can; I am working with the press, I am talking to the film producers. What else can I do?" I said.

That is true; what else can I or anyone else do? When we got back to the ranch I forced some money into Kathy's hand and told her to have some drinks in Aspen tonight on the horse. Kathy and her husband Orty were going up to Aspen to watch the parties and the costumed locals for Halloween. Aspen is well-known for having spectacular costumes at the bar parties and on the street.

I checked the mailbox before walking up the gravel drive to the house. As I pulled the bills out, there was a FedEx slip from the package of three CDs I sent to Debra Anderson in New Mexico. On those CDs were the tapes of last summer's COGCC hearings on the Divide Creek gas seep and testimonies from Chris Mobaldi, Laura Amos, Susan Haire, and Pepi Langegger.

"This is my FedEx slip from the CDs I sent to the film editor yesterday," I said to Kathy. When I looked down at that folded tracking slip of thin white, blue, and orange FedEx paper, it was like looking at a chance of hope. I continue to hope that somehow, somehow, we can bust on through to the other side with big media and get the general public to see and understand what is going on.

Kathy carefully backed the truck and trailer around at the end of the driveway and drove off. I was very lucky to have such a generous friend as Kathy, who had spent the better part of now two days helping me transport the horse for surgery.

I know that Kathy Miller now has a pretty good idea of what is going on, after our long conversation. Unfortunately, not all citizens of Garfield County or of the West will have the opportunity to be in such a conversation. That is what we need the media for.

I must remain hopeful, for what else is there except running away? I'm not saying that is not a wise choice—but in my case that involves the abandonment of two careers and the agreement of my husband. Right now, it is not going to happen. Not even a conversation about thinking of it.

Sunday November 5, 2006

It has been a very busy few weeks, and I have been too preoccupied with the horse's surgery and work issues to do much on gas and oil. Elections are around the corner, and last weekend I did spend a lot of time working on multiple drafts of the letter to the editor supporting John Gorman for assessor. The papers and the television are clogged with propaganda for the candidates, and as the November 7th voting day draws nearer and nearer the language of the attack ads has become downright ridiculous in some cases. We'll see what Tuesday brings…

So the horse came home from the Glenwood Veterinarian Clinic on Friday, and after a somewhat tenuous start to his six-week stalled recuperation time, things are going better now. Initially Buddy had little interest in being penned up in the corral-panel stall area I rigged up in the big hay barn. He pawed the ground and kicked at the metal corral poles with his newly incised bandaged leg where the two sections of nerve had been removed. I shuddered as he fought his strange confined surroundings and prayed that he would not damage his stitches…

Later that afternoon he settled down, although Doc Holliday, his equine partner, seemed fixated by either the nearby high-powered rifles blowing off at the neighboring ranch, or he sensed a mountain lion. Doc spent the next several days in remote sections of the pasture, staring up the mountain and snorting. Every so often he came to eat a little hay in the barn and visit Buddy.

Today I did my first bandage change on Buddy and thankfully there was no seepage from the incisions. Everything looked OK, and I laboriously and slowly cut the many layers of bandaging and thick white gauze off, then reapplied as best I could the pads, the gauze, the waterproof tape, the cotton batting layer, the vet wrap, and yet more waterproof tape. Thank God Buddy stood still as a statue during the process. In the vet clinic just several days ago he had put up quite a fuss when the veterinarian and his assistant were demonstrating the bandaging for me. I had dreaded a replay of that, but Buddy must have been a lot calmer now back at home in recognizable surroundings. I've always said that my huge bay gelding Buddy is like a golden retriever trapped in a horse's body.…If you have to have surgery done on a horse, it is a blessing to be dealing with a calm horse, not a high-strung bundle of equine nerves.

I sent the copies of the CDs to Deb Anderson in New Mexico for the film. I also had back up CDs made for myself. After an unsuccessful attempt to watch one of the CDs on my new laptop, with the help of Al I inserted one in the DVD player in the living room. It played just fine, and I watch the COGCC hearing from last summer with interest for about a half-an-hour.

How strange it was to see Laura Amos and Susan Haire, and hear them arguing their cases in front of the COGCC board. Nothing much productive was being achieved, and the carryings-on resembled a laborious legal maneuvering. I would have to watch the rest of the tapes later, and hear Chris Mobaldi's testimony. Chris was there, looking thin and ill, but determined and poised to speak as she stood next to Laura Amos.

A flier came in the mail a day or two ago about the Air Quality Control Commission's hearings being held in Denver on November 17th. I had planned to go, and hopefully still can make it.

Saturday November 11, 2006

It's been quite a week: mid-term elections were held this Tuesday and it was almost a complete Democratic landslide. On Wednesday morning I called Patrick Barker from Western Colorado Congress to get up to speed on who all was going to the air quality hearings November 17th in Denver. I asked him if he knew how the elections came out, and he said it was in the paper that Tresi Houpt won for county commissioner, John Gorman won for county assessor, and Bill Ritter won the governor's seat which I had seen on TV the previous evening. I was overjoyed, having worked closely with John and his wife Susan both on the ISOR initiative over surface owner rights, and then next with the assessor's campaign.

The other day out of the blue I got a call from long-lost Joe Brown, the guy working on the gas documentary this summer. He seemed to be getting back into it again after taking a multi-month break. Last night he called and we talked at length; he was on his way up to a cabin above the nuclear blast site in Battlement Mesa that one of the land owners had made available to him. I confessed to him that I did not discriminate with any form of press as to who I gave information to; I gave it to whoever was currently pursuing something actively.

It was up to each writer or film producer to take it from there, and at whatever speed or success level.

Ken Wonstolen from the Colorado Oil and Gas Association had written a lengthy letter to the editor blasting Patrick Barker's stand on the Air Pollution Control Division's proposed rules for cutting smog-forming emissions from oil and gas operations. He asserted that if western Colorado wanted to demand the same lower emission level as the Denver metro area (basically half of current emissions being proposed currently), then first western Colorado should impose the same regulations on auto emissions. Next, he asserted that EPA studies justified the emission levels.

After I read his letter I became increasingly angry as I thought about the EPA standards. How had the EPA handled Weston Wilson's whistle blowing action on the study done regarding the safety of hydraulic fracturing chemicals? They had swept it under the rug and ignored it until the current administration passed the recent Energy Bill that exempted those carcinogenic chemicals from the Safe Drinking Water Act. I did not need Ken Wonstolen of the Colorado Oil and Gas Association dictating to me his arguments using the EPA studies as his defense.

Folks from out here don't even call the EPA any more most of the time; they know that the Federal Government has put energy development on the fast track, and nothing else matters.

Basically, the fox is watching the hen house.

Today I went to a gathering to celebrate the recent local elections. It was out in Rulison, and eerily close to Chris and Steve Mobaldi's home that is still vacant and for sale. It was a wonderful afternoon at a beautiful Colorado classic ranch home. Jo Ann Savage, the hostess, had provided a sumptuous buffet, hors d'oeuvres and bar on the deck overlooking the jaw-droppingly beautiful surroundings (aside from the numerous gas rigs poking up here and there in the surrounding landscape).

Kim Phillips, the lovely and energetic Democratic organizer was tending the bar, and I took a beer and a cheese and cracker and began to mingle with friends both old and new. Shortly we were ushered inside to fill our dinner plates from the huge buffet.

I met a woman who looked familiar, I knew I had seen her somewhere before but I couldn't place where. We chatted for a while, asking about different committees, groups and meetings, but neither of us could figure where we both recognized each other from. As she and I continue to talk about gas-related issues, suddenly I realized where I had heard of her before. Nancy Jacobsen had told me about Lisa Bracken, a landowner up by the Divide Creek seep who had successfully settled with Encana.

Lisa was telling me about their property, their wilderness center and her father's dream of creating something for the future. Her family's heritage was mixed Native American, and she spoke with pride about her family and their work. Her mother was Cherokee, and her father was Comanche—their land was very precious to them, and part of their lives.

"We will never give up this land..." she said with certainty.

Then she began to talk about her father.

"He has pancreatic cancer; he has been dealt a full hand. I don't know what I am going to do without him..." she said.

The look on her face after she said these words I cannot describe.

When Lisa said these words, an immediate physical and mental sensation went through me. Another individual, close to gas and oil problems, coming up with another form of cancer...

I had felt this before, each successive time I hear of another person coming down with deadly symptoms living by the wells. I swear to God I am not lying.

I stared into Lisa's eyes and said,

"I don't know how to say this, but when you told me that I tingled all over; it hit me. I've had it happen before, when I've heard of other people's cancers and illnesses," I paused for a moment, not knowing how to say what I was feeling or thinking.

"I am not the kind of person who believes in auras and all that yoga kind of stuff…" I trailed off, trying to convey to her that for the most part I was pretty much just a regular basically grounded person.

"I could see the flush come over your face when I said that: I think that you are a person who is meant to be here to do this," said Lisa.

We talked more, and we said things that I cannot say here.

I drove home from the gathering as dusk descended over the mountains. I went the back roads to see the newest rigs along the river, despite the horrendous potholes on the frontage roads from the gas trucks. As usual, a big industry truck was backed up the Mamm Creek ditch with a hose dipped in along the Colorado River. The driver seemed to be washing the window of the truck. How I wished I could have the gumption to go over and talk to him, and ask what he was in the process of doing. Was he getting water to take up to a well? Was he dumping fluids? And if so, what was it, and what was in it?

Maybe one day I will stop, and I will have the right kind of person to talk to. So far, I have not been able to.

I got home and went to do ranch chores. Time to feed the animals, and attend to the horse still in bandages from the surgery to his leg. This morning the bandage removal went well, but after I rebandaged Buddy's leg I saw some suspicious discharge on the old bandage. I couldn't be sure it wasn't just moisture that leaked through from the red vet wrap; it might be a sign of an infected and seeping incision after the surgery.

I left messages with the emergency vet and got a call back later that evening. She will come tomorrow morning just to be sure that everything is going well. His stitches are to come out this week; I do not want to risk any setbacks.

COLLATERAL DAMAGE

I brought the horse back from the three-day stay at the vet clinic a week ago now. Much of my time and focus has been spent on his care and recovery. It has been both stressful and relaxing. The stress comes from concerns over complications from the surgery, the relaxation comes from the hours both morning and night (mostly night, after getting home from work) spent in the corral-panel stall situated in the back corner of the old hay barn.

I bring the horse fresh buckets of water numerous times a day, and clean his stall constantly when I am at home. At night, wearing cold-weather barn clothes and with the head lamp on over my wool winter hat, and a big flashlight for back-up, I fuss over the big bay horse and pick through the wood shavings with the apple picker for droppings that get tossed into the wheelbarrow. I brush the horse, and he cranes his neck with pleasure when I brush under his black mane. The well-fed barn cats Ginger, Mouse, and Boris wind about my ankles then settle into their hay houses in the stack. The lantern light plays dimly and beautifully over the scene in the weathered old wood barn, despite the cold temperature.

I read the Orion article for the first time completely today. The article was very well done, and the news was shocking. There were stories on numerous landowners and their health complaints. One would think that if people were becoming seriously ill from exposure to natural gas development something would happen to stop it. One would think…

Air Quality Hearings

November 14, 2006

It is late, very late. Got home from a GVCA meeting well after ten, then had to tend to the horse in the stall and cleaned the stall under lantern light for quite a while, and fed the barn cats and hauled more water for Buddy's bucket and fed the barn cats more food. Al did a hay feed and brought the horse some water, but not enough for the night.

I was glad for the time in the barn in spite of the cold, the company of the big horse contentedly munching hay with his nicely bandaged foot bright in new yellow vet-wrap progressing well, and the friendly barn cats brushing against me or watching me from their hay homes in the stack where I had built small, enclosed cubicles with the bales.

As I cleaned the stall and did the chores, information from tonight's meeting and plans for the upcoming air quality hearings in Denver two days from now filled my mind. There was too much to write tonight.

Dee Hoffmeister called me today, but I was busy at work and couldn't talk. She had a Rocky Mountain News reporter coming in an hour to interview her at her home. I told her about the air quality meeting, but it seemed she would be busy with a doctor's appointment and unable to attend. I wished her good luck with the interview, and said goodbye for now.

It is too late to write more. The film people are interested in the hearings, and I want to be sure that this time I get a copy that I can have. I will continue to pursue that tomorrow.

COLLATERAL DAMAGE

Too many times I have been on interviews and breezily and believably told that no problem I can get a copy, but after the piece is produced things are a bit different. It is just not a priority any more to call back people from that story any longer, or follow through on what they may have been promised.

Saturday November 18th

The last two days have gone quickly, and now I am back home after the long trip to Denver for the air quality hearings. I am very tired today; we got back late last night and I was unable to go to sleep as I thought about the hearings. The speech I had given was off the cuff, although I had written out a few pages beforehand and referred to those at times. As I always do after public speaking, I reran my recollection of what I had said before too much time passed and I would forget. Finally, after reviewing my testimony from recollection, I fell off to sleep.

The previous morning the alarm rang at an unwelcome 4:45 a.m. I got up and dragged myself from bed and fought the dangerous impulse to roll back over under the covers. That would be disastrous; I would miss the Denver air quality hearings.

I groggily pulled on barn clothes and prepared to do a hasty feeding of the animals before meeting Duke and Carol at 6 a.m. in the City Market parking lot. It was strangely reminiscent of the early morning departures Al and I made for our yearly vacations to Mexico. This time however, there was no lure of arrivals to tropical destinations. Instead long hours in the car and even more hours spent in a windowless hearing room at the Colorado Department of Health.

As I finished dressing it struck me that it was sadly ironic to be going to testify at a hearing when I already knew from very good sources that the decision was basically already made. It leaves one with the sense of participating in a charade; just going through the moves that had already been laid out. Nonetheless, I knew why I was going. It was to get the story out of what was happening to people's health that lived near the gas wells.

In autopilot I fed the animals in the barns and carried several buckets of water to the horse in the stall. It was pitch black in the corrals and I kept my eyes on the small circle of light before me from my headlamp. Strange to be out in the barns in what seems like the middle of the night.

As planned I met the others at the City Market parking lot then we were off for Denver. I knew there would probably be some down time, so I took the laptop along with a bag of "nicer" clothes to change into. Also, I had made copies the previous day of numerous articles of health impacts to pass out to the committee. Lastly, I brought a pillow in case I could sleep on the car trip.

When we got to the Department of Health, we soon learned that the anticipated one o'clock starting time was being pushed to three as a result of the current lengthy mercury hearings. We settled into the flourescent-lit and rather stark but large cafeteria to wait for others to arrive. Several people were hammering away on their laptops in the Formica booths. They were members of the press, and we knew some of them. David Frey from the Aspen Daily news drove down with Nick Isenberg, a media consultant who had generously offered to film the testimony for no charge.

Several other people from Garfield County were there, and we all greeted and had cafeteria breakfast at a large table by the windows. I pulled Beth Dardinsky aside and said,

"Hey, I heard your speech this week at the GVCA meeting, could I read you mine?"

Beth said sure, so we went into the corner of the room and pulled a few chairs close and I began to read directly from my notes. I was doing all right until I got to the part about Chris Mobaldi and her symptoms. I could not proceed, and I struggled to compose myself as tears filled my eyes and my jaw tightened. I began to leak tears and continued to read the speech in a choked voice. At points, I stumbled over words and choked up.

COLLATERAL DAMAGE

When I finished, Beth put her hand on my arm when I said I was sorry, and I didn't usually do that.

"It's OK, just read through it a few more times…it's hard to talk about these things…it's upsetting," said Beth.

Several of the group chose to go to lunch at a nice restaurant at Cherry Creek Mall. I would have loved to go, but fell compelled to stay as I had invited/asked some of the media people to come. Also, I feared others like Karen Trulove would show up and find no one around. Plus, what if they suddenly changed the hearing times and we all missed testimony opportunities?

I went back and forth from the cafeteria to the mercury hearings where quiet rows of people in dark suits listened intently to the proceedings. Who were they, I wondered? Lawyers? Industry representatives? The press? The people giving testimony to the committee talked about statistics of children damaged by mercury exposures. I all sounded so clinical, so distant. That, I knew, is how my own testimony might be received as I talked of the health impacts from natural gas development. The people on the Air Quality Committee heard things like this all the time…

At one point during the afternoon I saw a short man with a graying crew cut come in the cafeteria with some of "our people." I was introduced to him, and it was Lance Astrella, the renowned gas lawyer from the firm of Astrella and Rice. He looked different than I remembered from two years ago when I saw him from a distance as he spoke from a podium in a large auditorium in Rifle. His hair had looked darker back then… that night had been at the very beginning of my involvement with impacts of natural gas development…

Lance sat with us as we whiled three hours away waiting for four o'clock when we could testify. We talked about the issues at hand, and also learned more of each other's day-to-day lives, work, and families.

Finally the time came, and we all filed into the big hearing room. The two film crews were set up, Nick Isenberg and Joe Brown. The committee members were seated in a horseshoe pattern at the big square of table. Microphones were set for the people testifying at the other end.

When my turn came, I sat down and signed in, and began. I referred to my notes at times, but mostly just spoke off the cuff. I knew what I had to say. Thank God I did not cry, but I did definitely become emotional. My voice was shaking as I spoke of the symptoms of Susan Haire, Dee Hoffmeister, and Chris Mobaldi. I asked the committee members to please read the small packet of articles that described the ailments that these landowners suffered with—illnesses they believed were caused directly by exposure to toxins from the gas wells and production processes that surround their homes. I asked the committee to pass the same air quality standards for the entire state of Colorado as the standards that were being proposed for Denver. Why should those of us in the West have to breathe twice the amount of pollutants from the gas wells as the Denver population would breathe? Finally, I asked the committee to pass the kind of air quality protections they would for their own family members.

After I spoke in the hearings I made my way back to my seat by Patrick Barker. I was still shaking and my heart was pounding.

"Are you OK?" whispered Patrick.

I nodded yes, but I could feel the influence of adrenaline racing through me nonetheless. I sat in the chair and put my elbows on my knees and rested my forehead in my hands. My heart continued to pound away like a rapidly beating drum. I had literally given that testimony the very best and most forceful argument and presentation I could have. I was begging the committee to pay attention to those suffering from health impacts.

After the hearings a group of us went out to dinner. Oh, what a relief it was to have that over, for me at least. Others still had to speak tomorrow as the hearings had been much delayed, but I was done.

On the long drive back to Garfield County we talked, sang, listened to music, and finally fought sleep. Duke had to pull off the highway at one point for a catnap, and his snores filled the car as

COLLATERAL DAMAGE

I dozed on top of the piles of coats, notebooks, and my laptop in the back seat.

When we got to New Castle well after midnight, they dropped me off at my truck and I said goodnight.

"Another gas day," I said in parting.

It was a relief to get home. Now today, it is four and the sun is down behind the mountain. Time to go out and do chores in the barns. I need to sleep tonight. A lot.

Saturday November 18, 2006—evening

All day here at the ranch it was in the back of my mind that friends and neighbors from Garfield County were down there in Denver at the air quality hearings waiting it out to finally testify. Also, (nameless and faceless to me) others from the Front Range were going to testify also. As the Front Range air quality standards were being proposed in a separate ruling, the hearings were being held in two parts. First the Front Range rule, and then the statewide rule.

Lance Astrella was going to be there, as an observer, and the Rocky Mountain Clean Air Coalition, the Oil and Gas Accountability Project, Western Colorado Congress, and others would do their presentations, as would industry groups such as the Colorado Oil and Gas Association.

I called Patrick's cell phone after five; the hearings should be over by now. To my surprise, he said that a decision had not been made yet on the statewide rule. I was stunned; I had been prepped midweek to have little expectation of any ruling going in our favor, and I had heard the decision would be made by Friday at the end of the hearing.

"You people from Garfield County who came down here to testify really had an impact," said Patrick.

"I think you really got the committee's attentions with the stories of the health impacts, they probably didn't know about it except for the member from the Western Slope. I think you people got the ball rolling on this…there was so much discussion after the landowner testimony that things just went way too long to even get to the Front Range issue. They've continued it until December 17th," he said.

"I can't believe it! I was sure we had lost and that it was a done deal. Do you think we have a chance?" I asked as I stood in our old cellar, looking at the stored items we kept down there in the old-fashioned dirt walled underground area of our small farmhouse. Christmas boxes, furniture, the French horn, tools, canned goods and more lined the walls lit by the bare bulbs.

Suddenly, I felt overwhelmingly glad I had taken time off work and gone to testify in Denver. So what about lost pay? So what about everything else, and lost time from home and sleep? So what!

Maybe a few or more of the committee members actually listened to us, and now realize people are very likely getting seriously ill from living near natural gas development.

Thanksgiving Day November 2006

What a week! I still am catching up from the air quality hearings last week. There has been a lot going on with the articles that were written in almost every big Colorado paper, including the Denver Post. Also there have been many phone calls for me at least, as a result of some of the things that happened at the hearings.

This morning I went to the Murr's ranch to pick up a small load of hay. With this mild stretch of weather we're having, almost balmy and fairly dry except for the mud from the thawing ground, it is perfect hay-hauling weather. The rule with hay in the winter is get it when there is still some to get; it never gets easier to find the deeper you get into winter. It doesn't get cheaper by the bale, either.

Richard Murr met me in the drive, and waved me to come through the side entrance to the huge hay sheds. A large truck filled with beams and pipe blocked the regular entrance road. I got out of the truck at the barn, and thanked Richard for letting me come on a holiday.

"It don't make no difference to me," he said in his steady farmer's monotone.

Richard took a long pole and tipped a number of bales off the high stack. They thudded to the ground four and five at a time. In a short while we had the pick-up loaded down to full capacity. Gauge, the slightly graying older hunting dog nosed happily around the stack, obviously in his element.

"Gauge, that's an interesting name," I said.

"Gauge, like twelve-gauge. He'd like to be out hunting now, but he's too old," said Richard.

Just as I was writing out my check for the hay Richard spoke out.

"Well, the oil company finally got me," he said.

"Oh?" I asked. I knew full well that he had been refusing to sign a lease on his land, and that a force-pooled unit had been created there. His wife had asked me a few months ago for help, and I did what I could to provide information and then heard their family was taking care of it.

Then Richard told me that he had finally signed a lease, he had no choice because they were being force-pooled into it.

"We'd be worse off if we hadn't signed," he said.

"I know. The same thing happened up at the county property by the Rifle Airport. Not only can you get drilled on anyways, from what I understand, but you get whopped with expenses you wouldn't have if you signe," I replied.

We talked at some length, discussing the minute and complex details. Basically, the Murrs had balked at signing a lease to protect the surface of their hay farm. But, in the end they were forced to sign and allowed for directional drilling and by doing so they could keep them off the surface and avoid the "penalty fees" for lack of better words for not having signed. The Murrs didn't get the same monetary compensation as the county did, either. But with the assistance of an attorney they had protected themselves as best they could against the inevitable and unstoppable.

Richard picked up a scrap of wood from the ground and tapped his pen against it. He did the math for the hay on the truck and I finished making out the check.

"They've got you, like a cat has a mouse. They get you up against a wall or in a corner and just bat you around till they do you in," he said.

"Yes they do, that's why I called John Martin the county commissioner who didn't vote for the leasing of the airport land. I wanted to make the point that the system is inherently unfair and set up so that the gas company always can get what they want, ultimately," I said.

We talked more about that, and some of the statewide politics and the failed split-estate bill that was also so sadly unfair.

"Well, we're not giving up and we are getting press coverage. One would think that if they knew what was happening to people's health around these wells something would go differently," I said.

"One would think..." Richard agreed.

"How's that Laura Amos lady?" he asked.

"She's doing alright, they settled with the company and they are in Colbran now. They can't say how much their settlement was for with Encana; they had to sign a silence agreement. That's the way it is..." I said.

Richard looked square at me and said,

"That's not right that people can't talk about it."

"No, it's not," I agreed.

COLLATERAL DAMAGE

I got in the truck after we chatted about Thanksgiving plans and what family members were cooking what. His wife made pies the night before, and my husband was at home busy in the kitchen with the turkey now. He's quite a cook.

"Happy Thanksgiving," said Richard as I drove out of the barn past him and his dog Gauge.

"You too, and say hi to Linda!" I said as I waved goodbye.

I had a good feeling deep down having gone to their ranch today, and drove off happily knowing that the Murrs had gotten good guidance at the eleventh hour from an attorney. I was also honored that Richard, a true rancher, would talk to me about this. He knew I had been working on landowner information with GVCA over a year or now maybe two ago. He knew I wanted to see that landowners knew something of an inkling of their rights…

Nothing against people in suits, but if you have the confidence of a good rancher, you have something to be proud of. That's the way I feel anyhow.

It is Thanksgiving afternoon, a nice overcast but mild day. I want to go outside and forget about gas development, forget about work, and be thankful for my home, my animals, and my family.

My brother Tim just called earlier from Massachusetts and I talked to the extended family there: Uncle Carl, Mom and Dad, and nephew Nick. The turkey and pumpkin pie are in the oven, and the sheep in the neighbor's field across the creek complete the perfect scene as they contently graze at the brown stubble under their thick, woolly coats.

There is stuffing and gravy to be made, and I am off to clean barns, pet cats, visit horses and llamas, and unload hay. Happy Thanksgiving.

Sunday November 26, 2006

Regretfully, the four-day holiday weekend is drawing to a close. The sun has already slid behind the steep hill that shades the ranch in winter months. In the nick of time I got outside to do chores and feed in the last hour of warming rays. Now, right on time—3:30—the whole valley begins to gray and chill down for the night. Al is outside hanging Christmas lights, so I can do some typing in the quiet house now.

As usual, I fell fairly short of my lofty goals of achievement for that seemingly endless four-day long thanksgiving weekend. Nowhere near the amount of writing I had envisioned occurred, and the housekeeping and clothes sorting fell short too. On a good note, however, I dragged the virus- infected hard drive off the older computer up to Gary Gagne's to be fixed. Miracle of miracles, in spite of the "hall of shame record number of corruptions and viruses," as Gary said, he was able to salvage the computer. So I am up and running on that machine now at home, a huge achievement.

Driving up Dry Hollow yesterday I saw the newer pads and infrastructure going in close to the homes in Minneota Estates. More rigs loomed over the crest of the hill past Dee Hoffmeister's, the Amoses' old property, and Trulove's. I eased the picked slowly over the rutted road going into Nancy and Gary's, and drove along side the completed wellhead right off the main road. Everything looked tidy and innocuous enough, there at the sight of the well that caused the biggest finable violation in Colorado's history: the Divide Creek seep.

About four hours later I drove with my virus-free hard drive back down the hill, and I pulled into Chipperfield Lane just to see how things looked. I knew that 250 acres had recently been sold to Barrett, and that more wells had been going in and even more were slated. Word was they were producing very well, too. Oni Butterfly had just sold her home on Indian Hill, and basically fled to Carbondale.

I only drove down the road a short way, and saw rigs every which way I looked. Out by the main road a now complete sight which had once housed a huge assortment of fracking tanks and equipment had now changed to a somewhat larger than normal pad area. Cows and horses stood close to the pad, somehow looking strange next to the tanks and hardware.

COLLATERAL DAMAGE

Up at Nancy and Gary's that afternoon I had noticed a DVD about Corvallis, Oregon lying on the desk. Nancy started talking to me about Oregon, and she excitedly pulled out maps to ask questions. She knew I had lived there for quite a number of years, in Oregon, that is.

Today, I am sad that I may likely lose friends such as Gary and Nancy. I count them amongst my gas friends, as I call them. But my friendship with Nancy is not just about gas issues, it mainly focuses on animal issues and the work Nancy does with the Friends of the Rifle Animal Shelter.

And both Nancy and Gary have fabulous senses of humor, and they are smart, honest, and generous kind people.

I cannot be so small as to begrudge them a happy relocation away from gas and oil development, I will just be sorry to have them far away. As Andrew Nikiforuk, author of Saboteurs said, the people from Canada who fled the gas fields called themselves "the gas evacuees." Well, Garfield (or Gasfield) County, Colorado has them too, and their numbers are growing.

When I ran into the Mobaldi's neighbor Wilma from Red Apple subdivision we talked about Chris Mobaldi. I happened to be making Xerox copies of articles that included Chris and we discussed her severe health problems. Wilma and her husband and daughter moved well over a year ago, as did a number of others.

"Yes, after she [Chris] had her second brain tumor, we knew something was up," said Wilma.

She also spoke about the other family nearby experiencing serious health problems, who went door to door with results of blood tests containing high levels of toxins—thought to be a result of exposure to contaminants from the wells surrounding the homes and ranches in Red Apple sub-division. After receiving advisement from a medical specialist to get away from the gas wells, the family moved out of state. It is unofficially thought that a middleman bought the family's property for the gas company, and they received enough money to relocate. But the family cannot discuss the situation further for legal reasons...

Another nearby neighbor, who suffered a host of strange symptoms including irritations that cause her to speak in a high squeaky voice, was moved into a motel at the gas company's expense for the summer after she complained about the fumes and her maladies.

The Red Apple subdivision was the sight of a massive water-well explosion as work was done on a gas well nearby, and many of the residents believed there was contamination to their own water wells as a result. The company responsible trucked in potable water for a period of months but then stopped. They eventually deemed that the water was safe to drink.

I came to refer to the area as "The Old Hell Hole."

November 30, 2006

This afternoon I got a call on my cell phone from a number out of town I did not recognize; it was Judith Kohler the Associated Press reporter. She had a few questions for me and told me the article was out.

"What do you mean by out? Has it hit the papers?" I asked, confused because the Mobaldis had let me know it would be out next weekend.

Judith explained that no, it wasn't printed yet, but it was out in circulation on the Associated Press Western Wire which ran to fourteen states.

"Do you think it will go national?" I asked.

"I think so," she said.

We talked a bit more, then said goodbye.

I first met Judith when she and a number of other Colorado Press were covering a press release where Chris Mobaldi, Karen Trulove,

COLLATERAL DAMAGE

Susan Haire, and others spoke in Denver at the state Capitol about their illnesses and problems after living near the gas wells in Garfield County. Not long after that, she came up here from Denver to do a bigger follow-up story, as did David Frey from the Aspen Daily News. While David's extensive and excellent article came out almost immediately after the interviews, Judith Kohler's has been in the wings for a long time.

So next weekend it is.

I can hear Nancy Jacobsen's voice in my ears:

"So an article comes out, and then it's just another day..."

She should know, she as others have been interviewed and quoted in the New York and Los Angeles Times newspapers, and also has been in documentaries. Nancy once mentioned to me that after she took the French documentary film crew to the well pad by her home, she was no longer welcomed or allowed on that pad. Yet, in spite of the continuing press coverage, nothing changes as a result...not in any tangible way that the landowners can see, anyhow.

I called Steve Mobaldi, and he said that Chris was going to visit a sister for a Christmas and Thanksgiving combined holiday get-together.

I was surprised to hear she would be flying alone.

"She must be doing a lot better!" I said.

"Well..." said Steve in a cautionary tone.

"So people will be checking on her?" I said, meaning during her flight and transit.

"Yes, we have it all set up," said Steve.

I could well imagine that they had to get airport support like that needed for a disabled person or a child going on an airplane. There is no way the Chris I saw a number of months ago could handle a flight to California alone without some big risks. What on earth would she do if she went into one of her typical nonverbal and ill spells? What? Most of her time that I have seen lately seems to be that way: only a handful of hours of the day is she cognizant and barely verbal. I'm not saying I see her on a regular basis; but of the times I have seen her in the last year she has rarely been functional for very long, and sometimes not at all.

Friday night December 1, 2006

I called Steve and chatted about things, and he told me they had made all the arrangements for Chris's flight to Los Angeles for the family reunion, and they had wheelchair escorts there for her at the airports. That made me feel sad, imagining Chris being assisted by airport personnel as she might be having one of her bad times when she is intelligible and jerking with spasms, and ill. And being pushed around airport terminals in a wheel chair as an invalid...

Steve told me more about the attorney's decision to turn over copies of the law suit to Judith Kohler of the Associated Press, and David Frey of the Aspen Daily News. Both of them had been down at the air quality hearings on November 17th, and I saw David asking Lance Astrella for an interview. The two of them sat and talked for some while in one of the cafeteria booths as we waited for the mercury hearings to end before the gas and oil emissions hearings began.

"They say that there was radioactive material used years back in the fracking, and that's part of what made Chris sick," said Steve.

It was a bit later on Friday night when I was talking on the phone with Steve—discreetly in the laundry room and greenhouse area of the house so Al didn't have to listen to yet another gas phone call in our small house. He has already heard plenty...

As Steve spoke, I was stunned. I am not used to hearing the mention of "radioactive material" in these conversations; many other things, but not that.

All I could say was,

"Good God!"

I said goodbye to Steve, and still stunned, I walked back into the dark living room where the TV glow and the Christmas tree lights threw cheer about the cozy room as the logs snapped in the wood stove that burned hot.

"That was Steve, the lawyers turned the lawsuit over to the press. Steve said that radioactive materials were used a thousand feet from their house…that the companies some years ago used them when they fracked," I said to Al.

Al seemed shocked, and I know I was.

Sunday December 3, 2005

Last night I called Dee Hoffmeister and let her know that the Associated Press article in which she had been interviewed had been put out onto the Western Wire. She was happy to hear that, and we talked a bit.

She told me that someone from the Los Angeles Times had read the Rocky Mountain News article about her recently and he called and wanted to come up for an interview. He would be here Monday, and planned to stay a few days.

"Yes! That is great news! I am so happy they picked up on it! The Los Angeles Times has always followed these issues well!" I said excitedly.

We talked about how the interviews were going to go, and whom he wanted to see. He wanted to talk to other impacted landowners.

I said I would make some calls, and get back with her.

Then Dee told me that she would be at the doctor's office tomorrow having more tests.

"I have a tumor on my thyroid gland, and these tests they're going to do, these infra-red tests, I don't know what they are, are going to tell if it's active cancer or not," said Dee.

I was so sorry to hear the news about that, and we planned to touch base tomorrow. Poor Dee, in addition to being afflicted with constant health problems, and currently a bad case of bronchitis for which she was on antibiotics, now she had a thyroid tumor.

"I am saying goodbye to my grandkids right now, they've been over," said Dee.

I could hear the babbling of little voices, and Dee talking to them.

"That's what my family and I are worried about: what if this happens to the kids?" she said.

Then we said goodbye.

Earlier in the day I talked to both Tim and Karen Trulove. I was so sorry to have missed them at the air quality hearings in Denver, but they left after they found out they had to wait until the next day to testify. We talked about the Associated Press article coming out, and about what it had been like for all of us to testify at the hearings.

"Beth Dardinsky went before us, and it was pretty emotional by the time we got up there. So we just said it..." said Tim Trulove.

"I know, I read it in the Rocky Mountain News: you said, 'They're killing us.' When I spoke it was emotional too, I was shaking so hard that I could barely pick up my papers when I was finished. I was still shaking when I got back to my seat. But I think they listened to me, I barely read from my notes, I looked them in the eyes one at a time and I talked about all of you guys who are sick, and about the people who are not there because of the lawsuit settlements. And I gave them a big stack of copies of articles of the illnesses going on—not too much—but the best articles...Patrick Barker from Western Colorado Congress said that he thinks we opened their eyes, and that they had no idea what was going on. I hope that is true," I said.

COLLATERAL DAMAGE

By this time Tim and Karen were both on speakerphone and we were having a three-way conversation. It was so good to talk to them again; it's been a while.

Sunday night December 3, 2006

I called Nancy Jacobsen this morning at nine a.m., and she was still in bed. They had already seen an article in the Aspen Times from The Associate Press. Nancy told me about it.

Later, I found out from others that the Glenwood Springs Post Independent had an Associated Press article on page three, complete with the picture of Karen Trulove in her respirator mask. I couldn't even tell it was Karen.

No article today in the Denver Post, and I don't know about elsewhere.

It is freezing cold here, and has been all week.

It never went over twenty degrees today even thought the sun was shining bright in a cloudless blue Colorado winter sky.

I built a new hay-bale cat shelter late this afternoon for the cats in the big barn, with the sun warming us against the stack. The big three-sided old barn is situated so that the daytime sun shines in and warms the big space a good part of the day. The wise barn cats know this, and at night sleep in the blanket-filled spaces in the haystack and in the day sleep curled up in sun-warmed hay nests on the ground.

Buddy, the horse that had nervectomy surgery, is finally out of his stalled confinement and making his way around the corral in much less pain than before. After over a month of being locked into a fourteen-by-fourteen-foot enclosure that required pre-dawn and post-sunset hours of cleaning, feeding, and water hauling, it is a huge relief to have that time over.

He still limps in this sub-zero temperature, but he is so much better than before.

All this gas stuff has made me tired, as has the cold and the demands of work. So I sign off now.

Sunday Dec 10, 2006

After a lot of time and energy and expectation exerted over the last handful of weeks, between the gas issues and appointments and keeping up at work, I am wiped out. Yesterday I didn't even think about the computer or writing, and that is unusual for me on a weekend. It has been a very busy few weeks…

I had some calls with the Los Angeles Times reporter Nick Riccardi on Monday. I arranged to take off work in mid-afternoon so I could meet him that evening to discuss issues and talk about people for his upcoming interviews over the next few days when he would be up here. He worked out of the Denver office of the LA Times.

After the excitement over the much-anticipated Associated Press articles release the day before, I was both wound up and worn out. Monday afternoon I got home early enough to beat the early-setting winter sun so I could feed the horses, llamas, and barn cats before meeting Nick in town for dinner.

As I whiled away an hour-and-a-half in the small farmhouse here, I was very distracted. I pulled out some of the articles on the health impacts, and the *OnEarth magazine* article that had focused on Laura Amos's issues and the whole picture of gas development in western Colorado. I made a small pile of items to take with me to the restaurant. Unlike getting ready to give a speech before a committee, talking to the press was something that was harder to plan exactly for. You had to trust your gut and go with it. Finally it neared five o'clock and I got in the truck and drove to town to meet the reporter.

We had dinner and I gave him the information I had, and he made plans for his interviews the next few days. He said he was interested in how a community changes as a result of gas development, and he had spent the day mostly in downtown Rifle at city offices and the like.

"It seems like there are winners and losers in this," he said.

COLLATERAL DAMAGE

I agreed.

The people with the serious health impacts have to be the biggest losers of all, they literally are losing their lives. What more is there to lose?

A fellow from my office who moved here this summer told me at work this week that he got a packet from Galaxy Energy talking about pipelines, and what to expect. He was shocked to hear that the gas companies could come onto your land if certain conditions existed, and take over your surface. I had known that he and his wife purchased a home north of highways 6 and 24 where the gas development was headed. It was sad to learn that he would likely be in the thick of it, or very, very, near. I told him I would try and find out over the weekend what was up with the drilling sight. When I told him it would likely be near the Miller Lane area his eyes opened wide, and he said,

"Oh no! We're just one road away from Miller Lane on Antonelli!"

Today is Sunday, and the afternoon is winding down at 2:45. I went and hauled a load of hay from Murr's ranch yesterday.

As Richard knocked hay off the high stack with a long metal pole, and we loaded the truck with heavy green grass bales, we talked some about gas.

Richard figured the pad was going to go in down closer to Rifle, about where the racehorse-training farm is. He figured that from where the force-pooled area was, and the boundaries of that unit.

Today I stacked hay and cleaned barns. Al and I are going away for twelve days, to the Yucatan peninsula in Mexico. We go to Mexico every year, and enjoy traveling around and swimming.

I went shopping Friday evening and got a few gifts for my immediate family, and I have to wrap them and get them mailed off before we leave on Thursday at 4 a.m. I made cookies for co-workers last night and made a respectable mess in the kitchen. If I am going to get any cards out I need to get on that too; I always send cards to the special people and relatives in my life.

Last night I talked to Nancy Jacobsen. I called to see if she had any more room for new orders on dog bone wreaths for the Friends of the Rifle Animal Shelter fundraiser. She said she was already loaded down.

We chatted as we always do, then Nancy told me about a meeting she recently went to where she ran into another friend of ours and they got into discussions about gas development.

"I went to that bad place, and it ruined the evening," she said.

The friend she had been talking with had a husband who had a good-paying job with the gas industry, while Nancy and her husband Gary were being gassed out and finally were getting ready to move from their dream ranch to get away from it.

I knew exactly what Nancy was talking about, going to "the bad place"—when thinking or talking about the gas industry makes you so upset it overcomes you...

"One day this fall I was out in the yard, doing my thing and having a great afternoon. Then a car pulled up at the mailbox and I went down to talk to them because it was obvious they were lost. There were two ladies in the car, and they were looking for property on Silt Mesa that was for sale. They asked me if I like living here, and I told them. I told them the truth. I liked it before I knew about gas development, and I cautioned them about buying on Silt Mesa. Even though I didn't know these people, I had to let them know about it. But it ruined the afternoon, and I got upset," I said.

Nancy told me more about her plans to move away.

"I just want to go to Oregon now, then Gary can come after me. Did you read the article in the paper today where Barrett says they have the odor problem up here 'solved?'"

She laughed.

"Yeah, right, it smells like diesel for days on end. They have it solved…I need to get out of here and away from all of this," she said in disgust and resignation.

"Nancy, I am going to miss you when you move, but I am happy for you," I said honestly.

"Well, I think we'll put the ranch on the market in about February," she said.

We talked a bit more, then said goodbye.

Today when I was unloading hay off the truck I pulled open the Thursday local paper and scanned the articles. On the front page was a headline item entitled

" **LOCAL**

$90 Million County Budget is soaring with the oil and gas industry."

There was another article entitled

"Antero pooling aims to limit well pads,"

and also an article discussing

"Governor-elect also promises changes in Colorado Oil and Gas Conservation Commission.

"He also said he wants to revamp the Colorado Oil and Gas Conservation Commission, the state agency charged with promoting energy development, because it has gotten too cozy with the industry it regulates. Ritter says he will only appoint members to the commission who promise to protect the industry and the property owners affected by the industry."

TARA MEIXSELL

It is strange that when I read things like this, the revamping of the COGCC, it should make me happy. But it doesn't, it is too little and too late. Granted, it is better than nothing, but it is not enough.

And as Nancy Jacobsen said,

"It's great they are finally making changes, but what does it do for us? It's too late for us…"

January 1, 2007

It is New Year's Day, and a pretty cold one at that. We returned recently from a nice long vacation to the Yucatan Peninsula in Mexico. It was a very successful trip; the weather was quite good and we had plenty of time to swim, relax, explore the towns, and spend time together. I wisely did not bring anything related to this book, although I did think about taking along the external hard drive, and perhaps getting onto a computer at one of the hotels in one of those little white, cubical-filled rooms they make available for executives to do some editing. Thank goodness I did not!

It was strange, that even on the sunny beaches of Mexico so many miles away from gas development in Colorado's mountains, my mind strayed numerous times back to gas issues. I wondered how the statewide air quality initiative vote went just days after we flew out from Denver. I pondered the eventual outcome of the Mobaldi's legal case, I thought about the documentary film people and Debra Anderson's recently acquired grant from the State of New Mexico.

Even as the turquoise waves crashed against the powder white beaches and the warm Yucatan breeze riffled the edges of the palm-frond palapa under which I was sitting, all the issues going on with natural gas in Garfield County would not totally leave my mind.

A short lady walked by me on the beach wearing a white shade hat with Halliburton printed across the back of it. For some reason just seeing that name stirred up such negative emotions in me I had to remind myself that I had friends who were affiliated with the industry.

COLLATERAL DAMAGE

Nonetheless I stayed angry for a long time, just to have been reminded of the industry while so far away from it. And when I was trying to forget about it; at least for eleven days while in another country.

After returning home I spent the next day very quietly, unpacking from the trip and doing numerous loads of laundry. I pointedly did not turn on either of the computers, not the one I used for communications or the laptop I am writing this story on. As long as possible I was extending my vacation from gas issues. There were horse barns to be shoveled out and animals to reunite with. The cats were thrilled that we were home and they collapsed in grateful heaps in front of the glowing woodstove. They had no doubt missed that warm fire heat when we were gone. And judging by the way they nestled about us in bed, they missed their people too.

After returning to work for a few days, I was getting back into "normal" life again. I finally dialed Patrick Barker's number to find out how the air quality hearings went. I pretty much knew what the answer was going to be, but I held out an almost impossible hope that maybe by some miracle the commission had voted to uphold the more stringent metro-Denver standards of ten tons of emissions statewide. The only reason I held onto that almost certainly impossible hope was the memory of Patrick's exuberant phone call on Saturday of the second day of the air quality hearings in Denver when he said that the land owners testimony 'had really opened the committees eyes', and they had no idea that these problems were occurring to the landowners by the gas wells. The fact that he felt that the testimonies were powerful and effective gave me that impossible sliver of hope…

After greeting Patrick, I bluntly and not too politely asked,

"So, the air quality initiative; did we get screwed?"

Patrick paused, then replied,

"Well, that's one way of looking at it. But, it's a step in the right direction and it has opened up some doors for us. One member of the committee proposed that in a few years, emission levels statewide perhaps would be reduced, but the committee rejected that. They did say though, that they would leave open the ability to change the emission standards."

We chatted a bit more, and Patrick said that things had been pretty quiet over the holidays.

Last night on New Year's Eve I called my parents, and later I called a friend from Salida who I keep up with but haven't seen for some time. We got caught up with family events and goings-on, and then she asked me if I had done the second book yet. She was referring to historical fiction work such as my first book had been. I told her that I had put the almost-finished first copy aside to do a book on natural gas development impacts.

In retrospect a day later, I believe that after intentionally holding my tongue and not once, not ever, in Mexico talking about natural gas issues I let loose when I talked with my friend Bev.

"I decided to write about the gas issues, one of my friends is really sick from the wells that are by her house," I had said, and I then relayed more details about Chris Mobaldi's situation.

"Anyways, historical fiction novels don't make any money and it's a lot of work," I said. And that is true, after the thrill of getting my first book published and nicely acknowledged even, I still made very little money from it and spent huge amounts of time on it. It's not to say I don't take pride in it; I do, hugely. But it cannot take precedence over income-bearing pursuits, or in this case right now the gas issues in Garfield County.

"Well, I really liked your book," said Bev.

I thanked her, and knew that I liked my first book too. I wonder if that second historical fiction novel will ever get beyond the pile of notebooks that are in the bureau drawer. Time will tell…

COLLATERAL DAMAGE

Today, on New Year's Day, I had a goal to put my stacks of newspaper articles in files and box them up to go in the basement or at least get consolidated. For hours I worked backwards through the several stacks, and labeled manila folders appropriately. Some stacks of articles went together in big files; other got their own folders.

The living room is still strewn with piles of folders, discarded newspapers, boxes, and more. The folders are all labeled and ready to go into boxes.

As I read through one-and-a-half years of articles relating to the gas industry, it was like a time travel backwards.

I recalled different trips to Denver, to the Capitol, to hearings, and the like. Hours in meetings, hours of phone calls to lobby against the energy bill's exemption of hydraulic fracturing chemicals from the safe drinking water act. And hours and hours of working with the press and the film people.

There were articles about fires and spills, regulations, health studies, the Energy Bill of 2005 that I worked so hard on. There was a letter from Senator Ken Salazar, and articles from the Western Colorado Congress Clarion.

There was a health impact article by Rebecca Clarren, also the article from Natural Resources Defense Council's *OnEarth* magazine that featured Laura Amos's situation. There were articles from the split-estate bill issue going back for the two consecutive years the bill ran and failed. There were letters to the editor from irate citizens about spills, explosions, flares, and illnesses. There were copies of NOVA's (Notices of Alleged Violations) from the Divide Creek seep. There was a draft of Kemper Will's legal plan to rally citizen rights against the gas industry. More, and more, and more...

Thankfully, my sifting through the immense pile of folded papers ended with three articles about the hurricane Katrina animal rescue efforts we all launched from Colorado. Reading about saving displaced animals from a hurricane zone left a lighter feeling in my heart than did one-and-a-half years of press coverage of the struggles to preserve the environment and human health in the face of booming gas development. There was little cheer in that subject...

So, it is a new year. I make it my resolution to make it as good a year as I can; both for others and myself. I also want to finish writing about natural gas development this year. It is time.

Lawsuit Filed Over Underpaid Federal Oil Royalties

Sunday January 7, 2007

Yesterday I called the Gormans to wish them a Happy New Year, as they were traveling last weekend. John was getting sworn in this Tuesday as Garfield County Assessor, and it was going to be a huge change for both he and his wife Susan. During the campaign and prior to that while working with them on the "eye sore initiative" (which failed to get on the ballot), I came to know and greatly respect both the Gormans. They had literally worked as hard as physically possible on these two issues. The fact that John was willing to go to huge efforts on these hugely important gas-related issues in Colorado has earned my respect.

Susan answered the phone.

"Hi, it's Tara. Happy New Year!" I said.

"Hi, Happy New Year to you too!" she said.

We chatted for a while, and Susan said that John was under the weather and resting. I told her that I had called last weekend when they were out, and that a friend had given me an article from the New York Times about a whistleblower from the government regarding huge amounts of suspected royalties from the oil industry.

"Right after he reported this, his job was 'eliminated'. Not surprising," I said. Then I got the paper and began reading sections of the article about Bobby L. Maxwell, the twenty-two-year veteran auditor for the Interior Department. In fact, Edmund Andrews who wrote the Dec. 3, 2006 article writes:

COLLATERAL DAMAGE

"During a 22-year career, Bobby L. Maxwell routinely won accolades and awards as one of the Interior Department's best auditors in the nation's oil patch, snaring promotions that eventually had him supervising a staff of 120 people.

"He and his team scrutinized the books of major oil producers that collectively pumped billions of dollars worth of oil and gas every year from land and coastal waters owned by the public. Along the way, the auditors recovered hundreds of millions of dollars from companies that shortchanged the government on royalties."

In spite of a 2003 glowing citation from Gale Norton, then acting Secretary of the Interior, less than two years later his job was eliminated by the Interior Department. Coincidentally, this news came to Mr. Maxwell exactly "one week after a federal judge in Denver unsealed a lawsuit in which Mr. Maxwell contended that a major oil company had spent years cheating on royalty payments."

A few minutes into our discussion John joined in on speaker phone, and the three of us talked about Bobby L. Maxwell's currently pending (potentially huge) law suit against Kerr-McGee Corporation for committing fraud against the government by underpaying royalty payments. The suit also contends that the Interior Department ignored audits.

The trial is set to start on January 19th, but negotiations are already likely underway to reach a settlement before the trial begins. So chances are that yet once again, the story may well go away much more quietly than if it had gone to trial. In cases brought by private citizens under the False Claims Act, a company found guilty has to pay three times as much of the amount of fraudulent gains, and person who filed the suit can keep up to thirty percent of the money.

"Bobby Maxwell is going to be a very wealthy man…" said Susan.

"Yes, if things turn out well, he will be very lucky. The article says that between the years of 1997 and 2002 Kerr-McGee under-paid the government twelve million dollars according to Maxwell's calculations. That's a huge amount of money," I said.

We talked in general of all the unforeseen things to come, of the challenges that John had before him as county assessor, and of his unwavering commitment to pursue proper audits of the gas industries' royalty payments to Garfield County.

"It is why the people elected me, and I am going to do it," said John.

After we said goodbye I turned on the computer and began to do searches of the Maxwell article. I found a number of related articles, and the news was equally depressing about governmental officials intentionally committing what seems to be blatant fraud when it came to gas and oil business. One of the quotes from the New York Times December 3, 2006 article by Edmund L. Andrews rang so true:

"Mr. Maxwell says his frustrations with the Interior Department escalated after the Bush Administration took office in 2001. The Interior Department's top priorities became increasing domestic gas and oil production, offering more incentives to drillers in the Gulf of Mexico and pushing to open the Arctic National Wildlife Refuge and other wilderness areas to drilling. The department trimmed spending on enforcement and cut back on auditors, and sped up approvals for drilling applications.

"The agency's senior ranks also became more heavily populated with officials friendly to the energy industry."

This is just what is happening here in Garfield County, in so many ways. And all across Colorado and the West, too. The BLM appears to be turning into a drilling-permit factory, opening new offices and hiring new personnel just to handle gas and oil permits. The Roan Plateau situation is hopeless currently in spite of the huge public outcry to protect the top of the Plateau from drilling. The Colorado Oil and Gas Commission seems to fight the landowners and back the industry at every turn, as do politicians who are cozy with the gas industry—from Governor Owens to the majority of our county commissioners.

And why, why, I ask, did the state Air Quality Commission not pass the metro-Denver air emissions levels statewide? Why do we in western Colorado get twice the emissions? Do they not think it will soon become just as much of a problem here in the West as it currently is in Denver?

Patrick said that our landowner testimony of pollution and sickness from living near gas wells "opened the commissions eyes," but people getting their health and lives ruined don't matter enough it seems.

It has been very cold for three days, and right now the sun is beginning its last hour of warming the snow-covered ranch before it slides behind the mountain. I need to get outside and do some ranch chores when the sun is still out. It is much more pleasant to feed and throw hay when it is twenty degrees, not zero or lower when the sun goes down. It was ten below this morning when I looked out the frost-coated kitchen window at the thermometer hanging on the elm tree.

During my discussion with John and Susan yesterday I told them that after a two-week break it was odd and rather hard to come back to Garfield County and think about gas and oil, even though not a day passed in Mexico when I did not think numerous times about the issues.

I told them I was waiting; waiting to see what was going to happen. There are many political changes afoot, and the powers that be that were pro-gas industry are no longer solely in control.

But, when they were they did so much damage, and one of the things that bothers me most was the exemption of hydraulic fracturing chemicals from the Safe Drinking Water Act. And the COGCC has allowed the industry to operate pretty much as it pleases, with the exclusion of the Divide Creek seep that was discovered and reported by private citizens. As Laura Amos asked during a COGCC hearing over a year ago, was the commission looking out for our welfare? If so, how? What were they doing about contaminated water wells, dry water wells, and foul air? What did they tell Susan Haire about her health issues after she was knocked down and greatly sickened by exposure to gas-related fumes from a well?

The sad truth is the COGCC didn't do anything for these people who have health complaints. No one does.

Friday's January 5, 2007 local Glenwood Springs Post Independent ran a front page article entitled:

"Ritter passes on George"

The article reports that Governor-elect Bill Ritter replaced Russell George as Executive Director of the Department of Natural Resources. A Denver attorney Harris Sherman will be appointed to the post, a position he held from 1975-80 under Democratic then-Governor Richard Lamm.

The article goes on to talk about the committee that recommended the four candidates for the position, which was sponsored by State Representative Kathleen Curry.

Tresi Houpt, a Garfield County commissioner, served on the committee that screened candidates. She mentioned that one of the DNR executive director's responsibilities is to oversee the COGCC, and that she believes the COGCC "has failed to work to adequately protect public health and the environment. I would really want that to become as important a focus as moving energy development forward is. I think we're not doing anyone a service by not making that a top priority."

So, it is a new year, there is a change in the political make-up at hand. The nation has asked for a change of direction from the Bush administrations agenda, and environmental and energy issues are a part of it. Here in Garfield County it is a big part for so many, regardless of what "side" of the issue you are on.

Time will tell how effective at better protecting human health and welfare and the environment this group of newly elected and old officials will be. One can only hope that some changes for the better will be made for the future. It doesn't much help people like Chris Mobaldi, Susan Haire, Dee Hoffmeister, Karen Trulove, Rick Roles, and the countless others who suffer devastating health effects from the gas wells. But, a change for the better will not be worse, either.

COLLATERAL DAMAGE

The sun is almost behind the hill, so I go now out into the snowy pasture to do chores.

Oh, I just remembered: I got a return brief e-mail from Debra Anderson the film editor last night. She has another grant deadline for the film due Friday; she will be in touch soon. I was so happy to hear that bit of good news; that in addition to her newly acquired grant from the State of New Mexico is exciting. Perhaps one of these big projects will come to some kind of fruition, who knows what will become of my own computer work on this manuscript. I have no idea; it is not your typical book…

After the break of several weeks of writing, and the distance that I was able to achieve by going to Mexico in the middle of the Colorado winter, returning to the pattern of writing has been good. I have missed it. For me, writing is a kind of mental exercise and in a way an elevated escape from the mundane. It causes me to become secluded in my mind in a different and very personal place. I am not saying that at times it is not almost a forced task to take the time out of the short weekends or long workdays to find a quiet spot in which to write for a few hours. I guess what I am realizing is that it is a task and a regimen that I have grown accustomed to, and that if I avoid it I feel in some way lacking and even negligent.

While my second historical fiction novel "Book Two," (as I have simply titled it for now), lies temporarily abandoned in unfinished form in the bottom drawer of my bureau, I proceed with this book. I know I will go back to the historical novel one day, but right now "The Chris Mobaldi Story" as I have titled it in on my computer is what I am focused on.

In a way it is living history that we are living here in Western Garfield County every day with the natural gas industry boom going on around us. This book is perhaps more of a record of events, or a daily journal. Some years from now, I hope I will look back on these writings from a more positive outlook on the future of this area and the struggles of the landowners in the face of the energy giants.

It is hard to have faith in that idea. Instead, I think of the footage from the film "Land Out of Time" of Tweeti Blancett walking solemnly up the hill at her family ranch in New Mexico. I think of the dead cattle lying in the fields filled with waste from the evaporation pits, and of Tweeti talking about the extreme sadness of having to make the choice to abandon the ranch. She described the afternoon her husband came back from selling of their cattle herd, something they never thought he would ever do. It is heartbreaking.

Saturday January 13, 2007

On Monday at work I got a call from a co-worker.

"Tara, we went to a meeting last night in Debeque. The whole town came; it was about a waste pit they are going to build for the gas companies. It's about four miles from where we are buying the land for our house, do you know how far it is safe to live near one of those pits?" asked Lesa.

My heart fell in a now familiar way, another friend and co-worker being impacted by gas development right close to their home. Or in this case, their future home they were just now preparing to build. So far they had cleared a driveway and drilled a water well for the house.

"Those pits are not something you want to live near, there are fumes from the chemicals and people get sick..." I said, thinking of everything I had heard and read about human exposure to the air and chemicals released from the pits. Rebecca Clarren and the group who went up to check out some evaporation pits for the Orion article became so sickened they had to leave quickly, and some could not handle it at all.

We talked for a short while about it, but I had to go. I told Lesa to call Rick and ask him about it, and told her to check with Jim Rada in the Rifle office of the Health Department. The office was just across Railroad Avenue in Rifle from the City Market, in the new county building where the social services offices were. Lesa took some numbers down, and said she would make calls.

COLLATERAL DAMAGE

The next night I left work and came home briefly to feed the animals before going to a GVCA meeting in Rifle. I had to get up early that morning to cover a shift at work, and I knew I was going to be tired all week from this long day. I don't get to sleep well after these meetings; I never have been good at that. I always need to stay up and unwind for a few hours.

The big room at the Senior Center was all different than it usually was; it was set up for a formal event, with three long tables and chairs at the front of the room for the speakers, and about a hundred or so chairs in rows facing for the audience. Different than the more round table set up we were used to. I learned later from Marie, who works in the office at the Senior Center, that Shell Oil was supposed to have a public meeting that night, but they never showed for it. They didn't call either, so the help from the Senior Center not only put up all the chairs but also had to take them down again.

"We were going to have GVCA meet downstairs, because Shell would be up here," she told me after the meeting. Marie, bless her heart, always stayed from 7 to 10 p.m. or later, reading books in her office as GVCA met so we could utilize the Senior Center space. I doubt that she got compensated for that time; she stayed because she cared about community groups being able to meet.

How strange would that have been, I wondered, to be having a GVCA meeting below a Shell meeting. Most likely I would have not enjoyed the experience much, similar to many of the meetings where we share a room with industry. It is a very strange phenomenon, and I do not find it pleasant, as we are both on different sides of an issue. And it is a very personal issue.

This meeting was different for me than the last several years' meetings had been. The main reason I think is because of the change of course of the state and federal governments since the last election. We were getting ready for a new legislative session, and there were going to be issues on the upcoming agenda that we had been fighting as basically a lost cause for so long.

"We're going to have to make a lot more trips to Denver, but folks, this time we have to get used to getting our way. Things have changed," said Duke Cox, the President of GVCA.

What a novel concept that is, going down to Denver to the Capitol for a hearing and getting a positive outcome. It is such a novel concept that I cannot and will not believe it until I see it with my own eyes and hear it with my own ears.

Business was covered, and one of the agenda items was the Debeque evaporation pits. On a huge acreage near Debeque proper, a very small town with a tiny population, a 110-acre waste pit was being proposed. Companies would be able to dump their production water in this huge pit, and the residents of Debeque would have to live with the impacts of whatever effects the chemicals and byproducts in those fluids might have. Also, another group of individuals from out of town had bought a piece of property for another pit, and they were going to run their water trucks right through tiny downtown Debeque; likely a continual stream of tanker tuck traffic…

Understandably so, the residents were up in arms and furious over what was being planned. Why, people at the meeting asked, were these pits being proposed so close to the town and residential area? The most likely answer of course, is money.

I-70 and the county roads offered shortest and best access. To truck it out of the way where fewer people might be contaminated from the pits fumes, wastes, and seepage effects would obviously be more expensive.

The question came up of what was in these pits and how were they going to be monitored? As the group discussed the issue, I thought about the pit above the airport where the backfilling had occurred. What a joke that was: the liner had been all ripped up when they dozed it in still full of waste liquids, and the first liner had failed totally as it had been floating in the pit as observed by residents. The monitoring of these pits was totally lacking to the best of my knowledge. Certainly the slimy, black industrial mess I encountered that night after the pit had been filled in left me furious and disillusioned.

COLLATERAL DAMAGE

I felt that I had gotten if not a brushing off, at least a good deal of misinformation from the officials and offices I followed up with that week, from Garfield County to the Denver offices of the COGCC.

To my surprise, the random auger sample they tested after the pit was backfilled with dirt did still show industrial contamination. I had been pretty sure it would just be fill dirt.

So we will get more and more of that, and when these big waste pits go in they will be licensed as acceptable and the monitoring will probably be equally abysmal and non-existent.

Liz Chandler talked about the dumping of the fluids into ditches and streams that the companies had been routinely observed doing. It is said that they did it because it was cheaper to just pay the fines when they were caught on rare occasions then it was to do it across the board according to regulation.

Others talked about the condensate and wastewater trucks going down the roads leaking liquids all the way intentionally. The industry drivers had contests seeing who could dump the most before they got to the dumpsite, according to one person's report.

After the meeting, I ran into Beth Dardinsky in the ladies room. As we had both just come back from Mexico, we chatted for a bit about our recent vacations. I told her how after getting back it was hard to get back into the mindset of all of this gas stuff again.

"The first day I drove to Rifle, I thought that I could pretend that there was nothing going on and that all these rigs everywhere were fine. I could just drive by all these equipment yards and well pads on the hills and everything, and not think about it. Didn't last long..." I said.

"I know. I sat by a lady from Glenwood on the plane back from Mexico. When I told her I was from Silt she said, 'Oh, are you by all those horrible things happening around the gas wells?' I wanted to stick my fingers in my ears and say 'Lalalalalalalala'. I didn't want to think about it," said Beth.

Other things that I found out about Tuesday night at the meeting were that the Natural Resource Defense Council was opening an office in Boulder to address energy development issues in the Rocky Mountain Region.

The Peak and Prairie Rocky Mountain Sierra Club newsletter came out this week. It had the article written by Kirby Hughes, "Mog Log," about the Air Quality Commission's hearings in November. It also included the article Kirby asked me to write just after returning from the hearing. The article was titled, "Collateral Damage from Oil and Gas Development."

Now it is after 3 p.m. The snowstorm that began yesterday has let up, and the sun is shining though the kitchen window. We got about four inches, and before the sun again slides behind the hill to the south I am going to go out and feed the horses, llamas, and barn cats. It is so quiet and beautiful here after a snow, and a multitude of birds are busy at the feeders and around the grain bins and hay piles pecking for grains. Chickadees, rosy-headed finches, redheaded woodpeckers, and the huge and greedy magpies.

Monday night January 15, 2006

I just came in about 8 p.m. from doing extra ranch chores late due to the very severe cold. It is pushing down again near twenty-below according to our thermometer outside the kitchen window.

As I was trying to thaw my very frozen feet before going out again, Al mentioned to me that he had heard today on National Public Radio that the Maxwell case against the oil company's underpayment of royalties was going to court tomorrow in Denver.

After having spent yet another weekend afternoon perusing articles and doing Google searches on this, I could not believe it.

Going to court? These gas cases almost never seem to get all the way to court. I didn't really know what to make of it—was it good or bad news? My inclination was to think that it was good news, as settling out of court was never something that hit the headlines. Disclaimers were signed, and everything was tidily brushed out of sight. That is, after handsome cash settlements were agreed upon for that silence.

So, tomorrow is another day and how will this case begin? I am very curious, to say the least.

And, more importantly; how will it end?

As my fingers and feet finally thaw after going out to the barns, I need to prepare for tomorrow's chores. It is going to stay at these sub-zero temperatures until Friday, unfortunately. This morning I was very underdressed, and I believe I frostbit my nose and face somewhat. They hurt all day at work.

Tonight, before going out to feed I had put on two of everything. Two pairs of pants and gloves, two sweaters, two hats, my big barn coat and I still ended up freezing.

I went into the basement to get out the horse blankets I have not used for four years since we lived in Park County, at over 9,000 feet elevation where the winters were extremely cold due to almost constant wind chill. It is better for the horses to develop their maximum protection by naturally growing a thick hair coat each winter than to blanket them unnecessarily. That is my opinion, anyhow. But this morning, I saw that Buddy was shivering slightly.

So tonight, I carried the heavy rolled horse blankets that resembled high-tech sleeping bags out to the corral fence, along with some more wool blankets for the cat's hay houses in the barns.

As I unfolded the huge blanket, Buddy was not quite convinced he wanted to be part of this. I let him sniff the old scent of the horse that last wore it, which was Doc, his pasture mate.

Taking my time, I folded it inside out and held it up against his huge brown side, and I talked to him all the while.

I would say in about a half an hour I had slowly and calmly gotten the blanket on and the straps all extended to their maximum length then fastened to their hooks for that huge and gentle horse. He didn't ever move after the first fifteen-foot bolt, after which the blanket still stayed half on. He came to a halt in front of the three-sided barn, and I was able to slowly put it properly in place, and adjust each of the four straps one by one.

As you stand in front of a twelve-hundred-pound animal who allows you to again do strange-seeming things to him and he takes it calmly and in stride, it puts a good feeling in your heart. I think Buddy didn't take too many minutes to register the fact that he felt a lot warmer in his strange brown get-up that looked like a huge mummy bag.

In order to be able to write about gas issues, I need also to be able to write about my life here at the ranch. Somehow it helps; that is what it is all about for others and me. It's about our lives.

We just want to live on our ranches and in our homes, just like we thought we would be able to do before we ever thought, knew, or heard about gas development in Garfield County.

Saturday January 20, 2007

What a week, it has been so, so, cold here. Every morning this week the thermometer on the big elm tree outside the kitchen window registered just above twenty degrees below zero at 6:30 a.m., when we get up and ready for work. Some days the frost was so thick on the glass window that one could not even see out; you had to go outside and walk by the elm tree when feeding the animals to glimpse at the temperature reading.

Every night and morning it was a dreaded chore to get dressed again in two pairs of pants, big coats, two pairs of gloves, and wrap my face to go do the feeding. All the water bowls for the barn cats froze within hours, and they would not thaw as usual when left upside down for the day against the barn in the Colorado sun—if there was any. It was bitter, bitter cold…

COLLATERAL DAMAGE

Finally, toward the end of the week, a warm-up came. This morning it was a balmy (by comparison) zero when I got up. What a relief!

Then, this afternoon when I went out to feed and do chores before the sun went behind the mountain, it was near twenty degrees. For once, after seven days of barely being able to stand to be outside for the minimum necessary time to hastily feed and hay all the animals, I could stay out for a longer time.

Two mornings ago I found the outside water pump frozen and no longer functional. Today Al and I ferried buckets of water from the kitchen sink to the horse trough. If it keeps up we need to run a hose from the basement or greenhouse; carrying buckets of water doesn't cut it.

This cold snap has kept me focused on just getting through day by day, we have burned all the wood in the woodbin and Al is out there now filling the bin in the mudroom from the log pile on the hill. We are burning a lot of wood these days.

After the almost two week vacation break in tropical Mexico, it is hard to get back in the regular groove of things. Plus, the holidays came and everyone took a break from the routine of the daily grind.

Yesterday morning before work I called Patrick Barker from Western Colorado Congress. He was on his way to Grand Junction where they were going to have a meeting.

I told him if he needed help with any of the film presentations of *Land Out of Time,* I would be glad to help if my work schedule allowed. We caught up on a few things, and we talked about the Bobby L. Maxwell lawsuit for underpayment of royalties.

"If it's happening on that level with companies based in Denver, it seems very likely that it is happening here in Garfield County," I said to Patrick.

"I have no doubt it is happening in Garfield County, it is just a case of proving it..." said Patrick, almost laughing as he said it.

"I spent a lot of time the last two weekends doing searches on the case, and I got some contact information on it, mostly the lawyers. It seems to me that if there are other cases that have been proved successfully for underpayments then the experts who worked on those would be good contacts for Garfield County," I said over the cell phone as I sat parked in my truck on the snowy on-ramp of I-70 on my way to work in Glenwood Springs.

"Yes, those contacts are always valuable to have," said Patrick.

We talked for a while more about the proposed huge Debeque waste pit for production water and Patrick groaned when I told him that a co-worker of mine had called and asked me about the implications of living within four miles of the pit where she and her husband had just bought land to build there new home. They had a two-year-old and another one on the way.

Patrick said today WCC was going to plan and discuss strategy for the upcoming legislative issues this season. We agreed to stay posted on what was going on with that, and I brought up the issue of hydraulic fracturing chemical exemption from the current form of the Energy Bill.

"I heard on the news yesterday that they took away billions of dollars of subsidies for gas and oil companies. Lewis Libby is in court, and the Valerie Plame case is under investigation. With this change in politics, things are happening.

"I still want to address the exemption of fracking chemicals. We need to have that changed," I said.

Tonight, as I sit here in my kitchen typing on the laptop at the table I recall the GVCA meeting some months back after the recent elections when I said,

"We need to fix the Energy Bill! We need to get rid of the exemption of hydraulic fracking chemicals from the Safe Drinking Water Act!"

COLLATERAL DAMAGE

Some of the folks there laughed. They did not laugh at me, I know, but they laughed at the idea of us small and piddly people from Garfield County trying to change national policy.

I talked to Steve Mobaldi briefly yesterday morning, and I asked him how Chris was doing.

"Not so good," said Steve in his ever-cheerful voice. Nothing seemed to ever get him down.

"How isn't she doing so good?" I asked.

"Oh, she's puking a lot," he said matter-of-factly.

"Has she put on any weight?" I asked, knowing that she had plunged from about 165 lbs. to under 100 lbs. in the past few years.

"Yes, she's about 115 lbs. now," said Steve.

"That's good, when we did the TV interview she was about 100," I said.

Then Steve got a call from a client and had to go.

Today, I was going through things on my desk, and I came across a photograph of Chris from about four years ago when we had worked together. She had a big smile on her face, and she was glowing, robust, and healthy. I stuck the photo behind some of my family pictures. I had wondered where that picture had gone to.

Tuesday night January 23, 2007

On my way home from work I called Nancy Jacobsen. I haven't talked to her since coming back from Mexico.

"Computer support and sales," said Nancy when she answered the phone in her signature style of answering for her husband Gary's computer business.

"Hi Nancy, it's Tara," I said.

As usually, when she heard it was me she burst into her exuberant voice.

"Well Tara, how are you! I was beginning to wonder if you decided to stay in Mexico after all! I haven't heard from you for so long!" she said.

"Well, reentry was kind of hard after we got back, and there were a few issues like all the animals here at the ranch, and the money needed to stay at the hotels in Mexico, so we came back," I said.

"I almost called you a few days ago; they're going to have a showing of *Land Out of Time* this Thursday in Rifle, and I wondered if you were going to go to see it again," she asked.

"Yes, it is showing three times here this week. I might go if Patrick wants help, but it makes me sad to see it. It makes me want to get up and walk out," I said.

It was true. I sat through the first showing, and in a strange way it was like the meetings, when you heard and saw all these things that were going on but in reality they were so incomprehensibly awful you didn't want to think about them. It was painful.

Then Nancy said,

"We had the first showing of our house today. Well, it wasn't really a showing but someone came to see it."

"Oh no!" I said.

My heart dropped. This was the first real step of Nancy and Gary going away. It was a real step, too. I was extremely sad, for I could envision the day when they would be gone, and slowly we would drift out of one another's lives. Except for Christmas cards, and maybe a stray phone call here and there.

COLLATERAL DAMAGE

Nancy has become a good and important friend to me and I don't look forward to losing that.

Nancy told me about an article in the Sentinel today, where John Salazar our state representative talked about not putting waste pits near where people lived, like they were proposing to do in Debeque.

"They're doing some things better now, but too slowly. What about us, and the pits up here? What does that do for us?" she asked somewhat angrily.

"The people who looked at the house asked about the gas impacts; they had heard about the Deitrich's well getting contaminated. So we discussed that. Of course I'm not going to lead anybody on about what goes on up here. I wish my well was contaminated so Encana would buy my place and I could leave!" she said.

Tonight, as I drove home I knew I wanted to at least turn on the computer for a minute or two. After I filled the grain bucket and loaded up the barn-cat food in the coffee can, I went out and fed the animals. When I came in the house, I plugged in the laptop for a rare midweek writing session.

People argue and talk about pollution levels, emissions, contaminations, gravel pits, trucks, illness and more from the gas industry here in Garfield County.

What impacted me tonight personally, was hearing that Nancy is now ready to sell her house and leave, and I will soon be without a good friend. The community will take a loss, as Nancy is a huge support and advocate for community programs, including the Rifle Animal Shelter.

I also, as a friend, will take a loss. Those are the things that never make it into reports, or articles that the few out there even bother to read about gas development impacts. I guess I am just not in a good mood tonight.

Wednesday Jan 24th

At lunch break I was in Copy Copy in Glenwood Springs making copies of the Bobby L. Maxwell case articles. I had heard the previous day that the judge ruled favorably in his case against Kerr-McGee. I printed out the latest articles by Edmund L. Andrews of the New York Times, and I made ten copies to pass around at a meeting I was going to that day.

As I ran the stack of papers through the copy machine feed tray, I saw a familiar looking person with short dark hair coming through the door of the shop.

"Oni Butterfly!" I called.

She turned in surprise, then smiled as she saw who it was.

We gave each other a hug, and then caught up a bit. I knew that Oni had sold her place up Dry Hollow in the thick of gas development country after many years of struggling. That's how I knew her, from GVCA meetings and such.

"How are you, how is it now that you're up in Carbondale?" I asked.

Unlike the others who have relocated and seemed immediately happy about being gone from the industry effects and overwhelming pressure on a daily basis, Oni did not seem happy. She had a sad look on her face as she spoke;

"I miss my place, my place in the country. It's not the same where I am now. I have to take the dogs to parks to take walks, and they get sick all the time. I miss my land," she said.

We caught up a bit, then said goodbye again. I had to get to a meeting and back to work.

Throughout the next days, as I followed the developments in the Bobby L. Maxwell case, I continued to think about my meeting with Oni. She, like many others, made the choice to leave her country dream home because she could no longer deal with, or stand, what was going on with gas development around her.

Some lose money on their investments in their properties, but regain a healthier frame of mind in a new and different environment.

Others like Oni, in spite of moving away, lose something for good. Although I have never been to her former ranch home, Indian Hill, I have heard about it. The vast property is said to be gorgeous and have Native American petroglyphs and artifact sites upon the land. I am sure it was a hard place to leave, and a new location within the town limits of Carbondale would never be the same.

January 24, 2007 Headlines

New York Times

"Kerr-McGee is found Liable in Lawsuit Over Oil Royalties" by Edmund L. Andrews

"Washington, Jan. 23—A federal jury in Denver agreed Tuesday with a former top auditor for the Interior Department that the Kerr-McGee Corporation had cheated the government out of millions of dollars of royalties on oil it produced in publicly owned coastal waters.

The decision, reached by the jury after deliberations of about four hours, is a vindication for the auditor, Bobby L. Maxwell. He became a whistleblower and sued Kerr-McGee as a private citizen after top officials at the Interior Department ordered him to drop his audit findings."

This is just the beginning of another of the articles on the Bobby L. Maxwell lawsuit. The implications of how deep the federal government's involvement in the underpayment of royalties went are astounding.

For me, as I read back through some of the articles, the whole picture of how the last four or five years of the federal government's handling of energy development has been so blatantly industry-biased and perhaps criminal.

Aside from the public lands being used for exploration and development, and royalty payments being perhaps manipulated and underpaid, what about public health and safety, environmental health, and the rights of those being impacted by this industry?

Today I called a transcriber from New Mexico who had been recommended to me by Debra Anderson. I had some audiotaped interviews I was interested in getting transcribed if it would not be too expensive to do so.

Although I put my best effort forward when I write, it is hard after the fact to get the words exactly correct of the discussions that happened with people the day before.

After spending many, many hours talking with landowners and members of the press and some filmmakers, I began to make a few audiotapes of my own during interviews. I wanted to get the words of the people being interviewed exactly, and not incorrectly rehashed later.

The transcriber and I talked for quite a while. She had done a lot of Debra's work already, and she was familiar with the issue.

"People need to hear about what's going on with this," she said.

"Yes, they do. So hopefully this film will get done, and maybe my manuscript will get done," I said.

We talked at length about the impacts here in Garfield County, and the realities of what might change if any greater public awareness of the problems was achieved.

She told me her timeline of getting audio interviews transcribed, and the price was doable for me and I was in no great rush. But, I didn't want it to go on for too long. The first book I wrote took five years to get done, but that was a totally different story.

I told her I would call in a few months, and that I had a few more interviews I wanted to get done.

COLLATERAL DAMAGE

In closing, here is what is on my mind. When I reread the article by Edmund L. Andrews about the royalties case, it hits the nail on the head. All the way to what is happening to the people here in Garfield County.

Dec 3, 2006

the New York Times:

"Mr. Maxwell said his frustrations with the Interior Department escalated after the Bush Administration took office in 2001. The Interior Department's top priorities became increasingly domestic oil and gas production, offering more incentives to drillers in the Gulf of Mexico and pushing to open the Arctic National Wildlife Refuge and other wilderness areas to drilling. The department trimmed spending on enforcement and cut back on auditors, and sped up approvals for drilling applications.

The agency's senior ranks also became more heavily populated with officials friendly to the energy industry. For example, its new deputy secretary, J. Steven Griles, worked as an oil industry lobbyist before joining the department, and Chevron and Shell had paid him as an expert witness on their behalf in the Benji Case."

That is what is happening here in the West: the push is on at the federal and local levels to expedite energy extraction and development, and to remove any barriers to it.

Sunday January 28, 2007

It is a beautiful, sunny day today. Although it was zero this morning at 7:30, once the sun came up the temperature steadily climbed. By ten a.m. it was twenty degrees above zero, which in the full sun felt positively balmy. I did not even put a hat on when I did the feed and chores, for the first time in probably months...

The pump out back where we draw water for the stock tank is still frozen though, so I fought with hoses for a while and hooked them up in the basement. All the indoor cats went out for extended play times, and they rolled in the patches of snowless earth that showed through, frolicking as though it were spring. One of the hoses that went all the way out to the back tank broke when I tried to free it from the ice and snow it was buried under, so I had to ferry big buckets out to the back stock tank for the male llamas Scout and Willie. For once in a long while, it was pleasant to be outside doing chores.

When I came back in I called the Mobaldis. Steve answered; we had not talked in a while, so we got caught up.

The lawsuit was still rolling along in slow negotiations in Denver.

"Chris and I are not really involved directly in it; we get the updates, letters, e-mails, and some phone calls. The lawyers are handling everything," said Steve.

"It seems like such a long time now since it was filed," I said.

And it has been a long time. It makes one wonder what and when an outcome may ever be decided.

"The house is going into foreclosure in March," said Steve.

I was surprised.

"Wow, what does that mean exactly?" I asked.

"The lawyers are handling it," said Steve.

"Would it be different if the lawsuit wasn't going on?" I asked.

"I don't really know…but we can't sell the house knowing what the contamination problems are there," said Steve.

I laughed.

"Wait, you're telling me that you can't sell the house without disclosing contamination problems, but the COGCC claims there was nothing wrong with your water? That is ridiculous! Is it because of tests you did and things that got found out for evidence for the case?"

"Probably," said Steve.

I told Steve about Nancy and Gary getting ready to sell their house, and about Oni moving. He already knew that Peggy Utesch was gone too, farther west in Colorado.

"You know, Steve, when I started getting involved in this stuff, I had a more hopeful attitude. I guess I thought that some things could change as a result of bringing things forward. Now, it just seems like an ongoing depressing story, and more and more things keep happening but little changes. I want to finish this book, because otherwise I could keep on writing it forever, as more and more people keep getting sick near the wells. It's not going to stop," I said.

"I know, we can tell that the smog in Garfield County is getting much worse. We were up there last week to see a doctor because Chris is having trouble with her liver," said Steve.

He described their visit, and said that the doctor was puzzled by Chris's strange accent, and suggested she see a psychiatrist.

"Then he wanted to turn her into a pin cushion and do all kinds of blood tests on her. We told him that she was seeing a specialist already, and had all those tests done. We specifically wanted to see him about her liver functions, and that was it. We told him that she had been exposed to all these different chemicals, and we thought that was why she was sick. Also, we thought there was some type of DNA change as a result, and maybe that is why she doesn't have fingerprints anymore," said Steve.

"Did he recommend that she see a psychiatrist before or after you told him she didn't have fingerprints?" I asked curiously.

"Before," said Steve, chuckling a bit.

"Did he seem surprised that she had no fingerprints anymore?" I asked.

"He looked a bit shocked…" said Steve.

All of the sudden Steve began talking about the chemicals that the lawyers had figured that Chris had been exposed to in the water from their well. He was rattling of the names rapidly.

"Wait, wait! You're going to fast!" I said as I hastily left the greenhouse where I had been sitting and went inside the kitchen to search for a pen.

"Are you writing this down?" Steve asked.

"Can I?" I asked in reply.

"Yes," he said, and then continued talking.

Apparently, in 1996 at least three gas wells in close proximity to the Mobaldi's water well were involved in experimental radioactive fracking tests. Or, perhaps better put, the fracking mix was different from what had traditionally been used up to that time. There were two periods of time when these tests were conducted by the Department of Industry in the Mobaldi's neighborhood. The studies were to see the performance of the frac—to see where the materials go.

Steve read me a direct quote from the actual language of the lawsuit:

"The harmful substances contained in the fracking fluids which made their way into the plaintiff's water well which Elizabeth Mobaldi ingested day after day included:

scandium
zirconium
iridium
antimony
arsenic
barium
cadmium
chromium
and lead

COLLATERAL DAMAGE

Some of the heavy metals used in the fracking fluids were radioactive isotopes used to detect the flow of the fracking fluids through the oil and gas formations."

I was scribbling the notes furiously on the back of a homemade thank-you card written on red construction paper. Because my nephew had created this unique piece of art, I planned to put it in my keepsake box of family photos and card I had begun keeping about a year ago. It briefly crossed my mind that one day, years down the road, if I flipped the card over, I would see these scribbled notes and lists of chemicals, and comments on radioactive fracking mixes.....

We talked a bit more about the chemicals, and then I asked Steve how Chris was doing.

"Well, about the same, she has problems. She still struggles, and she can't work mainly because of the communication issues. That gets her really upset," said Steve.

"Is she still getting sick every day, and does she still have days when she has to sleep all the time?" I asked.

"No, she doesn't have to sleep all the time, and she's pretty much over getting sick. She still has days when she has diarrhea all day, though. But she's gained weight, even though she doesn't eat very much at all," he said.

"Yes, you told me that last time we talked. That's really good news! Last week I found a picture of her I had been looking for, from when we used to work together. We had just finished moving a client into a new apartment; Chris looks like a redheaded chipmunk!" I said.

It is true, the photo from just four years back shows a merry-eyed redheaded robust woman with round cheeks dressed in her plaid shirt and blue jeans. So different from the photos recently of Chris that appear in articles in papers and magazines, showing a frail, gaunt, and stooped woman with not a visible ounce of body fat.

It is now four o'clock, and the last of the day's cheery warm sun still fills the kitchen. The days are getting longer, and if I get outside within minutes I might be able to catch the last rays.

Yesterday, when I talked to the transcriber from New Mexico who does work turning audiotapes into written word documents, she asked me about my involvement in these gas issues.

"It is so sad what is happening up there to these people! It is important that their stories get told; then something can be done to change it," she said.

I guess my involvement in gas issues is that I have friends who are sick from living near wells, friends who are driven from their homes to get away from it, and I feel driven enough to spend hours each week typing into this computer.

What, I suppose, would I have been doing if it were not for gas and oil development coming into my life and my community?

Would I be working on another historical fiction novel, would I do more volunteer work or be a better housekeeper?

Somehow I doubt any of the above, but I suppose that is part of the mystery of life. You cannot wind the clock backwards and change the setting of the stage. We are where we are, experiencing what is going on around us at that time.

When I think of the venues I have been in, and some of the people I have talked to and met due to oil and gas development in Garfield County, it makes me realize that it has somehow become a part of my life in a bigger way than I ever would have imagined.

That is why this book needs to close. Life is finite, and although this is a hugely important issue, as my favorite literary quote from Walden by the author Henry David Thoreau says:

"I went to the woods because I wished to live deliberately, to front only the essential facts of life, and see if I could not learn what it had to teach, and not, when I come to die, discover that I had not lived."

The time has come for me when I want to get back to the woods.

Feb 1, 2006

This morning about ten a.m., I got a call on my cell phone. I didn't recognize the voice immediately, and something was said in a garbled accent.

"Ees these Tah-ra?" the voice asked.

"Chris? Did you try to call me?" I asked in return. From the way she answered the phone, I couldn't tell if it was me she was trying to call, or if she got me by mistake.

"Yais, I ahm calling yuh. Ah want tah come say yuh toe-day, ah cahn lev rot now," she said in her strange accent.

I couldn't believe it; Chris was making the first trip to Garfield County in almost two years, all by herself.

"Are you sure you can do it?" I asked, seriously concerned. That last trip she later confessed she had a difficult time making it back, and that she overdid it.

"Yas, Ah huv gud dyes ahnd bad dyes, toe-dye ess ah good dye. Toe-morrdah mah beh ah bad dye, ah neh-va know," she said.

We talked for a minute or more, and I said for her to have the front desk receptionists at work call me when she got there in an hour and a half. I could not believe she was coming; it was so sudden and unexpected.

At work when I heard over the intercom about 12:30 that Chris was downstairs, I ran down and gave her a big hug. She looked good, and her eyes were bright and she was happy. She had gained a small bit of weight and did not seem quite as frail as last time I saw her.

It is very hard to describe what the next several hours were like, as Chris went to visit the people in the offices at the agency where she had worked just three and a half years ago or so.

Some of the folks probably thought the language issue might have been cognitive, and some were able to understand what she was actually saying, or most of it.

Chris spent some time in her old supervisor's office, and a number of people came in and out to see her. And Chris and I made the rounds of some of the people who would have been there long enough to remember her.

We went into the executive director's office, and Bruce rose from his desk with a warm smile on his face, and reached to shake her hand as he said,

"Hi Chris, I'm glad to see you."

"Yuh rey-mem-buh mey nam!" she said, with a big smile on her face.

"Of course I do!" said Bruce as he shook her hand.

Suddenly, Chris stepped back and winced in pain and held out her hand.

"Oh, I'm sorry!" said Bruce.

There it was, one of the common symptoms shared by so many who live near the wells have: severe pain in the hands and feet.

We said our goodbyes and went to grab a fast late lunch at 2:30.

In the Quizno's sandwich shop assisting Chris with her order at the counter, I could see how the people around us didn't know what to make of her.

I had to interpret for her to get her order in, and I don't think others can tell if she is either ill, disabled, or foreign.

COLLATERAL DAMAGE

Chris barely ate anything; she took some bites of the small cup of soup and lifted the sandwich only once or twice.

As we ate at the small table in the corner of the fast-food restaurant, I asked Chris how things were going.

Chris began discussing their strained financial situation due to her medical bills, the foreclosure of their Rulison home, and the need to give up her car.

"Thees ees meh last beeg fling..." she said, with a wan smile on her face as she held her sandwich with shaking hands.

At that moment I realized that Chris had most likely over pushed herself to drive to Glenwood today, as they were giving up her car tomorrow.

The struggles that the Mobaldis faced not only included huge issues with Chris's health, but their overall ability to remain financially functional and solvent.

As I stared down at the plastic trays of food on the small Formica table where we sat, and Chris continued to shake and look unsteady across from me, I had the fleeting and seemingly impossible hope pass through my mind that perhaps, perhaps their lawsuit would be successful.

This woman has suffered unduly and her life has been about ruined physically. And she and her husband have paid a huge financial price also.

This moment of realization as we sat in Quizno's restaurant was strong.

There is the knowledge of the filed legal charges against the gas companies for the damages brought upon the Mobaldis from natural gas development; then there was the real life reality of today. There is no lawsuit settlement yet, Chris is not well and barely functional, and there are money problems.

Who is paying for the profits of the natural gas industry in Garfield County?

I believe that Chris Mobaldi, and others, are paying the biggest price, and very few will ever know of it, or acknowledge it either.

When we got up to leave, Chris dropped her drink on the floor and was upset. Everyone was very friendly about it, and we went out through the bitter cold parking lot with Chris holding my hand and leaning heavily against me. Chris cannot handle cold at all.

She lost her ability to speak intelligibly, and she began to shake.

I helped her into the car, and I was not happy to leave her, but I had to go to an appointment. We talked for a bit, but I couldn't really understand her much.

As I got into my truck next to her vehicle, I looked over, and saw her slumped over the wheel, with her hands on the dashboard. I panicked and ran out of the truck and over to her door.

"Chris! Are you OK?" I asked in a frightened voice.

"Yas, ah weel bay fahn, don't wo-rey," she said and smiled at me with an exhausted and pained look on her face.

"I'll call you when you get home," I said.

As I drove away I was worried, worried that she had pushed herself too hard and was going to collapse as she so often did after periods of exertion.

The one thing that stays in my mind tonight is the clarity with which she herself understands the peaks and valleys in her functioning level.

There are good days and bad.

COLLATERAL DAMAGE

And there she is, in the bad times, stuck in a world where she cannot interact with others, and where people react to her as though she is incompetent.

Steve and Chris called me tonight, and told me that she got home just fine. But when I talked to Chris, her speech was beginning to get a bit too hard for me too understand.

How I hope that the legal case finally settles sometime and the Mobaldis can have some relief from the pressures of what is going on now, for whatever money can do in a situation like this.

The bottom line is, no amount of money in the world can ever replace for her or anyone like her what they have lost when they have been through what she has been through.

You cannot buy your health or your life back, ever. You can just move on.

The Mobaldi's case is still slowly winding its way through the legal process, with the firm of Astrella and Rice.

Hopefully one day soon a conclusion will be arrived at, and that settlement will help ease the pain and stress of Chris Mobaldi and her husband's life.

About one year ago the Mobaldis asked me to write Chris's story. I have done my best, and now that they are deeply involved in a suit my information has become limited on their particulars, so as a result my story expanded into further areas, and other people's situations involving their impacts from gas development near their homes in Garfield County.

As gas and other energy development continues at a furious pace in Garfield County, I know I could continue to write about this subject for endless years.

I feel the need to get back to my own life now.

TARA MEIXSELL

Sunday, March 11, 2007

The past few weeks have been busy. So much preparation for the trip to Denver to testify before the House Committee at the Capitol for House Bill 1223 with people from the Western Colorado Congress, Grand Valley Citizens Alliance as well as other groups and landowners from across Colorado. Then finally, after endless meetings, phone calls, last minute plans and the inevitable expected changes, the sizeable group assembled in Denver at noon for a briefing at the cafeteria of the Inspector General's Office Building just across the street from the golden-domed state Capitol building.

Matt Sura, the Energy Organizer for the Western Colorado Congress out of Grand Junction ran the briefing, and everyone gave a short synopsis of what they were expected to speak about. We were cautioned to keep our testimonies brief—three to five minutes—so as not to wear out the patience of the Agriculture, Livestock and Natural Services Committee members. We were then warned by Gunnison State Representative Kathleen Curry, the sponsor of House Bill 1223, that committee members would lose focus if testimonies ran long or repeated information. She said they'd begin to turn on their laptop computers and work on other things instead of listening to testimonies or possibly even vote against the bill as a result.

A group of us had carpooled early that morning from the Western Slope in Laura Amos's company's SUV with a logo and photograph of a snarling mountain lion across the back window. I'm not a hunter and, in fact, am opposed to hunting. That's never been a stumbling block for me in my interactions with Laura though. The reason we've worked together for a number of years relates to the health impacts of natural gas development.

Laura picked up Rick Roles and Dee Hoffmeister at the Kum-and-Go gas station in Silt, then met me in New Castle, where I left my truck by City Market. Next, we picked up film producer Nick Isenberg and his huge camera in at the Ramada Inn in Glenwood Springs right off I-70. Nick had again very generously offered to film the hearing free of charge so that we could send tapes to Debra Anderson. When

COLLATERAL DAMAGE

I had called him a few days prior to the trip to get a quote for his film services to cover the hearing, he said it would be well over $1,000. "But for you," he said. "I'll do it for free." Finally, we were off to Denver.

As usual, the conversation on the way to a gas hearing evolved into a story swap with people recounting symptoms and events such as exposure to noxious fumes near the wells and ignited waste pits. Laura, Rick and Dee began discussing the strange high-pitched ringing in their ears they had in common. They also talked about the pains in their joints, muscles, hands and, especially, feet.

As we drove through the spectacular Glenwood Canyon, with its dramatic cliffs alongside the Colorado River, and then beyond, through the flatlands past Dotsero into Eagle County, we took turns reviewing prepared speeches and taking notes to add to testimonies. Dee had hers typed out, and it was excellent. Her past experience as an editor showed in the smooth, flowing and concise testimony. She just jotted down a few things in the margins to add as we gave her suggestions of how to tie it directly to the bill.

Rick, the quintessential roughneck cowboy, planned to speak off-the-cuff and he gave us a version of what he was going to say. Dee took some notes and everyone in the car commented on what they thought would be most effective. When Rick started going, he could go. After having lived in the gas field for ten years on his family's ranch and also with his past experience of working for the oil industry, his knowledge of the technical aspects of the operations left me far in the dust.

I had written the beginnings of a testimony in a spiral notebook while waiting for the group to pick me up that morning in the parking lot. It was similar to several other speeches I had written for other hearings, but every speech is always different. The others in the car listened as I read what I had written and I wrapped up the few hand-written pages confident that I had something adequate to read if I was asked to. The main reason I was going on the trip was to offer support and encouragement to Rick and Dee and to ensure their safe arrival. Their testimonies were more critical than mine.

There was an impressive line up of people who testified at the hearing, including Lance Astrella, an attorney from the firm representing Chris Mobaldi in her ongoing litigation against the gas companies. Lance spoke about the need to get on top of these health issues now before they became an even bigger problem and the huge expenses that the gas companies would face if they got into class action lawsuits. He also talked about the number of individuals he has represented due to damage to their health as a result of exposure to toxins from the gas industry. He presented himself very clearly and effectively. He never came across as being against the industry, instead pointing to the need for solutions to current problems that will only increase with time.

Dr. Gerdes of Colorado Springs was there to testify about Chris Mobaldi's symptoms in addition to her neighbor Ted's symptoms. Ted has since sold his home and left the state after becoming ill from living in close proximity to the gas wells. The levels of heavy metals in his blood were reportedly off the charts—200 times the allowable EPA levels, and that's just where the tests capped out. He followed the doctor's advice to move away from the gas wells and the contaminants they produce. Dr. Gerdes said Chris suffered from debilitating skin and joint pain, respiratory problems, severe skin rashes, massive nosebleeds, headaches, nausea, two pituitary brain tumors and acquired foreign accent syndrome. Chris and her husband Steve also moved away from the gas wells. Dr. Gerdes described the link between exposure to toxins known to exist around gas wells and infrastructure and the health problems that these two landowners from Rulison had experienced.

Dr. Wezensky from Grand Junction spoke about his patient, Susan Haire, and the dramatic effects that she had suffered after her exposure to airborne toxins while irrigating near her home. She collapsed after breathing the foul air and became violently ill. The extreme exposure to toxins left her permanently sensitive to industry chemicals and she was unable to breathe the air around her ranch home. She had three air scrubbers installed in her home and wore a respirator while outside. Susan also left Colorado to escape exposure to the toxins emitted from the gas wells and she now resides in Texas. Away from the gas fields, her health has improved. Doctors diagnosed her with symptoms likely developed from exposure to hydrogen sulfide or similar chemical mixes.

Mary Bachran of the TDEX Institute in Paonia, Colo., gave an excellent PowerPoint presentation discussing the toxins known to be involved with natural gas exploration and the symptoms that accompany exposure to these chemicals. The research being done at The Endocrine Disruption Exchange by Dr. Theo Colborn investigates the link between the toxins and the symptoms they may trigger. Many of the symptoms linked to the chemicals associated with natural gas extraction and development were ones that landowners living near gas wells from across Colorado were reporting.

But, as always, it was the landowners' testimonies that took center stage.

Morgan Bear testified about her husband. He'd been welding on trucks transporting what he was told was water for the industry and suffered nerve loss in both legs that spread into both of his arms. The young wife and mother began to cry before the house committee as she continued her testimony.

She told how they were in danger of losing their home and that she was her family's sole source of income. What she earned was not enough and the only way they were getting by was with the help of church members and their extended family.

When Dee Hoffmeister testified, she also barely made it through a few sentences describing her symptoms before she was overcome with emotion.

"There was a big, gray cloud of benzene," Dee told lawmakers with tears in her eyes. "I just passed out from the fumes in our home. I just couldn't breathe."

She described how her family had carried her from the house and driven her away to escape the fumes that had knocked her unconscious—and how she had been overcome by severe and lingering symptoms including severe headaches, intense dizziness, nausea and pain that traveled around her body. She said she couldn't stand without holding onto furniture and had trouble walking. When she attempted to go home again, time after time she stepped out of the car at the Silt country property only to become immediately sickened by the odors and fumes. And she would have to leave her home and go back again to her daughter's home to stay some twenty miles east in Glenwood Springs, further away from the gas fields and their wells and fumes. For over half a year she was unable to live in her own home due to her inability to breathe the tainted air.

(GWS Post Independent March 8, 07—A P Steven K Paulson)

The testimonies were running long and Kathleen Curry turned the floor over to the opposition.

To my surprise, although by this time it was nearly five o'clock and the hearing had been going on since 1:30, all those who wanted to testify were allowed to do so. They began calling names off a list and Rick Roles was next.

Rick stood before the assembly in his tweed suit coat with his thin, long braid and grizzled beard. He spoke angrily and passionately about how a person who knowingly spread the AIDS virus could be tried for murder in a court of law, and that the gas companies should be tried for murder for knowingly harming the people in western Colorado. He pulled a well-worn document from his pocket and said it had his blood test results that showed benzene, toluene and other toxic and carcinogenic chemicals.

After Rick said his piece, state representative Wes McKinley asked about his livestock.

Rick told the committee that half his kids from his goatherd were stillborn last year, a live colt hadn't been born on the ranch for ten years and he stopped breeding mares after one had dropped an eleven-month colt dead in the yard the previous year.

The committee members looked stunned and horrified. Even the two of the Republican representatives who had seemed to be ignoring the proceedings—at times snickering rudely and working on their laptops during the landowner testimonies—glanced up with slightly shocked looks on their faces.

He said his goats that had large growths on their necks that burst open and exuded black goo, leaving them with stained and lumpy necks for life.

"I'm a walkin', talkin' condensate tank," Rick said before he took his seat. "[My stock and I] are all going to die. It's just a question of when."

Several other landowners spoke, including a woman named Gopa Ross. She approached the podium with a jug of yellowed, discolored water in her hand. She held the jug high for the committee members to see as she explained that this was what had happened to her once crystal-clear well water after the gas company drilled on her country property. She, like the other landowners who testified before her, described how her quality of life and her property had been ruined by the gas industry. The committee members appeared to be shocked by what they had heard in the long session and their sobering closing statements reflected as much. House Bill 1223 passed 9-4.

After the hearings ended we made our way through the ornate Capitol building, its marble and gold railings polished and gleaming. I couldn't actually imagine working there every day myself; too much political maneuvering.

As it was after hours, we had a bit of trouble getting out because the Capitol doors were power- locked and we spent some time reconnecting with Nick, who tried to leave by the lower exit with his heavy gear. Finally, we all assembled and departed from the Capitol. As the sun set behind the Denver skyline and the neon colors of downtown began to glow against the graying light, it suddenly seemed beautiful, in an urban way. Part of the good humor of the group was the relief that the bill passed and the other part was the fact that the long day was finally over.

We went to the Warwick Hotel for a post-testimony reception. After mingling with representatives and environmental folks and downing some snacks and beverages, we left the swanky affair for our long drive home.

I got home close to midnight. Although exhausted, I could not get to sleep until after 3 a.m. I kept replaying the testimonies in my mind from earlier in the day: Lance Astrella, the two doctors, Mary from TDEX and the landowners. Over and over again.

The next morning I groggily awoke to a snowy and gray day at about 8:30. I had planned to take a day off work and use a vacation day knowing I would be coming home very late and very tired from Denver. The weather matched my mood. I had absolutely no get-up-and-go. None. It was either exhaustion or the lingering effects of a stubborn cold.

The mud was wet and deep in the churned-up corrals when I went out to feed the animals. It was hard to find good places to put out the hay since the big barn had water from the ice melt water running from up the hill saturating the dirt floor.

I was feeling quite unproductive although I did have a long call with Amy Mall from the Natural Resources Defense Council office in Boulder. She had been at the hearings, too, and apparently had been sitting right behind me for a good part of the day although I didn't know.

Yesterday, just after I turned on the computer to edit, Debra Anderson called. We talked for two hours about the hearing, the successful vote in the House of Representatives and about her progress on the film and my efforts at the writing.

Well, today's work is done. It's nearly fifty degrees outside and I have been sitting here in the small greenhouse on a tall plastic bucket with the laptop in front of me on a folding stool. Our flowers are beginning to bloom in here, the rosemary shrub is frilly with light purple blooms and the parsley and mint are bright green. The chard shoots are brilliant red and the microscopic baby-lettuce leaves are poking their tiny heads out of the soil in the pots that Al has planted. The cats inside the house have come in the greenhouse as I have been typing to sniff at the plants and investigate the thick bed of dense, green catnip growing in the corner.

It is time to go out and enjoy the rest of this day. The gray, long-haired barn cat, Mouse, has been meowing at the open screen windows of the greenhouse and Mister, the huge tiger cat, was prancing on the clear roof of the greenhouse. The horses napped in the warm dirt at the top of the hill in the sun. The mud is drying up, Al cleaned the stock tank after fiddling with his new vintage tractor and I am going out to have some fun myself!

Saturday, March 17, 2007

This week we had a Grand Valley Citizens Alliance meeting. There were updates on the progress of the state legislative bills and recent hearings in Denver, which a number of us had attended. There was more discussion about Presco's expedited plans to pursue drilling on the moratorium site in Parachute where the 1969 nuclear test blast was detonated. The 43-kiloton device was three times as powerful as the bomb dropped on Hiroshima. GVCA members emitted audible groans when the subject of drilling close to the nuclear blast site was brought up.

Liz Chandler, a large animal veterinarian, reported on a goatherd whose success rate in producing live kids had rebounded from last year's low numbers. She said that the air quality was much better around that ranch this year than it had been last year when there were a large number of stillbirths.

Then she talked about a sheep operation near Rifle that reported large numbers of stillbirths this year. The gas wells have gone in very close to the fields where the sheep are and the air quality is being affected. The wells are just across the road.

Today is Saturday and it is an incredibly warm and sunny spring day. The mud from the last snows and rains is dried up and the churned-up soil in the corral with hay bits sticking out is hardening like stucco. The first red-breasted robins of spring are hopping around on the greening grass, dragging their wingtips across the ground as they look for food.

It is far too nice to stay indoors, so I am sitting on the haystack in the open-sided barn with the laptop. The barn cats are snoozing and grooming their coats. The barn is blissfully fly-free this early in the spring, though if this warm weather keeps up that won't be for long.

This week I wrote a letter to the editor supporting Kathleen Curry's work on behalf of the landowners and the environment in regards to gas and oil issues. I e-mailed it out to the local papers and to the Denver Post and then today forwarded it to a few more papers including the Grand Junction Sentinel, the Gazette and The Rocky Mountain News. Curry has been a landowners' champion for the last three years and has bravely sponsored another big gas and oil bill. We'll see if any of those letters get printed. Hopefully some will.

In just a few weeks Debra Anderson will return to Colorado to finish up filming on her documentary.

Mike McKibben, a reporter from the Grand Junction Sentinel, interviewed Rick Roles a few days ago.

"Usually I just let them come up to the ranch and I talk to them," Rick said. "But this time I took a different approach. I drove him up where he could see it up close and personal. I showed him the well pad by my place where there are no secondary burners on the production tanks."

The afterburners serve to burn off emissions from the tanks that include methane, volatile organic compounds—including BTEX chemicals, which are carcinogenic. Rick said that despite years of requests for afterburners, there still were none.

"I guess we just don't count, way up here in the oil gas fields," he said.

Monday, April 2, 2007

Today I spoke to Dee Hoffmeister. We haven't been in touch much since the hearings in Denver in March. I asked her how she was.

"Not so good," she replied.

COLLATERAL DAMAGE

What she told me left me stunned. And five hours later, I am still stunned. I'm also disillusioned and not too hopeful about any chance that things will improve much when it comes to regulations dealing with health impacts from natural gas development. But the irony and the timing of what happened to Dee Hoffmeister is uncanny, if not dumbfounding. I feel at a loss for words here.

Dee told me that the night she got back late from testifying in Denver on the health impacts of gas development she couldn't move her body, head or neck even an inch without becoming incredibly dizzy and experiencing spinning sensations.

What she told me next I could not believe. Three nights later, very early on Saturday morning before 3 a.m., she said that the well by their property exploded into flames not 750 from their house. The flames went over 175 feet into the air. Their son, who lived just up the hill, called them to tell them that the well was on fire.

"We heard some loud popping noises and that woke us up," said Dee. "It was really hot when they went out to take pictures."

I asked her if they had been worried about the house catching on fire and if they were thinking about evacuating.

Dee gave a little sad laugh. "I was so sick from the fumes that I couldn't have gotten up out of bed even if the house was on fire. They would have had to carry me out of there."

She went to the emergency room at Valley View Hospital in Glenwood Springs after the fire, and the doctors, as she said, got a glazed look in their eyes and ran a battery of tests on her.

"I told them they wouldn't be able to find out what was wrong with me because I was poisoned by the fumes," she said with a chuckle. "And I was right, they couldn't figure out the cause. They never can."

Dee said that investigators had been sent out. I asked her by whom and she wasn't sure. She said that Frankie Carver from the COGCC had seemed to want to help and had given her the number of someone from the Colorado Department of Health and they gave her a toxicologist's number and he needed to talk directly with her physician.

This was strange because the COGCC and not the Colorado Department of Health oversaw all health issues relating to gas development. At least that's my understanding of it. I asked her what the reported cause of the explosion was.

"The investigators said it was vandalism," Dee said.

"What! Vandalism?" I replied. I was having trouble believing this.

"They said that someone had opened the valve on the condensate tank and that's what caused it to explode," she said.

"How do they know it wasn't human error, or as they say, 'silly things go wrong sometimes'," I asked.

Dee didn't know.

Dee was going to Grand Junction the next day to see her acupuncturist. Coincidentally or not, it was the same one that Chris Mobaldi saw. Both women were prescribed a regimen of Chinese herbs and chemical detoxifications. I urged her to go see a medical specialist who could do the right kind of blood test on her to find out what exactly what she was being contaminated with.

Home tonight tired after a long and event filled day at work, this news of Dee's increasing health problems aggravated me even more after the well fire. What's happening to people like Dee, Karen, Chris, Rick and so many others is so very unfair, tragic and heartbreakingly sad.

Things are happening so fast with gas development around here and in my view, the oversight of the industry is pitiful and nonexistent. The irony is astounding: Just three nights after a woman already suffering from severe health problems as a result of the wells just 700 feet from her home returns from testifying for the first time to a House of Representatives committee at the state Capitol for the health impact bill, she ends up being sickened by fumes to the point that she is rushed to the hospital.

And if that is not enough, the well by her place explodes and flames towering over 175 feet high burn over her home. If this isn't sad, what is?

The next day I talked to Dee briefly and she said that the doctor told her that her thyroid had totally shut down. He had put her on a regimen of detoxification herbs and had given her a very specific diet that included no white flour and sugar. For now, she was going to stick with this doctor and not pursue other avenues.

Saturday, April 7, 2007

What a whirlwind of a week. There was a lot of gas activity for me. Almost every night I had some phone calls and some of them were long.

On Wednesday I went by Dee's children's house in West Glenwood to pick up a copy of the videotape of the well fire. Once again, because of the fumes from the wells and her illnesses and symptoms she had gone away from home. I rang the doorbell of the suburban two-story house on a cul-de-sac and Dee opened the door. She was steadying herself with a cane that had four prongs on the bottom. Her eyes looked very tired.

She handed me the video and also a few pictures and a short article that had been printed in the Glenwood Springs Post Independent. We talked briefly and then I looked at one of the photos.

"Oh my God," I exclaimed at the photo of yellow-orange flames billowing against the black sky. The flames were higher than the illuminated gas rig that stood alongside.

Even though Dee had described what had happened, I was not prepared for the shock of that image. And this happened 750 feet from their home. Unbelievable.

I dropped the film off with a person who could transfer it to DVD. The next afternoon I picked it up and took a copy over to Patrick Barker of the Western Colorado Congress.

By this time Matt Sura was aware of what had happened. I got a call from him saying that Judith Kohler of the Associated Press and Nancy Lofholm of the Denver Post were interested in talking to Dee and doing interviews and articles. As soon as I talked to Dee and found out about the explosion, I had called Judith. She had always done such a great job on covering the issues both here and in Denver. It was great to hear that both reporters had taken an interest in our cause.

There were so many questions. Who did the investigation that concluded it was vandalism? Why did the Glenwood Springs Post say it was an overactive heater on the condensate tank? What role did COGCC play in looking into the incident? Dee seemed to indicate they were involved but she was unsure of the details.

Late in the afternoon I stopped at the grocery store to pick up a few things on my way to Rifle. Lance Astrella returned my call about Dee as I sat in my truck. I was hoping he might have some suggestions for Dee and her family. As we discussed details of the explosion and fire, water and air samples, and contaminations the Hoffmeisters had proof of already, I saw a lady walking out of the grocery store toward a minivan.

"I think I see Dee Hoffmeister," I said to Lance.

I ran out of the truck and across the parking lot.

"Dee," I yelled.

She turned and seemed surprised to see me. She still looked wan and leaned heavily on her cane.

"Dee, I'm on the phone with Lance Astrella. Can you talk to him?" I asked.

COLLATERAL DAMAGE

She took the phone and I listened to her end of the conversation as Lance questioned her about the contamination and the nature of the explosions. He gave her the name of a water testing company to call. If the explosions that caused the fire were underground then they could very possibly have caused a rupture that would lead to water contamination from the gas well.

When Dee was done talking to Lance, she handed the phone back to me. I hugged Dee goodbye and went to my truck as they drove off to a soccer game in Rifle.

"I just hope something good can happen for Dee," I said.

"Yes," Lance replied. "She's in a tough spot."

"She needs to get out of here."

He agreed. "Yes, she does."

I have now known Dee for over a year and this was the first time I've ever run into her in public. Ever. How fortuitous to see her when I was on the phone with Lance Astrella making inquiries on her behalf. I could count the number of times I've talked to Lance in the last two years on one hand since I first called him with specifics regarding the Mobaldis.

Thursday night I was watching the news from Denver and at the end there was a blurb with a number for CBS 4 Investigates.

I spontaneously picked up the phone, called the number and left a brief message about the fire at the Hoffmeisters and Dee's health issues. I said that Dee had just come back from testifying at the state Capitol in Denver on bills regarding health impacts from gas development and then the well in her yard blew sky high and burned just 750 feet from her home.

A little while later, one of Dee's neighbor's, Carol Bell, called and said she had just found out from another neighbor, Dan, that the COGCC never even knew about the explosion until he called the following day about a yet different problem involving an open condensate tank. Carol had been away at the time of the explosion and fire.

"When he called COGCC, Dan said, 'I'm not calling about the fire, I'm calling about the something else.' The individual at COGCC then indicated they didn't know what fire he was talking about," Carol said.

How, I wondered, could the COGCC not have known about the fire? The Hoffmeisters had called 911, five fire trucks reportedly appeared on the scene, although they stayed down the road for over half an hour. People from the gas company had to come shut down the well, according to reports I heard about why the fire trucks didn't get to the Hoffmeister's house earlier.

Friday, yesterday, I got a call from Libby Smith from the CBS affiliate in Denver, Channel 4, when I was at work. I left my desk and went outside to talk to her for a few minutes. I gave her Dee's number and told her that the video of the explosion was being posted on YouTube and said that I could overnight a DVD of the film if she wanted.

She said she had to run it by her boss and then she would know if they were going to do a story on it.

It is Saturday afternoon. This morning it was a relief to sleep in at least until 7:30 and to not have to rush off to work. After visiting with Al for a while, drinking coffee and then feeding the animals in the pasture, I still felt restless. I felt the urge to do something regarding the gas issues but I didn't know what. Just knowing reporters from two major media outlets and maybe even a major network were considering Dee's story made me hopeful, but also nervous. There was always the chance there would be no coverage.

Finally I just left the house and went to town for a hair trim and pre-Easter shopping trip. On the way in I called Patrick and asked him a few things. I was overjoyed to hear that Raj Chohan of Channel 4 was doing a story on gas issues and the bills and that he would be here in Garfield County this week. Also, he wanted to talk to Dee.

COLLATERAL DAMAGE

Patrick said he would call Dee and tell her and also send her the link to the video of the well explosion and fire that he posted on YouTube yesterday.

I tried to call Dee earlier but she didn't answer her cell phone.

Well, it is already 2:30 on what started as a warm and sunny spring day. Now some clouds are blowing in and the wind is picking up a little. It is time to get outside and re-cover the haystack with the big tarp before it blows off. I loosened one end this morning to aerate the hay a little in the sun. Some narcissus are blooming in the small garden behind the house and the tulips Al planted below the vegetable garden are budded out. Any time now the sheep and their lambs will come down to the long pasture we see from our place.

The tulips and the lambs always come every year at the same time in the six years we've been here. I remember that because the first year we were here before the back pasture was fenced in the sheep came up around the house and feasted on the tulip blooms just as they were beginning to open. That was the same spring when I planted lots and lots of flowers one weekend, then came home from work and found all the blooms eaten off. The sheep were gone by then, rotated into a different pasture. I never knew who the happy culprit was, perhaps a deer or some marmots. Who ever feasted on my flowerbeds had a belly full of several hundred dollars worth of blooms. After that, I switched to hanging baskets.

One night, looking out the large bedroom window into the darkening night, I saw a raccoon sitting on his rear end and swinging happily in one of the petunia baskets hanging from the elm tree. I called to Al and we both watched the coon as he swung and feasted on sunflower seed bits the birds had dropped in the flower basket. Despite some nibbled-on shrubs, trees and flowers, we enjoy the wildlife here. When Al plants trees he fences them and we know that any gardens that aren't fenced might sustain some damage. But on the whole, we all coexist pretty well together here, and except during hunting season, the wildlife is quite approachable. It is one of the reasons we chose to live in the country, to be around wildlife.

I bought a nice bouquet of flowers today, something I do not often do. There are yellow and white lilies, purple mums, some greenery and a big white, delicate, puffy hydrangea. As I type about gas issues, they cheer me up in the small farmhouse kitchen.

Tomorrow we go to Emily and Dave's on Silt Mesa for Easter dinner and an egg hunt at 1 p.m. We have gone there for a number of holidays and it will be fun to be with friends. Al is making a cake in the shape of a bunny's head and I am making hors d'oeuvres.

Tonight I plan to dye Easter eggs. That's why I went to town today, to buy white eggs. Even though we don't have kids I still enjoy the holiday traditions, foods, smells and rituals. It brings childhood back, in a way. And for fifteen years when I was a teacher, we always dyed eggs at school and if I didn't do it now, I'd miss it. It's the same way I feel about carving pumpkins at Halloween, although I have missed a few years here and there.

Monday, April 9, 2007

Today I had a scheduled call with Gilad Wilkenfeld from Henry Waxman's office in Washington, DC. Mr. Waxman chaired the Committee on Oversight and Governmental Reform and it had been recommended that I contact him regarding concerns over hydraulic fracturing chemical exemptions from the Safe Drinking Water Act.

I had called Weston Wilson from the EPA shortly after the air quality hearing and asked about the possibility of bringing back his whistleblower action regarding hydraulic fracturing chemical exemption. Mr. Wilson had given me a number of people to follow up with, and I did.

After a good conversation with Greg Dodsen from Henry Waxman's office, my name and number were turned over to Gilad Wilkenfeld.

Today Patrick Barker and I spoke with Mr. Wilkenfeld together. I talked as succinctly and honestly as possible about the issues, as did Patrick. Our call to Washington lasted about an hour.

COLLATERAL DAMAGE

I tried as hard as I could to say nothing but what I knew to be the truth and at times I felt like I was a tape recorder rolling out the words I have said many times before. With the specific issues I am comfortable with, the words roll off my tongue with no effort.

Again, as so many times before, I felt as Mr. Wilkenfeld was hearing for the first time some of the finer, or more horrific, details of what really happens to people who live close to areas of natural gas development.

While he listened attentively, asked excellent questions and sought our best contacts for real scientific and expert advice on these issues, it made my heart sink a bit nonetheless when he said they had so many issues being presented to them at the Oversight and Governmental Reform Committee.

For the hour we sat in the second-story apartment with a view of Glenwood Springs and the Roaring Fork River emptying into the Colorado, talking on a speakerphone to an aide in DC regarding the health impact issues in Garfield County, I felt I was on a mission. I was asking myself to do the very best that I could for everyone that lives here. No wonder I feel tired tonight. I believe I used up all of my adrenaline during that call.

It is pouring and the wind blew hard tonight just before the rain starting falling after sunset. Today was gray and dreary and my mood seems to match.

As during the lead-up to the 2005 Energy Bill that exempted hydraulic fracturing chemicals from the Safe Drinking Water Act, it's very hard to have meaningful conversations with those in DC who caution that they may not have the ability to help.

On the other hand, we had a good, productive conversation with the Governmental Oversight Committee and gave them our best information on health and hydraulic fracturing chemical concerns.

I am sure the most critical parts of the conversations were those spent discussing the experts available to contact on the issue. They can get to the nuts and bolts of the issue.

In the greenhouse here, surrounded by Al's thriving plants—the chard, the herbs, the green lettuce shoots—I sit in my bathrobe with wet hair after a shower and type away. The greenhouse is a nice place to type in the winter when it's not too cold. It's a quiet, small plant-filled extension off the house that provides privacy in a semi-tropical setting. The scent of catnip, mint, chives, oregano and other herbs float in the air and at times I pick small pieces off and crush them in my hands or chew them.

I don't usually type on weeknights but I also don't usually have a call with a senior staffer from a congressman's office, either. I usually now only type on weeknights if I found out that something blew up. Or something like that.

Saturday, April 15, 2007

What a week. So many things are going on that I don't know where to start. On Tuesday night we had a GVCA meeting and among other things, we discussed the latest developments on the gas bills. The senator sponsoring the bill was reportedly trying to insert language that would be detrimental to our cause, such as changing the definition of "waste."

We discussed lobbying options and Duke Cox described what might happen in the near future with both the health impact bill and the COGCC reform bill. There might be a need to rally again and go down to Denver with a group of impacted landowners to testify before a senate committee. It all depended on the fate of House Bill 1223. Would it get rolled into the COGCC bill or would it stand on its own all the way through the senate? Time would tell.

We heard a report on the nuclear-blast-site drilling plan in Battlement Mesa. An undisclosed entity or individual had filed a lawsuit to stop the potential drilling for natural gas just half a mile from the nuclear blast site. Finally! Something was being done to stop the unbelievably risky plan to drill there. For a change, when I drove home from the meeting I was not depressed by the topics we had covered. That's not to say that there were no reasons for concern. Quite the opposite… What was different was that we had a little ray of hope.

COLLATERAL DAMAGE

Late yesterday afternoon I got a call from Matt Sura of the Western Colorado Congress in Grand Junction. He gave an up-to-the-minute report on the bills. Apparently, it looked like the Health Bill was being rolled into the COGCC bill and would keep in language that would allow the Colorado Health Department to remain involved in the permitting process.

Matt also gave a positive update on the COGCC bill and said that it was basically a done deal and Monday should be a fast-moving hearing at the Capitol. He said he and some others were going down to Denver for it, although he didn't think that much testimony would be heard.

Then I called Patrick Barker, our GVCA appointee from the Western Colorado Congress, and we talked at length about the bills. I thanked Patrick for all the help he'd given me and the other landowners as we worked on the bills. He fielded so many phone calls and worked out the details for our testimony trip before the House Committee on Natural Resources and Energy. He has also been a huge help with copying numerous DVDs and posting film footage on the Internet.

Dee Hoffmeister confirmed last night that the crew from Channel 4 in Denver was, in fact, coming to interview her today. They planned a stop in Grand Junction before going to Silt. Raj Chohan had apparently already interviewed Kathleen Curry and he was going to use the interview with Dee and the others in conjunction with coverage of the legislation.

Last night I spent a lot of time thinking about the gas bills and their hopeful outcomes. I am so pessimistic when thinking about anything good happening since so many times over the last few years all the gas impact legislation has ended up failing or falling short of the mark. Until it's a done deal and I get the word on Monday that we were actually successful, I cannot truly believe it.

Nonetheless, I have some cause for joy since Channel 4 is coming to Silt to interview Dee about the well fire and her health issues, plus the Denver Post interview is scheduled for Monday. Late last night I was very quiet as I sat alone thinking about how all these events have unfolded. Usually Al and I relax on Friday nights, watching a movie in the living room and savoring the freedom of the upcoming weekend. This particular night, however, I spent time alone contemplating all the little pieces that have gone into bringing attention to the health concerns related to gas development. Without the landowners going public with their stories, none of this would have happened. I washed my hair and sat on the bed in the little spare bedroom wrapped in my old green bathrobe. I thought about it all for a long time. On the antique bureau below the large mirror there was a cluttered pile of miscellaneous items that had made their way there throughout the last week or so.

There was an article about the health bill hearing and I could see the photo of Dee Hoffmeister as she leaned forward to answer State Representative Roses's questions about her health issues. She sat beside Kathleen Curry, who was biting her lip. I wondered if she was doing that to compose herself. I remember when Dee testified about her failing health and she had broken down and cried, tears had fallen out of my eyes, too, as I sat in the hearing room before the committee at the Capitol. The room had been packed with angry and sickened landowners, environmentalists, oil and gas employees and lobbyist and members of the press. Lance Astrella had come, too, and for most of the hearings, I sat by him and Morgan Bear. The work her husband had done on trucks that carried production fluids from the pits, most likely containing condensate fluids and hydraulic fracturing fluids known to contain highly toxic and carcinogenic chemicals, made him so ill he suffered from seizures and depression and lost the use of his arms and legs.

On the bureau there was also the usual assortment of everything that I had emptied from my pockets over the last week. There were piles of change, paperclips, a few sticky notes, a FedEx slip with Debra Anderson's address from when I mailed the last batch of DVDs of the well fire at the Hoffmeisters. Next to it was the small external drive that contained a copy of the computer manuscript of the gas book. In a way, the items on the bureau were a kind of time capsule of my recent past.

Finally, I went to bed.

COLLATERAL DAMAGE

This morning I called Kathleen Curry at home. I had her number from a legislative contact flyer we got at the Capitol. It has all the senators' and representatives' contact information and lists all the governmental offices and committees at the Capitol. Kathleen probably knows my name and I know she recognizes me, but I don't speak directly with her often.

I thanked her for working so hard on the health bill and told her I knew what a hard job it's been. We talked for quite a while until Kathleen revealed she was in the grocery store. I must have called her cell phone.

As we discussed the bill, Kathleen said that during recent a meeting at the Capitol in Denver, an industry lawyer had accused her of fabricating the information from the testimony on the health bill.

I was stunned.

"What! They're saying we lied? That everyone made up their testimonies? What? You didn't even know those people," I said.

"Yes, they said I made it all up, that it was all lies," she said.

Kathleen described how she was being pressured to drop the health bill as a compromise to let HB 1341 go through, but that she refused to do so.

I told her that here on the Western Slope we were waiting to hear each of the latest developments and that we knew it was politics and it was out of our hands.

She asked if I had seen her quote in yesterday's Rocky Mountain News. It had gotten her into some hot water and the governor's office had called and asked her to apologize to the industry. Again, she refused.

"You know, when we were there in person testifying, no one accused us of lying," I said. "That testimony was powerful and we got the nine votes to prove it. But when we're not there, the industry says it's all lies. I think we should go have a meeting with the governor and let him hear it from us."

Kathleen agreed.

I told her about Dee's well blowing up and about her recent hospitalization. I could tell she was shocked. I also told her that Channel 4 was up here today to interview Dee, the Denver Post was coming Monday and documentaries were being done on gas development. I let her know that Debra Anderson was almost done, but she had at least one more trip up here planned.

Kathleen was happy to hear about the continued press exposure. She had to go and get her son to a game, so we said goodbye. I am really glad I called to thank her for her work. I know she was happy to hear from the grassroots people she was doing such a good job of representing.

After I got of the phone, I was still furious that the landowners had been branded liars, and even worse, that there was a suggestion that she was behind it all.

"I just talked to Kathleen Curry and she said the industry's attorney accused her of fabricating the testimonies from the hearing. They said it was all lies," I told Rick Roles.

"What? Lies! Are my blood tests lies?" Rick asked furiously.

"Yes, apparently so. It's all lies, we made it all up," I said sarcastically.

Rick and I spoke a bit longer and he told me that Channel 4 had wanted to interview him after the "See Red" guy, Lamb from Debeque, told them about Rick's blood tests.

"Well, Channel 4 is coming to see Dee today, so call her up and find out when," I suggested.

I then called Dee and told her what we just found out. She thought it was ridiculous but she wasn't surprised.

COLLATERAL DAMAGE

"They don't care about us. It's all about the money," she said.

I agreed. "Yes, it is."

Tomorrow is a GVCA retreat. We're going to Beth Dardinsky's house up Dry Hollow. Hopefully the fumes won't be bad because I don't want to get headaches or whatever else it is that one gets from being there at a bad time.

Al went to town to get firewood where someone cut down apple trees. He's a good hunter and gatherer. I've taken advantage of the time alone and typed for a few hours in the quiet house. My writing isn't quite up to par; I'm not feeling that good today. But at least I got something down.

It is a lovely spring day today and the elm trees are beginning to uncurl their tiny green leaves. The cats are lounging in the sun on the wooden deck, soaking it up after this past week's cold and wet snap. We actually had some snow, the temperature dipped down to ten degrees and I had to break the ice in the horse's stock tank several mornings. Previously it had been warm enough to take out the electric stock-tank heaters that had been plugged in all winter.

Maybe if I go outside in the sun I'll feel better. If I do some chores in the barn or walk around the pastures surely I can get my mind off all this gas stuff. Yet, it's strange knowing that the Channel 4 crew is perhaps over at Dee's house doing their interview right now. It's so important for people to know what's happening to those like Dee. I hope the interview and the final piece go well. Another drop in the bucket, another step along the way.

Monday, April 16, 2007

Getting in a quick type here. Al isn't home yet even though it's 7:20 p.m.

The Channel 4 crew ended up being delayed. They were still filming with the gas companies on Saturday and I got a call on Saturday night saying they would be going to the Hoffmeisters at 11 a.m. on Sunday.

Sunday morning was the GVCA annual planning retreat. This time it was at a private home just a bit up the rode from the Hoffmeister's place in Dry Hollow.

I stopped in and talked to Dee on my way to the meeting and we arranged for her to call me when Channel 4 finished their interview. I was going up with them over Jenkins Cut Off road into the heart of the gas field to then meet up with Rick Roles and tour more wells.

When I got the call at the meeting and was told to go down to the Hoffmeister's, I left the GVCA meeting. Dee and her husband Hoff introduced Raj Chohan and his cameraman to me and then we went up the hill and over the gravel Jenkins Cut Off road at a slow speed so as not to get any flat tires.

As we came down the crest of the hill and first saw the dramatic view of the once lovely valley now densely dotted with gas pads and infrastructure, I wondered what the two fellows from the Front Range were thinking. I know the first time I saw it I was shocked. There is something so incongruous about the high mountain rural landscape of cattle farms, rolling pastures and a haystack inundated with well pads and all the massive industrial equipment and impacts everywhere. It was reminiscent of a science fiction movie set. It just doesn't seem real. Welcome to Colorful Colorado.

I glanced into the rearview mirror and saw the Channel 4 vehicle flashing their headlights at me. I knew that meant they wanted to get a shot.

After that, we met up with Rick where he was waiting in his white pickup truck. His dog Rowdy, a Rottweiler mix, was chained up in the back. We went up to a few different wells and Rick first showed them a well site that had the proper afterburner installed, then showed one of the pads at his ranch that did not. Of all the pads on Hunter Mesa that Rick knew of, there was only one that had an afterburner installed to burn off the methane instead of venting it directly into the air.

"I guess we just don't count up here," Rick said.

COLLATERAL DAMAGE

As the cameras and recorders were positioned at Rick's pad, I stayed out of the way and watched and listened. About fifteen minutes into the shoot, a increasingly loud noise began coming from one of the huge tanks. It sounded like a huge pressure cooker blowing off steam. The noise must have been interfering with the recording, but it took a few minutes for Raj and his cameraman to turn their heads and look at the huge tank emitting this now deafening noise. It just about drowned out our voices.

"What is it?" they yelled above the noise.

Rick explained that the tank was blowing off methane, as it always did up there. He said that we better get out of there pretty soon or they would begin to feel sick. I believe they looked a bit nervous when he said that. I know I was in a tenuous frame of mind. I was excited that the television people were there to film it, but I don't like to be up and around wells that are blowing off fumes or up around pits that are emitting chemical evaporations. There are others who have gotten sick on their first visits to certain evaporation pits that are filled with various and unknown production fluids.

The well blew off two more times before we completed our stop there.

The Channel 4 crew was pressed for time to get back to Denver and we finished up after about an hour or so of their shoot. They got directions from Rick for a last location where they could film an active drilling rig.

Before we shook hands and said goodbye, Raj said that when he does his investigative reports, he ensures that the stories are based on facts. He said he won't pursue pieces if people talk about conspiracies. I could understand that, and to the best of my knowledge no one on our side had said anything about a conspiracy, so I assume he found our report to be credible.

I told him that others were actively working to address the fact portion of our story regarding human and environmental health by collecting air grab samples, blood and water tests from contaminated wells. He said he would be interested in seeing the information. He wrote his e-mail address on a piece of notepaper and I folded it and put it in my pocket. Now it sits on my computer desk and I need to follow up.

After all the preparation and anticipation, Channel 4 will air a three-minute piece in late April or early May. We'll wait and see what happens.

On my way home from Rifle tonight I called Matt Sura from the Western Colorado Congress. They were in Denver just finishing up after the senate committee hearings of HB 1341.

He said it went very well—a 5-0 vote in our favor. He sounded exuberant, yet he said we had work to do. He gave huge amounts of credit to Harris Sherman, the newly appointed chairman of the Department of Natural Resources, for the work he did on pushing for this bill. He said that Harris Sherman was totally exhausted now.

I told Matt that I am so pessimistic I have a hard time getting excited even when good things appear to be happening. He commiserated.

As people are here on the front lines of gas development issues, living in their homes as they endure fumes, illnesses, contamination and more, little is getting done to help them. That's because the infrastructure and current regulatory agency are not there to do it, nor are there adequate state or federal policies.

We're working so hard on these legislative efforts to get sound policies in place and even though there seems to be some hope for positive change on the horizon, a resolution still seems far off.

Perhaps it's because we're still living here and watching and experiencing it going on around us every day. It's hard to have hope in the face of all this suffering and it's even more painful to have hope when there's a chance it may all go up in flames.

Sunday, April 22, 2007

I decided to call Nancy Jacobsen. I knew they were moving to Oregon and I didn't want her to disappear before I had a chance to talk to her. They'd had it with being surrounded by gas development up Dry Hollow in Silt.

Nancy picked up the phone as soon as it rang and she was exuberant as she told me about her recent trip to Oregon. They leased a house on one acre on the Willamette River outside Corvallis, and the closing on their lovely ranch in Silt was either on May 10 or 17. The date was tentative due to the buyer's financial issues.

I had lived in Oregon for twelve years and listening to Nancy's glowing reports of her future home and state made me nostalgic. Yes, it was lush, the people were friendly and the communities were well-balanced, especially in the university towns. Everything was so much better than it was here.

In my mind, the main difference is that, to my knowledge, there is no natural gas development in Oregon.

The impacts of natural gas development are my only problem with living in Garfield County. I can deal with any other potential negative factors.

Nancy described how they had so much to do in a short time. Gary was selling his computer business and needed to spend time showing the ropes to the buyer. They had to pack all their belongings from their large home and prepare for the movers.

I told Nancy that Debra Anderson was coming back for her last visit to finish filming and asked if they would consider giving her some time. Nancy was hesitant and said they had to be careful. She said that she would talk to Gary about it and that I should e-mail her details.

During the last minutes of our conversation, my eyes welled with tears as we talked. I would likely not see Nancy again before she left Colorado, or for that matter, ever again. Even if I did see her again, the fact was simple: Nancy and Gary had sold the ranch in Silt to get away from the horrors of gas development and in a matter of weeks they would leave Colorado. They would be gone.

TARA MEIXSELL

Wednesday, April 25, 2007

Yesterday morning just before I was leaving the house for work, I got a call from my friend Kathy Miller who lives on the north side of Silt Mesa. She told me that there was a fire and huge plume of black smoke going up in the air across the river.

"Something is going on over there, something big," she said.

"Maybe they're lighting an evaporation pit off. That creates a lot of black smoke," I said. Then I explained what burning off the pits was and we talked about flaring.

"Is the smoke roiling," I asked.

"Yes, it is, and there's a huge black cloud over it, kind of spreading out like a mushroom shape," Kathy replied.

I called a few people and learned at 9 a.m. that the fire was under control. When I called the sheriff's office earlier, they said the fire department was out there.

I called Jaime Atkins from the COGCC to ask what was going on and left a message on a machine asking him to call me back. I haven't heard from him yet.

Today a co-worker who lives in Silt left a picture of the Glenwood Springs Post Independent of the well fire explosion by my keyboard. The photo shows flames leaping out of the blackened equipment on the pad.

I got an e-mail today from Matt Sura, the Energy Organizer from the Western Colorado Congress, about the Senate vote on the COGCC reform bill, which also now includes the health impacts bill.

The bill passed the senate 29-6 in our favor.

I wrote Matt back, "Are we happy now?"

COLLATERAL DAMAGE

Last night I talked to Lisa Bracken up Divide Creek near where the fire was. She told me that her father recently passed away from pancreatic cancer. Their place was near the Divide Creek seep, the sight of the largest finable violation of natural gas contamination in Colorado.

She talked about her Native American father's passion for the land and told how he had gone down into Divide Creek where the gas seep had been discovered and lit the toxins on fire as he stood in the water. She said that they filmed it. How strange it all seems.

Although I am tired tonight after a long day at work, so much has been going on I thought I better get a bit of typing in before it all fades away.

It's ironic that at the same time there's another well fire disaster with unknown toxins burning in our air, a milestone is reached on our gas bills. And it wasn't long ago that Dee Hoffmeister's well blew up and she became so sickened from the fumes that she was hospitalized, just three days after the hearings she testified at for the health impacts bill.

Nancy Loftholm from The Denver Post came to interview Dee last Tuesday for an article. She said she was finishing up the piece last night and it would likely in the paper this Sunday.

Al is on a weeklong raft trip on the San Juan River and it's quiet here at the house. I called Dee Hoffmeister and Rick Roles to see if they heard about the senate vote. Rick hadn't but Dee had. Dee asked if I was getting a lot of typing done with Al gone. I said no, as it was a weeknight. But at least I managed to pull out the computer tonight for a short while.

It's hard to believe that anything really significant will occur from legislation in regard to the impacts of gas development in Garfield County. It seems so disconnected and far away from what's happening here every day. Then again, I realize that without improved legislation protecting landowners and environmental health concerns, nothing can be done at all.

Front Page News

April 28th 2007

The day after the latest well fire, the Glenwood Springs Post Independent had a front-page article entitled,

"COGCC REFORMS CLEAR SENATE"

It read:

"Bill would increase the board's size from seven to nine

"Landmark regulatory reform of the oil and gas industry could be headed to Gov. Bill Ritter's desk for his signature by the end of the week.

" The Colorado Senate on Wednesday voted 29-6 to pass a measure that changes the mission and make up of the Colorado Oil and gas Conservation Commission.

"The bill has already passed the House of Representatives, which now will consider changes contained in the Senate version. The Western Colorado Congress, an environmental organization that supports the measure, is predicting that the final version could be delivered to Ritter before the end of the week to be signed into law.

"Passage of the bill has been a top priority for Ritter. Harris Sherman, Ritter's Executive Director for the Department of Natural Resources, drafted it.

"The bill would increase the COGCC board size from seven to nine people, while reducing the number of oil and gas industry representatives from five to three. The industry has said that its representation is needed for reasons of technical expertise, but critics long have contended its domination of the commission resulted in failure to adequately address complaints related to oil and gas development.

"The measure would also change the COGCC's mission to put more emphasis on protecting public health and the environment. One of the COGCC's board members would be the Executive Director of the State Department of Public Health and Environment.

"Others would include the DNR chief, and people representing the interests of the environmental or wildlife protection; local government; agriculture/royalty owners; and soil conservation or reclamation.

"Duke Cox, president of the Grand Valley Citizens Alliance, which looks out for landowners' interests on gas development issues said that the bill would be 'very significant' in helping balance energy development against other interests.

"However, he is concerned by statements made by state Sen. Josh Penry, R-Fruita, that Ritter's appointments to the commission would be subject to significant scrutiny.

"'If they have to go through Senate approval, you can look for a fight'," said Cox.

"Penry said the appointments are subject to Senate confirmation, but Cox need not worry.

'I think there are some in the environmental community that need to know when to claim victory. This is a sweeping rewrite (of oil and gas regulation) and they are still grousing', he said…. 'I doubt very many other states would change their oil and gas laws in one fell swoop as we just did.'

"Elise Jones, executive director of the Colorado Environmental Coalition, praised Penry along with Senate President Joan Fitzgerald and the bill's sponsors, Rep. Kathleen Curry, D-Gunnison, and Sen. Jim Isgar, D-Hesperus, for brokering a compromise on the issue.

'Colorado deserves a balanced approach to regulating the many impacts of the gas and oil industry'," she said in a news release.

Penry said he doesn't know many people who think the COGCC should be dominated by the industry, but it is also important that it not become anti-industry. Cox also said that balance is important.

'We don't need radicals on this commission. We need people who are willing to put aside their personal feelings and consider the best interests of the public', Cox said.

"The bill originally sought to eliminate current laws that call for oil and gas to be efficiently extracted and not wasted. Opponents of that law say that such efficiency can come at the expense of the landowners and the environment.

"Industry groups had opposed the bill but took a neutral stance after it was amended to retain the efficiency clause. Sherman agreed to the change because he believes the bill still will provide adequate consideration of environmental and health impacts of drilling.

"Curry had sponsored a bill that sought to deal with health concerns, but Penry said that the bill was dropped because the concerns will be addressed via the COGCC reforms. He said another bill aimed at protecting wildlife will move forward. However, the concerns raised by all the bills will be addressed in a single rule-making process that will result from the passage of the COGCC bill and will be subject to legislative review, he said.

"He believes that the unified process should help alleviate industry concerns about the large number of bills that have been brought forward this year."

(Article written by Dennis Webb, Glenwood Springs Post Independent)

COLLATERAL DAMAGE

On page two of the paper in the Local News Briefs section, there was a four-paragraph article about the gas well fire from the day before. It was entitled,

"Lightning a possible cause of oil well fire"

That seemed rather odd, as to the best of my knowledge lighting rarely occurs in the early morning; typically it is later in the day when we have strikes. I am not saying I don't believe it; it just seems odd.

Yesterday I got a call from Kirk Siegler at Aspen Public Radio. He was doing a story on the current gas and oil legislation, and he wanted to interview some of the landowners who had health complaints they believed were cause by nearby gas development.

We agreed to meet next Wednesday morning at one of the Silt gas stations then go up into the gas fields to various people's properties to do interviews. I was going to e-mail him phone numbers after I called people to ask if they were interested in taking part in the radio show. Kirk said it would definitely be broadcast locally through KJAX from Aspen, and that it also might go to National Public Radio.

As we talked and I gave him background information, he asked me if I thought that changing the GOGCC and passing these gas bills would make any difference to people's health issues near the development.

I answered that as it currently seemed to be, no one was responding effectively to any health complaints. When people call with complaints, and they are typically told that they "might be sensitive" or "nothing has ever been proven to be harmful to health" from gas development impacts.

Any change in a more proactive direction is very welcome and very much needed.

I told Kirk that there were a good number of people who could no longer speak to the issue, and for a variety of reasons. Some with serious health impacts gave up in disgust and fled the area. Others, though not very many, had settled monetarily with gas companies and could no longer discuss their grievances against the company.

And then there are those who are desperately trying to sell their homes and get out. Although they may have complained and come forward to various forms of the press and government to speak about their health issues before, now that they had homes in the heart of the gas fields up for sale, they couldn't discuss it. Or perhaps better put; they declined to do so. And absolutely no one can blame them for that. If you cannot prove that anything is being done negatively to impact your health by the gas industry, why on earth would you come forward with concerns when you had finally made the hard choice to leave? And for most, the once lovely country homes they are leaving were their hard-worked-for retirement properties.

Many of these people have spoken out and fought valiantly to improve the situation around their homes to no avail. After years of hopes that the situation might improve by speaking to the press, hiring lawyers, going to countless meetings, even the most dedicated finally choose to leave the area. For many, it is just not worth the constant struggle and fears of health impacts to stay.

Last night I talked to Chris Mobaldi for a long time. At first, Chris was very hard to understand. But, after a few minutes we settled into a lengthy conversation that was comprehensible in spite of her foreign accent syndrome. As we talked about the gas bills currently underway and more, I was constantly surprised by the different words that peppered her speech; they ranged from Spanish to German and French and I am sure more languages that I have no experience with.

Chris seemed happy to hear that some improvements were hopefully underway, and when I told her that the first draft of the book I was writing about her was almost done, I could hear excitement in her voice. And perhaps the disbelief as well…

Something that Chris said stays with me today; and probably will for a long time.

COLLATERAL DAMAGE

She said that her friends don't talk to her anymore; that because of her illness people are forgetting about her. She made an exception for me, but I must say that aside from some phone calls and my work on this book and the issue I am not a very present friend for Chris. We live too far apart now.

This morning I watched more of Joe Brown's documentary *National Sacrifice Zone*. I got a good way through it, then in the middle of an interview up Dry Hollow with Karen Trulove at their ranch the film stopped. I had adjusted the volume, and the DVD cut off. I will have to start over, and go through the first hour of the DVD again to see the rest of it.

Some of the interviews in the film I had been present at, and it brings back memories of those events.

During the footage at the crest of the Trulove's property, as Karen described the fumes from the wells above her property I had been unable to stop some tears from leaking out of my eyes. Karen was being brave, but what she was saying about what was happening to her home and to her health was heartbreaking to listen to. I had heard Karen speak at the state Capitol, and she had been close to tears at times during her testimonies.

This time, on her own ranch, she was not tearful. But I was.

Tomorrow Al is coming back from the raft trip. It is now midafternoon, and it is absolutely gorgeous outside. I have a house to clean, and several loads of laundry to deal with.

The thermometer outside the kitchen window is pushing eighty degrees; the trees are leafing out and blooming, as are the spring bulbs. Last night when I was out in the barn I could smell the sweet scent of the apricot tree drifting on the breeze.

The blackflies have also hatched this week, and this morning I put my first round of bug wipe on the grateful horses. Their underbellies, ears, and throats were beginning to get chewed upon.

Enough of this typing now, I am going outside to the barns and to enjoy this beautiful day. I am happy that I got some writing in though. That is what you have to do; keep diligently at it and soon you have a big piece of work. There is no other way to do it…

Sunday April 29, 2007

This morning I watched the rest of Joe Brown's documentary film *The National Sacrifice Zone*.

As I have been going about my routine on this Sunday, the film stays in my mind. The voices and faces of the people impacted by gas development, from the ranchers to the landowners, environmentalists, and those living literally on top of the Rulison nuclear blast site where Presco Corporation wants to drill for natural gas.

Kim Weber from Debeque who lives close by the Black Mountain water-waste pits said she was available for an interview with Kirk Siegler from Aspen Public Radio on Wednesday. We talked at length about her symptoms, and the operations at the pit. The last time I had spoken with her she had had friends over, and we were going to get together at a different time.

"How was your visit with your friends?" I asked, thinking that she and her husband had had out-of-town visitors or family there.

"It was good, they are people I know from CRED," said Kim

"Oh, they are gas friends? Or did you know them before?' I asked.

"No, I know them from gas."

"I have gas friends too, but a lot of them are leaving…" I said.

Then we talked about that for a while.

Earlier this morning I had spoken with Dee Hoffmeister who was feeling quite ill after being exposed yet again to fumes up at her Dry Hollow home from the gas wells by her property.

She said that she was dizzy and had headaches, and that doing these press interviews was wearing.

The article she was interviewed for was supposed to be in today's Sunday Denver Post, but it didn't come out yet.

Today it is above eighty, and it feels hot. The plants around the yard and in the gardens look dry. Hopefully we will get rain soon.

Al is not home yet, nor have I gotten a phone call. He usually calls when they come in cell phone range after they take off the river.

I can imagine them all down there; breaking down the rafts and huge amounts of gear at the take-out under the broiling sun after the long trip. Then they have a shuttle drive to pick up the other vehicles before they get on the highway to head home.

I likely will be asleep when the rafters finally get here.

Wednesday May 2, 2007

I just got back from a long session of interviews with Kirk Siegler of Aspen Public Radio. We met at new gas station in Silt, and then we drove up in his car to see Dee Hoffmeister up Dry Hollow.

The interview was very casual and quite long, and initially he taped the conversation out behind the house where there was an excellent vista of the surrounding landscape and the industry infrastructure. Gas trucks of various shapes and sizes and colors roared up and down the hill almost constantly. Later, shade was sought under the spacious roof above the front deck.

As Dee talked about her discovery last year that "she was not the only one who was sick" when she ran into her neighbor Karen at an Associated Press interview, she began to choke up and cry. I was standing below the porch at this point, watching the unusually colorful black and yellow and black and red and yellow blackbirds and smaller finches singing about the juniper tree in front of the house. I had tears on my cheeks as she tried to speak, and they were not just prompted by her, but also by a flashback to the time well over a year ago when Joe Brown had been filming Karen at her ranch just across the road and her story had brought me to tears then. She herself had been very calm and stoic, but her descriptions of the events that unfolded around their ranch and her horrific deterioration in health had been heartbreaking.

Karen Trulove and her husband Tim will always be heroes to me, along with all the others who have spoken out about their health impacts from natural gas development. I don't think I will ever forget her powerful testimonies both at the Capitol and in the interview with Judith Kohler of the Associated Press. On both occasions when Karen described her life, her deterioration, and the scope of her symptoms she came to tearful halts in her speech. It was just too much...

The interview at Hoffmeister's lasted about two hours, then we drove up to Hunter Mesa and met Rick at a juncture in the road where Jenkins Cut Off ends. Up the very steep gravel road we went, and drove slowly through the gates onto his well pad that had five wellheads.

Again, Kirk conducted a lengthy and detailed interview and Rick explained the ins and outs of the equipment on the pad, and the substances in the various tanks, in addition to discussions about the companies, his health and symptoms, and more.

I drifted away a bit, but continued to listen to the interview. At one point, as I stood behind the condensate and produce water tanks, I squinted hard to read the small print on the industrial warning labels pasted to the tanks.

I made out the words hydrogen sulfide. Hmm, that was interesting as I had often been under the impression that there was no hydrogen sulfide associated with the production in this area; although there was in Canada and elsewhere....I pointed it out to the others and they also found it interesting.

As Rick talked, I gazed about the surrounding landscape and mountain ranges, and admired the herd of fat appaloosa horses grazing below, with small white sheep moving back and forth between them in the pasture. A pretty gully that must have contained a stream was just downhill of the well pad, thick with lush green vegetation sprinkled with yellow flowers. On the roads behind, the industry trucks moved back and forth, and within view a number of well pads and several newly erected rigs were visible. It was a busy place...

We said goodbye to Rick after we stopped at another pad and looked at the afterburner; the only one Rick claimed that existed on Hunter Mesa. It served to burn off the excess methane gas as opposed to letting it flow directly into the air. Rick said he had asked a good while back about where the afterburners were for his property, but no one had gotten back to him ever.

When Kirk dropped me off back at my truck in the park and ride near I-70, I asked him how he felt. I myself felt a bit sick, had a headache, and felt lightheaded.

"Well, I wondered if it was the red wine from last night..." he said.

"Yes, that's what Karen said it's like when you breathe the fumes. It's like having a really bad hangover that never goes away," I said.

Kirk was off to Rifle to the Energy Expo, and I gave him directions and suggested a restaurant that was nearby that had fast and excellent pizza and sandwiches. He said he would let me know when the piece would come out, and indicated it would not be months by any means, probably a few weeks. He said we'd be in touch.

I drove home feeling a bit woozy from the time up on the pad around the fumes. I know I had smelled fumes a few times up on Rick Roles's well pad, and once at Hoffmeister's as well. I didn't feel much like driving into town even though I had planned to go to City Market to get a few things. That could wait; I just wanted to go home.

Now that I have been home for about an hour, I feel better. I brought the laptop out into the big barn (in spite of the hay and dust) so I could enjoy the newly leafing green elm trees and listen to the wonderful sound of the sheep and lambs baaing in the field across the creek. The birds and crickets are also active, and the whole combination of animal, bird and insect noises is like a lovely symphony to my ears. Out here, smelling the fresher breeze than was up in the gas fields, plus the grounding scent of green grass hay and the horses, I feel less sick although still lightheaded.

I remember talking to Carol Bell, the retired chemistry engineer who lived right across from Dee's place and saying,

"You know, I don't know if I'm imagining it or not, but after I come up here I swear I have bad headaches, and sometimes nosebleeds."

Carol looked straight back at me with her serious blue-eyed gaze and said,

"You're not imagining it."

Yesterday Jaime Atkins called me from the COGCC to answer a call I left last week regarding a well fire. We talked briefly about it, and then I asked him about what had happened at Hoffmeister's when their well exploded into towering flames. I asked if any violations had been filed, and said that Dee had heard nothing back.

Jaime said that an NOAV (a Notice Of Alleged Violation) had been filed against Barrett, but he hadn't seen their response yet that was required within a fifteen-day period.

Jaime said he went out and looked at the site and he concluded that the seals had been broken on the valves of the condensate tank, and it was likely vandalism.

I told him that Dee was not doing well, and said she had asked for an air quality monitor and had been told there were none to be had.

Jaime said to call Jim Rada at the Department of Health, and to call the State Air Quality Commission.

We had a cordial phone call.

COLLATERAL DAMAGE

Sunday May 6, 2007

Today's Denver Post front-page headline read:

"Well-ness checkup

Ailing residents believe nearby oil and gas drilling is to blame for their conditions, but proof of the link remains elusive."

I was at the grocery store when I got the paper. As I deposited my quarters for the paper, I glimpsed a picture on the front page through the newspaper machine glass window. It was Dee Hoffmeister, standing outside in front of the well pad where the fire occurred. She was holding her jean jacket over her mouth and nose, as if to block fumes.

I could not believe it: Nancy Loftholm's article had made the front page of the Sunday Denver Post! It was an excellent article that covered a good part of two pages and included numerous photos of the Hoffmeisters.

When I got home, I made a few phone calls.

I wanted to have a day that was just a good day, and not to think too much about gas issues.

So I went out to the barns and the animals, and I saw a coyote running along the fence line through the scrub, then he jumped into the thicket by the creek. He was heading of toward where the sheep and the lambs were in the large pasture below. A bit later, a very fluffy-tailed second coyote came into view and followed his path.

Worried for the lambs, I climbed the hillside thick with prickly cactus and cheat grass to the fence line. I let off a few loud whistles, and watched the herd below. They were moving around and the ewes were bellowing to their small lambs. The herd dog was barking furiously, and I could just hope that the pair of coyotes would kill no lambs during the mayhem.

They are my neighbor's sheep, but I watch over them. I would not, however, go any further today.

I spent the next few hours in my own barn, deep in thought, as some of my own animals were close by. I brushed the big horse's mane and neck, and he craned his neck with pleasure. I thought back to the more stressful times we had spent together in that same space: Buddy penned up while his stitches healed beneath the huge bandages, for four weeks last winter after his leg surgery. How wonderful it was that he is doing so well now six months later. He is the most placid and friendly huge and beautiful bay horse.

The two barn cats Mouse and Ginger lay curled in hay nests side by side and groomed one another.

For me, being with my animals and watching the birds fly by and call, and seeing the clouds float above the mountainside beyond is what I cherish in my life. The work I am doing with gas development impacts is all about that. Just trying my best to see that those of us who live here in the epicenter of gas development have a chance of keeping what is precious to us about our lives and where we live.

Sunday May 16, 2007

Another very busy week, I have been swamped at work. There were deadlines for projects that were due immediately at the State, plus my boss was out on leave for a second week in a row and I was doing my best to fill her shoes. That alone was way more than enough, so of course on top of that we had to have a time-consuming emergency that took a huge chunk out of two days for me. By Friday I was pretty much trash material, and by the time I left work my mood was low.

Saturday morning I woke up and was barely excited about a potential day off because of everything I knew that needed to be done that was sitting in orderly piles in my office in front of the computer. For the first time in four going on five years I went into the office on Saturday to get a jump on that pile. It was nice and quiet at the office, and I managed to be somewhat productive before I drove back home.

COLLATERAL DAMAGE

Once at home I logged onto my e-mail computer and began a series of frustrating and lengthy phone calls to technical support MSN employees in various countries. I spent over two hours getting nowhere, and in frustration gave up.

This morning I continued the battle and after at least another hour with technical support, (this time from Malaysia), I finally fixed the problem. After a nice long Mother's Day phone call to both Mom and Dad, I began getting my thirty-five backed up e-mails I had been unable to retrieve for over a week.

Amy Mall from the Natural Resources Defense Council office in Boulder was collecting data for a report she was going to make in Washington regarding health and environmental impacts from gas development. She had sent me a number of e-mails, including the information that she had seen on the Channel 4 series that Dee and Rick were interviewed in. She said they were very good, and that the piece had been fairly balanced, as they also interviewed industry people. A few days later she came across the Denver Post article, and she said that was excellent also.

As I scanned through the lengthy list of unopened e-mails and deleted the unwanted ones, a return address from Jaime Atkins of the COGCC caught my eye.

I clicked the e-mail open, and read his lengthy letter. It was a report on the Notice Of Alleged Violation's filed at the Hoffmeister's well fire—the "stone well pad"—I believe is what it is called. Additionally, there was information about Bill Barrett Corporation's required response to the NOAV, plus information about a sheriff's investigation and an FBI investigation of the fire.

"Oh, my god!" I said in stunned disbelief. The FBI? I had known the Sheriff's Department was doing work on it, but not the FBI!

I read down further. Jaime said that both the Sheriff Department's investigation and the FBI investigation concluded that leaked condensate had been the cause for the fire, and that vandalism or arson had been the cause of the leak.

"Oh, my god!" I said aloud again as I stared at the computer screen.

Al, who was cooking lasagna in the kitchen, said in a somewhat frustrated voice,

"What?"

I hadn't realized that I was talking out loud to myself, and I explained to him what was in the e-mail about the FBI.

That seemed to satisfy his questions about my "Oh my gods!"

I am in the barn typing on this warm day, and the battery is about to go...

Making More Movies

Tuesday May 22, 2007

It is after 9 p.m., and I have been on the phone for most of the night since I got home. One work call, and two lengthy calls regarding gas issues. I just got off the phone with Amy Mall from the Boulder office of NRDC.

Before it gets too late, I need to write down a few things about the past few days' goings-on...

Sunday I drove down to Denver and spent the night near I-70 in a hotel so as to be close to downtown Denver for the 8:30 a.m. film set-up time at the law office of Astrella and Rice.

Debra and another film professional would be arriving there at the same time; Deb spent the night at her family's place in southern Colorado after driving up from New Mexico the previous day.

First Lance Astrella was going to do his interview, and then Weston Wilson from the Environmental Protection Agency would do his. For me, this was a big deal.

After I arrived in Denver about four in the afternoon on Sunday, I headed over to a store in the huge Colorado Mills complex just a few blocks from the hotel to buy an extension cord. I knew that tomorrow the film equipment would take precedence over my tiny tape recorder, and I might need to get creative to get a plug in. Deb had told me that she needed to have at least six electrical outlets for her equipment. Although I was near a lot of stores I typically didn't get access to, I wasn't too motivated to go shopping or poke around once I made the long drive from Garfield County to the Front Range. I had left home at 11:15 and got to Denver about 3:30, with a brief stop in Vail to eat lunch. The highway was packed with Front Rangers returning to Denver from their weekend in the mountains, and some of them drove like maniacs, tailing and passing on all sides as the traffic roared along at more than seventy-five miles an hour. In Idaho Springs it began to rain heavily, just to add to the highway driving fun.

After getting settled into my hotel room, I took my files of gas articles and information on the recent gas and oil bills and went to a restaurant near the hotel and ordered dinner. It was a nice mid-range restaurant that was also a small microbrewery. I got a big corner booth and read the *OnEarth* article that featured Laura Amos and Weston Wilson to prep myself for any questions I might want to raise the next day in the interview. The waitress was friendly, the lighting was subdued in the rustic-themed setting, and good music played as I read and sipped my drink. I filled two pages with notes and questions for the following day. I called Al and got an update from home, and let him know I didn't know if I left the back water hose on when I was filling the llama stock tank. I was relieved when Al said that I hadn't left the hose running.

After dinner I walked back to the hotel and took a long bath. Usually when I came down to Denver I was preparing to testify on some legislation, and I often would spend the night before writing or editing my speech as I sat in a hotel bed. This time was a bit more relaxed; there would be no legislative committee, no imposing room at the Capitol packed with landowners and gas and oil people, and the press. Nonetheless, it was going to be a very important day filming Lance Astrella, the noted attorney, and Weston Wilson, the EPA whistleblower. I finally fell asleep before 11:00 p.m., my mind filled with anticipation.

The next morning I woke up early in the dark hotel room—at about 4:40 a.m.—and I sleepily pictured Deb driving through the dark from southern Colorado to get to Denver in time. Then I rolled over and slept until the wake-up call from front desk came at 6 a.m.

I got out of the huge hotel bed and flipped on the radio, trying to get the NPR station. I found it and turned it up. This morning the first part of the two-day series we worked on was being broadcast at KJAX, Aspen Public Radio. Maybe by some slim chance it would come on down here, but I doubted it. At the same time, I turned on the television and watched the local news for traffic and weather reports.

I packed up most of my things, dressed, drank the tiny pot of hotel coffee I had brewed, and then set out to the nearby coffee shop to get another cup to go. After I filled my large cup, I went outside the shop and sat at the small ironwork table by a spouting fountain. There were big box stores nearby, and I-70 was just behind them. A small brown sparrow flew down and landed next to me, cocking its head toward me. I'm sure he was waiting for the day's supply of dropped bagel bits to begin. How, I wondered, could birds live in an urban setting like this? It was mainly a sea of parking lots. I could hear the birds chattering from the immaculately manicured shrubs around the chain restaurants that flanked the big stores.

Map in hand, and with good directions from the hotel front desk, I drove east on busy Colfax Avenue into the heart of the Denver business district. I almost made it exactly to the law office—I could see the landmarks the secretary had told me about—but I was on a one-way street in traffic and couldn't find the parking garage. I pulled to the curb a block past the address and called the office. Elisa, the pleasant receptionist, gave me exact directions to get to a lot just behind the office. As soon as I parked and got out of the truck I heard someone call my name. It was Deb, and with her was the cameraman she had arranged for to help with filming that day.

We hauled the piles of film equipment into the building through the back entrance, and then the building staff helped us get an elevator to bring it up all in one load.

Up on floor fifteen, we found the office and went in. It was immaculate and very high-end. Beautiful art, spacious offices and meeting rooms, and vistas of the downtown and mountains that took your breath away. Although it was an office, I noticed that every surface was gleaming and uncluttered; not a piece of paper was in sight except for at the front desk where a tidy stack sat by the phone.

COLLATERAL DAMAGE

Lance showed us a large room conference room where we could set up, and also showed us his office where we could do an interview. Lance's office was breathtaking; he must have a fondness of fine Japanese art, for there was a collection of very expensive-looking sculptures and figurines in glass cases in the entryway to his inner office. Deb and Dave went back to the larger room, and I talked to Lance for a few minutes about the state legislation that was going to be signed next week in Colorado. He asked me if I was going to the signing, and I said that I had been asked to go to the signing of Bill 1341 which was what we refer to as the Colorado Oil and Gas Conservation Commission reform bill. The health impact bill had basically been rolled into that, and that was the bill I had worked most directly on. The signings were going to be held in three different cities on the Western Slope; a different bill would be signed at each location.

As I talked to him in this incredible office, with two walls almost completely glass on the corner edge of the high building from which I saw not only the sprawl of Denver but the snow- capped Rocky Mountains in the West, I knew it was a once-in-a-lifetime moment.

I am sure I will talk with Lance Astrella again about issues, or see him at a big meeting at the Capitol or something, but this was different. Today we had a film crew setting up in his office, and right now he and I were discussing potentially important and effective legislation that was going to be signed by Governor Ritter next week. For me, it was a huge honor to be here.

The laborious and technical process of setting up the cameras and all the lighting equipment began in one of the large conference rooms. I had no idea what Deb and Dave were doing, but after what felt like longer than an hour unrolling electrical cords, positioning and repositioning some very strange-looking lighting effects devices, testing, and retesting the huge cameras, they finally got it all done and then began to roll film. The small black mike was carefully snaked into Lance's pressed suit and clipped on. Lance is always impeccably dressed, but today he looked especially well turned-out.

Deb described how the interview was going to proceed—we were only going to hear Lance's side of the conversation—not Deb's prompting questions. She asked him to speak in statements, and from time to time she would ask him to go back and restate something in a slightly different manner. Lance had it down, and I am in awe of how exactly he words everything: so carefully, so succinctly…I imagine after being a high-end energy attorney for thirty years (and now a "Super Lawyer") one learns to pick one's words with extreme care. I sat on the floor behind the camera and listened intently, occasionally crawling on my hands and knees out of camera range to change or flip a tape in the small audio recorder I had running.

Lance began by addressing the importance of America's need to develop the mineral resources, including natural gas. I learned a valuable lesson as I listened—it is so important when one is a spokesperson for a side of a cause (or as I have been, a supporter of the need for better health and environmental protections in natural gas development)—to not come off as being "against" gas and oil development. If one does, you immediately create a situation of alienation and shutdown in the face of those listening to you. You paint yourself as black or white on an issue, not reasonable…

As Deb quietly staged the questions, Lance began to get into the need for change in the industry. He talked about his frustration with an industry that was slow to embrace change, and said he had been disappointed that the industry failed to take advantage of new technologies and methods that would greatly lesson the damage to the environments, aquifers, and people's health. Lance himself had been an attorney working for energy companies for the first twenty years of his career, until he switched to representing those who believe they had been damaged by energy development in one way or another. One issue that he returned to over and over again was the dangers of the VOC emissions from the wells and their infrastructure—adding that relatively inexpensive equipment was available to burn off and greatly reduce those dangerous emissions—but the industry typically failed to install burners on most wells. They needed to pushed into doing so by regulations—either state or federal—which currently do not exist.

Lance stated that he unequivocally believed that many individuals living in proximity to gas and oil development across the country had serious health problems resulting from their exposure to chemicals—either airborne, waterborne, or both. The striking thing was that the symptoms were the same in all the different areas: people suffered from skin rashes, burning eyes, sore throats, respiratory difficulties, dizziness, fatigue, depression and more. Lance believed that it was imperative for the state to get more directly involved in conducting comprehensive health studies on the effect of chemical exposure from the gas and oil industry operations on nearby populations. Currently, there is almost no data available; few, if any, studies have been done. When a landowner is sickened and believes that they have been contaminated from nearby gas and oil operations, it is impossible for them to prove it in a court of law. No landowner can afford to conduct a study on the subject—it must be taken on by a governmental entity.

The cases that are brought against gas and oil companies are more typically based on property damage issues, which might include water contamination. The very hard part of bringing forward a successful case is producing the expert scientific (and very expensive) data that stands up in a court of law. Few landowners (if any) could ever afford to pay for that level of defense against the extremely well-heeled energy companies. In order for a case to be plausible, it must be with a group of clients, or a well-organized community in a class action situation. But what often happens in cases where damages done by an energy company are apparent, is that the energy company negotiates with the most glaringly provable individuals privately—and after compensation has been dispensed a confidentiality agreement is signed which prohibits the plaintiff from discussing the details of the settlement, including the amount of the compensation. Therefore, the story goes silent. Also, and importantly, if that individual had been part of a cluster of landowners with damages in common, they are now splintered from the group. The industry is well known for the "conquer and divide" method, and for good reason. It is a technique that works very, very, well.

The topic of hydraulic fracturing was discussed, and Lance emphasized the necessity for in- depth research of an area's geologic nature before fracturing operations proceeded. He described the dangers of product communicating with water supplies if the geology was of a fragile geologic nature. He also discussed that fifty-percent of the liquids injected in the fracturing process were typically bought back to the surface, and in most cases placed in pits that were dug when the well was put in. Any dangerous chemicals or VOCs then evaporated into the air, or had the potential to make contact with water supplies. Lance urged for the need to utilize "green" non-toxic fracturing mixes to best ensure that the practice was safety done, with a minimal risk for environmental or health damages. He also talked about technology that eliminated open-air pits—closed loop system technology that contained all the liquids so they made no communication with the air or water on the site of extraction.

An issue that came up twice was a conversation that Lance referenced regarding a meeting held between an energy company and a Native American tribe. The energy company wanted to develop for energy on the reservation, and discussions were underway. As the company spokespeople were discussing the processes involved in the extraction, they referred to the produced water (the water that was brought back up after the drilling had been completed) as a waste product. One of the Native American elders stated that the tribe's people did not consider that waste—that everything was a resource.

Lance closed up his interview by talking about ways to more efficiently reduce the amount of produced contaminated water the industry left behind after its operations were completed. He described a process of separating the contaminant from the water, purifying the water and making it again available to communities for usable purposes, and then dealing with properly and safely disposing of the concentrated waste product.

Deb asked Lance if he thought we had gotten it all, and he said he thought we did. A few closing statements were made about solutions, and the need for more effective and responsible regulations. The examples of the new regulations both in Colorado and New Mexico were presented as examples of progress in the right direction.

The cameras and lights were turned off, Lance went of to a meeting, and we began to prepare for the next interview with Weston Wilson, the senior EPA official and whistleblower. Another one of my heroes…

We did not stop for lunch, even though it must have been midday. I think Deb had a bag of jerky, and I had a bag of sparkling water. We just powered on—they had to set up the lighting, create the second staging in the same room (so it appeared to be a different location) and both Deb and I had long ways to drive that night after the filming session. We rearranged furniture, and created a backdrop for Weston's interview—and I had to put a new tape in my audio recorder. After the preliminary testing, the film began to roll. What Weston relayed in the next hour or so had me stunned, shocked, saddened and horrified as to what our government had allowed to happen regarding hydraulic fracturing exemption regulations—and the blatant lack of response to the large mishaps that had occurred in Garfield County.

Weston Wilson is an environmental engineer for the US Environmental Protection Agency. Part of the function of the EPA is to review all governmental environmental impact statements as they refer to the Clean Air Act, and to make public their findings on the report. The EPA also has the authority to sign off on the environmental impact statements, or to make a conclusion that the reports are faulty. Then the EPA has the authority to bring their conclusions to the President's Council on Environmental Quality if they feel that a study is flawed, and inconsistent with the law that you have failed to control pollutants that are necessary. Then, the President's Council is asked to broker a dispute between the agencies involved.

Weston described that for the most part, the way the Congress set up the administration of the oil and gas industry, the EPA is not included. The EPA has provisions to control hazardous waste—but Congress exempted them with respect to gas and oil. In the late 1980s the new technology of hydraulic fracturing was introduced to expedite the extraction of natural gas from tight sand and rock formations. After a landowner in Alabama brought forward a claim that their water well had been contaminated by injected hydraulic fracturing materials, this brought about the decision of the 11th Circuit Court of Appeals that says that the EPA should be regulating hydraulic fracturing and all injection to protect drinking water supplies. So over the years the EPA has banned the injection of hazardous waste into the ground, and regulated other practices that inject. For the most part in the oil and gas business, this was related to bringing brine and oily water to the surface, and then the liquid is re-injected into the same formation. Then as long as the well tube itself was engineered well, then the brines would go down and not harm any drinking water sources along the way. The gas companies can get permits to utilize class-two disposal wells to dispose of quantities of produced water—it is re-injected into a tested well—into a salt-water aquifer. The overseeing regulatory entities are required to approve the construction of the well prior to the injection of the liquid waste product. The 11th Circuit decision said that fracking should be treated like any other waste disposa—again, if there is any oil field waste, the industry can request of the EPA a permit for disposal of any such waste. The 11th Circuit Court says fracking should be handled in the same way. In other words, the oil companies need to get a permit to show it is safe to inject fracking fluid.

He went on to describe how hydraulic fracturing was quite different than the more traditional older method of extracts. Hydraulic fracturing involves injecting into the same well that becomes the producing well. The well is injected with a slimy liquid material that carries sand or glass beads into the rock, then it's sort of like putting jello into there—you use the jello to carry the sand into the rock, then over a short period of time, about an hour or so, the liquid is no longer jello—it leaves the sand, propping up the rock and they pull most of it back out. Then they're said to acquire more gas. The problem is, many of these fracking materials themselves are toxic—and therefore they risk contaminating underground sources of drinking water. Or, they run the risk of contaminating the air when they are brought to the surface and put into the reserve pit, and allowed to evaporate. Wilson made a conclusive statement that the EPA failed to regulate injection of fracturing fluids.

Debra then asked what has happened politically regarding regulation of hydraulic fracturing injection. Weston answered the question bluntly, saying that what has happened politically has been quite a disaster. After the 11th Circuit case in Alabama that said the EPA should regulate the injection of fracturing material under the Safe Drinking Water Act, the EPA prepared a technical study, and it began under administrator Whitman—shortly after the Bush/Cheney first election in 2000. That technical study was going along, and the EPA quickly determined that the most toxic part of it was diesel fluid; diesel enabled the fracturing mixture to be thicker and thereby carry more sand. The EPA did work out with the three major companies, which included

Halliburton, a ban on using diesel. But the EPA's technical report went on to say that there were other constituents of the fracking materials that could be toxic. Wilson said that conclusion was what led him, in the fall of 2004, to object on technical grounds that the EPA had actually failed to say what the toxic routes were and the need to regulate them. Weston then became a whistleblower on the issue; he objected to the EPA's conclusions, and he asked the United States Congress and the Office of the Inspector General to look into that matter, and see if there was abuse of agency discretion—after all, if it were correct that this leads to some health risks, it jeopardizes the public welfare. After Wilson published his report to Congress, there was a summary of that information in the LA Times in October of 2004. That was the first time the amazing political part of the story came forward—that the former chairman of Halliburton, Dick Cheney, within a few months of coming into office—was holding meetings with energy companies on this subject. As vice president, he was pressing the then administrator of the EPA, Christie Todd Whitman, to exempt hydraulic fracturing from the Safe Drinking Water Act regulations. At the time, Whitman was punching back and saying the report was not finished, but Wilson said it was surprising to him to learn the role of the vice president in trying to get this practice exempted for the very industry he was involved in prior to becoming vice president. Next, Nicki Tinsley, inspector general of the EPA began an investigation of Wilson's complaints. Several months into that, Congress took the final report from the EPA saying fracking did not present a risk, along with other information, and exempted hydraulic fracturing from regulation under the Safe Drinking Water Act, in August of 2005 as part of the Energy Policy Act. It was a blanket exemption in the Energy Policy Act: the practice of injecting materials for the purpose of hydraulic fracturing is exempt from reporting to any federal agency, specifically exempt under the Safe Drinking Water Act. Two weeks prior to resigning as the inspector general of the EPA, Nicki Tinsley sent Henry Waxman's office a letter saying that there was no point in pursuing Waxman and Wilson's request for an investigation in the whistleblower complaints, as a result of the exemption allowing for hydraulic fracturing due to the passage of the recent Energy Policy Act.

When asked if this exemption happened as a result of pressure from above, in both Congress and the EPA, Wilson replied that he couldn't exactly pinpoint from where the pressure came, but he was surprised about the direct role played by Vice President Cheney, as reported in the LA Times. What Wilson found most alarming as a technician was that the EPA had described how toxic the materials are—toxic at the point of injection—and still came out with a summary that they don't need to be reported or regulated. That leaves you and me as the American public in this position, he explained:

"We cannot know what the industry injects in our land—it is exempt from being reported. The industry can claim that the ingredients are confidential business information with respect to what the recipe is."

Wilson said that, of course, what he and the other technicians want to know is if the ingredients are toxic. If no such reporting occurs, because the Energy Policy Act exempts such reporting from the Safe Drinking Water Act requirements, then, in order to correct this, the law would have to be changed, either on the national level or on the state level—each state would have to do it on their own. He went on to say that he was heartened by the fact that in 2004 after his report was presented to Congress, Mr. Waxman wrote to the Office of the Inspector General and said that the allegations made by Mr. Wilson are proven in the body of the EPA's own report. So now, one hopes that Congress can see this as the risk that it is, and that certainly you make the requirement to report and let the public have access to this information—as, after all, that is the hallmark of democracy. After all, even if it is safe, the citizens should not be denied access to this information. Wilson's complaint is that Congress was not given the full information available on the risks of contamination from hydraulic fracturing. Therefore, they cast their votes without having the necessary information.

Next, the issue was raised that last year the EPA got several requests from citizens from Garfield County. The citizens were saying that "my drinking water might be contaminated from this practice, or the air we breathe might be affected—EPA, can you look into it?" Wilson went on to state that the EPA should have looked into those citizen requests. He and another staff person had prepared the letters and were ready to write to the COGCC stating that we felt that this practice caused imminent substantial risk to the public drinking water source and that the EPA was going to take over the investigation. However, as soon as they got their letter to the political appointee, supervisors cancelled the investigation. Therefore, the EPA did not investigate the legitimate complaints from citizens in Garfield County.

Debra asked Wilson why he thought it was cancelled, and Wilson laughed a short laugh and said that he could only guess that it did not meet the vice president's or other parties' interest in maintaining the exemption in this practice. He then went on to say that he didn't really know, but what he did know was that the agency has the authority to look into ground water contamination; it has the duty to respond to citizens' complaints—to bring in the scientists, to bring in the toxicologists to help us understand what occurred in that particular case,

"But we did not," he said.

Then specifics were discussed about Laura Amos's request to the EPA to investigate the contamination of her water well. Wilson had participated in a study of the Laura Amos case in Garfield County, and her well water was clearly affected by nearby drilling. The question came up in Wilson's technical team whether this was a case of a good gas well gone bad, or a bad gas well gone bad. Wilson said,

"And we never got the answer to that. By a bad well gone bad I mean one that was drilled with leaks, and therefore could have leaked high up in the drilling column to contaminate the near-surface water supply. But we never got to the bottom of it because the EPA management failed to allow us to investigate. Whether another circumstance could have occurred in that case: whether it could have been a mechanically sound [gas] well that caused fracking laterally and then these materials moved up vertically to her [water] well—that was never resolved. Had it been the latter it might be a bigger problem for the industry, because the oil and gas commission is checking the mechanical integrity of these wells—they use various techniques, basically like pressuring a tire and making sure that the tire still holds air, and then you know it doesn't leak. As long as the well is sound, then one would expect that the fracturing materials are delivered so deep into the ground that it's not likely to have much chance of contaminating a much shallower surface-water well."

Weston continued, saying that Laura Amos has asked the EPA to investigate her case, and it was a legitimate request. He said it presented a real mystery—is this woman ill from what she claims or is it from something else? And the EPA should have investigated. At the time the EPA staff prepared a letter that said, in our professional opinion, this practice by Encana of drilling a well adjacent to a domestic well represents a substantial risk to underground sources of drinking water. The way they wrote it was actually "imminent and substantial"....However, our supervisors stopped that letter and the investigation, and nothing took place. He added that for him, that was very hard to take, because this complicated arena of good well gone bad, or bad well gone bad, the complicated pathway, and then the human response, required the kind of talent the EPA employs. The EPA employs petroleum engineers, groundwater specialists, toxicologists, medical PhDs— people with a broader interdisciplinary experience to understand this mystery than those employed by the Oil and Gas Commission. So, Wilson said, we not only had the responsibility, we had the talent to investigate. And it did not occur.

Wilson described how difficult it was to bring forward the sort of complaint that he did in his whistleblower action; that one constantly worries about the future of their job and their family's security. He also explained that he had been pulled off of all gas and oil work at the EPA, and now he worked on sewage treatment and other important issues, but he was not working in the field he used to be in. His income has not been affected, but his knowledge of what goes on in the industry has been cut off.

As the interview began to wind down, Wilson described his initial feelings over reading the EPA's preliminary report in June of 2004, and saw that its conclusions were inconsistent with the body of the report itself—and he smelled some sort of political interference. So, he said his only recourse was to report outside of EPA channels to the US Congress and the Office of Inspector General. He said he was stunned by the report for two reasons: because it was inaccurate—actually the EPA itself was showing that the injection of these materials was toxic. He also felt that as a civil servant, regardless of what one's technical position was on this, there was no reason to keep this information from the public. It is after all the public's gas that these companies are acquiring, typically on public land or the subsurface is owned by the public. "There is no reason," he stated firmly, "to keep this secret from citizens. If it is truly benign, and some of the formulas can be benign—show me—show me it is benign. Don't exempt them because you believe they might be."

Next the make-up of the external review panel was discussed. There were seven members of the review committee reviewing this report for the EPA, technical people. Contrary to EPA's own provisions about blocking actual conflict of interest and trying to prevent the appearance of conflict of interest, the members of this committee didn't fulfill those requirements—because they were employed by members of the industry who would benefit financially from being exempt from the Safe Drinking Water Act and not reporting. One member of the committee was an employee of Halliburton; others came out of academic interests and had previously been employed by the oil and gas industry. Wilson said that regardless on your view on that, it didn't present a proper analysis that avoided that appearance of a conflict of interest. The EPA's rules require that when it does a scientific study it include independent experts. In this case, in a seven member panel, five of them have some type of appearance of a conflict of interest. They currently work for the industry, or had worked for the industry. So, this doesn't meet the requirements of the EPA's procedures to have experts who are truly independent—financially independent—from the industry.

As closing statements were made, Weston addressed the fact that we are a nation that's afraid of the security risk of how we get our oil, so it is an awful strong position to come in and say that additional environmental protections may slow down domestic gas and oil production. Wilson stated that in his view it doesn't slow it down, but it does add cost—it costs more to prove that you are not harming citizens, but that seems only justified in a democracy.

The last thing that Weston Wilson stated in the interview was that it is time to reverse this. "What more egregious practice can there be than keeping secrets?" he asked. "This isn't a secret about war—it's secrets about getting gas. There is no reason for this to be secret. The industry needs to prove it's benign and it's not harming people. And the only way that can occur is for the information on the chemistry and toxicity of these hydraulic fracking materials to be reported to a state body or to a body like the EPA, so that citizens can have access to it and make their own judgments. It should not be a secret."

As I sat quietly on the floor, behind the filming equipment and out of sight from the camera, I was overcome with an indescribable sense of horror over what our governmental entities, including the EPA and the federal government, were allowing to take place. It was a feeling of utter betrayal and lack of trust in what these entities were and were not doing to protect the public and environment from the dangers of gas development. I felt as if I had just listened to secret Watergate tapes—this information was seriously disturbing. It was as if a chunk of what I had perceived to be responsible oversight in our government had now fallen away, and I was seeing the ugly and corrupt and damaging side of the inner workings of our governmental entities bent on maximizing profits for gas and oil companies. Equally disturbing was the fact that the government turned their back on seeking the facts in the egregious cases of contamination in Garfield County. They just turned their backs and walked away. What could we—this small group of individuals hearing this information this afternoon in this Denver skyscraper—what could we do to change what was happening? In the words of Shakespeare, "Something was rotten in the country of Denmark." At that instant I knew that something was rotten in the United States that involved the underhanded maneuvers that took place to get exemptions for hydraulic fracturing chemicals in the 2005 Energy Policy. I was stunned, and felt totally disillusioned, and quite powerless. It was both frightening and deeply saddening at the same time.

It is a day after the filming, and I have been thinking about what went on.

There were certain specific moments when what was being said was so poignant, so powerful, and so unbelievably incomprehensible that it could possibly be true just stunned me.

It wasn't that anything I heard yesterday came as a surprise completely; it was just that at that time it had a huge impact.

COLLATERAL DAMAGE

I guess it boils down to this: in spite of regulatory entities, the EPA or COGCC or any other, no one is fully looking into these health issues. Even when they know they are concerns. Even when the agencies' policies dictate they should conduct studies or investigations, they are stopped.

Help, help, help!

What about the people?

After the filming was done, Deb, Dave and I went across the street to eat a late lunch at the beautiful historic Brown Hotel. I was totally exhausted, as was Deb. We had both driven a fair way to get here; but also the interviews and filming drained energy—from me at least, even though I mostly just sat in the background and listened. Although I had started the morning energized and looking as professional as I ever manage to, in a tan summer suit coat and nice capris and new black sandals, by the end of the long afternoon I was sitting on the carpeted floor as the film rolled, my suit coat and shoes long-since tossed aside. As each audiotape in my little tape deck clicked off, I crawled on all fours out of camera range to flip the tape or insert a fresh one.

I dreaded the long drive back over the mountains and was so tired I wished I could just stay another night in Denver. It would have been so easy and so wonderful to just go to a hotel in the outskirts of Denver and book a room, then take a long bath and relax and go to bed. But, I had to work the next day and staying one more night was not an option. Also, there were the ranch animals to feed and Al was getting ready to leave on his raft trip. Not to mention, everything costs money. Every day off costs money, every hotel room, every trip to Denver, and the extra expenses that I incur just by being in Denver and spending more than I would have if I had stayed at home.

Before any trip for gas issues, I know ahead of time about the cost it will take on my allotted vacation time. That doesn't bother me; I count the days carefully and save what I need for my vacations with Al. I have to watch the other costs, however. It is easy to spend money while on a trip out of town.

TARA MEIXSELL

Lucky, I finally made my way out of the business district after only getting slightly lost in the confusing grid of one way and diagonal streets at rush hour. Finally I was on I-70 headed west and I put the pedal to the metal and sped my way back home, anxious to be back at the ranch. As I drove up and over the mountains, the hours passed so quickly compared to the day before on my drive down to Denver. This time, I had the whole incredible experience behind me and I was processing the events of the day, and the content of what had been said in those two very powerful interviews by Lance Astrella and Weston Wilson.

I got home about 7:30 and the sun was just setting. How good it was to be home! I went out and fed the horses and llamas and barn cats, and sat in barn and watched the ewes and lambs across the creek in the expansive green meadow below the mountain. The baa-ing and maa-ing filled the evening air, backgrounded by the beautiful sounds of birds getting ready for the evening.

Now back home, very tired, and still with my mind on the events of the day in Denver, I know why I go there for these things.

It is for this; these peaceful and beautiful moments at the ranch that I work for.

So many others have lost their peace already.

Many are gone already, and others who are holding on are having what is left of the rug of their lives pulled out from beneath their feet as impacts from gas and oil surround and overtake them.

This morning I heard the second half of the Aspen Public Radio piece where Rick Roles and Dee Hoffmeister were interviewed, along with several others. I heard it at 7:50 a.m., as I got ready to leave for work.

Sunday May 27th, 2007

Yesterday morning I woke at six a.m., just beating my radio alarm by a minute or two. That morning I was going to meet Debra and drive with her to do the interview and filming at the Mobaldi's house in Grand Junction. During a phone call the previous evening Steve said that Chris was hoping I would come. Although I had not planned to previously, I decided I would go. Deb said that I might think of some good points to address in the interview. Anyhow, it was always nice to see them, and over the last two years I only saw them when we were doing interviews or taking trips somewhere for testifying at gas-related hearings. This would be much more of a personal visit at their home, without all the pressure of television cameras, reporters, or governmental officials and industry representatives in the audience. And Debra is so relaxed while behind her huge camera; she interacts with and talks to whomever she is interviewing in a way that makes it seem that the filming is barely going on at all.

I got out of bed wishing I had been able to sleep in more after the long work week, but not really minding that too much. I made a pot of coffee, then ran a quick bath. As I was finishing up bathing I heard a horrible and very loud sound coming from somewhere in the back pasture or up the mountain. It sounded repeatedly; a kind of a screaming or howling noise. Amidst the screams was a continued series of loud barking from a dog. The worst type of thoughts crossed my mind; one of the most horrible memories I keep hidden in the very back of my mind is the memory of walking in on the demise of my young llama as it was being torn apart by two dogs at the ranch up in Fairplay where we used to live. It was beyond awful.

Frightened for my animals, I jumped out of the tub and grabbed my old green robe and wrapped it around me as I slid into my clogs and ran out back through the pasture. The llama and horses were acting strangely; usually they would rush as closely as possible to the site of a commotion, but this time they held back on the top of the pasture slope, just watching the steep canyon from which the awful noise was coming. Again and again, the terrible howling and screaming sounded. It was just beyond the pasture fence, where our property took a steep climb up the mountain to the north.

Soon I spotted the large white herd dog, one of the Akbash dogs that guard the neighbor's sheep and lambs in the adjoining property to ours.

He was standing on the rocky hillside wagging his feather tail slowly as he continued to bark, and his gaze was focused on a spot deep in the crevasse beneath the oak scrub trees and juniper that I could not see. Suddenly, to my surprise, I saw a tall man standing on the steep and rocky mountainside above him. He was dressed from head to toe in brown Carhartt clothing, probably overalls and a coat. I could tell he saw me from his posture.

I was at the gate between the upper and lower pastures, a fair distance away from where the ruckus was going on between the dogs and whatever was out of my view. All I had on was my robe and clogs, and I held the robe closed with just my hand.

Then I heard the man say the word,

"Coyotes."

It almost seemed like he was talking to the dogs, but perhaps he was saying it for my benefit. He was on our property, but I knew and liked my neighbors and their hired hands. They had the large herd of sheep and lambs to care for, and I had many times assisted with interfering when wayward dogs were causing trouble by chasing and stampeding the sheep. Usually some sheep were killed when that happened. One time several years ago I went out to pitch hay in the back pasture and all the llamas were staring intently down at something going on in the creek bed. I walked all the way to the back of the pasture to see what had them so captivated. It was a small lamb, wet and scared, stuck in the narrow creek bed. With my work shoes and clothes on, I stepped into the water and lifted him out. Frightened, wet, but now safe, he bleated and scrambled up the sandy bank to the pasture. Another time, I called the neighbors when a ewe was stuck upside-down in an irrigation ditch that ran through the field. I thought it was dead, but they pulled her out and she was O.K. I always like to keep an eye out for the sheep or whatever stock are in the field so close by.

I couldn't tell if the man on the hill was our neighbor Warren, or one of his sons or hands from the distance between us.

As the animal's yaps and screams continued from the crevasse I slowly turned away and walked back through the dusty corral to the house. It was the best thing to do…

COLLATERAL DAMAGE

Back in the house, I continued my bath and braced for the sound of a gunshot as I scrubbed. Thankfully, I heard none and soon the animal noise ceased.

I met Debra at the Base Camp Café in Rifle where we both had breakfast before going in her car to Grand Junction. Just as we were leaving I saw Rick Roles at the counter, as usual with his coffee in hand. I hadn't noticed him earlier to my surprise, as he was often at the café in the morning.

We all visited for a while, and discussed the barbeque celebration that was going to be held at the Savage's ranch in Rulison. It was an event that was planned to close out the day of the bill signings with Governor Ritter. I told Rick that he had to come, or he would be in big trouble. Rick, along with many others, gave important personal testimony at the Capitol that helped to get the gas and oil legislation we have worked so hard for finally passed. Both Kathleen Curry, our state representative who had carried the health impacts bill, and Governor Ritter would be there. This was going to be a once-in-a-lifetime event—things like this do not happen every day in Garfield County.

The drive to Grand Junction passed quickly; we talked about the film, the book, and more. Once we found the Mobaldi's new home that I had never seen, we pulled in the drive and began to unload the camera gear. This time, there was much less equipment than in Denver at the law offices of Astrella and Rice.

After introductions Debra toured the lovely house and chose to film in the sunny dining room with plenty of natural light pouring in from the large windows and sliding glass door that lead out to the back yard. Unfortunately, not too long after Debra began filming, Chris became cold. When this happens she loses her ability to speak intelligibly, she experiences pain, and she convulses. Steve turned off the air conditioner and wrapped Chris in a blanket and helped her to the huge sofa in the living room to rest and warm up.

As Chris recovered, Debra interviewed Steve by himself. Even though Debra knew a fair amount of details about the Mobaldi's situation, the answers that Steve gave to some of her questions left her stunned, with a dropped jaw and a shaking head. She literally could not believe some of the things that Steve said.

After a while, Chris felt warm enough to come in and have her time in the interview chair. Today she was speaking in a very strong English-type accent I thought, although Deb felt it was Irish. I cannot say I would even know the difference between the two accents.

The majority of what Chris said was quite intelligible, though a few phrases were too hard for me to make out. So often I can interpret what some words she is saying are because of their context in the stream of conversation she is carrying on. Also, there are some key words and pronunciations she has that I am now familiar with.

For instance, "nyetta" is how she often says yes. Come to think of it, she has a number of words she uses to say yes. Along with "si" as in Spanish, I also recall her saying "oy" (like it rhymes with toy), and "oyee" almost like the French "oui." Chris's language syndrome, what ever it is, continues to baffle me. Maybe one day there will be a clear, understandable explanation for it, but for now just listening to her strange and changing accents is just bizarre. And oddly enough, unless she is just speaking gibberish, I swear that if I try very hard to grasp what she is saying and ask her to repeat things several times, I can usually eventually understand her. Like the one day when she was doing quite poorly, but we talked on the phone and she kept saying, "Mall ghut…" Finally I understood she was referring to the flowers that Deb had given her, and she was saying that they "smell good."

It is strange, or perhaps funny, that as I sit here trying to think of the key words that I have come to recognize and understand as Chris's own language, I cannot recall many. I have to hear them in a conversation; then I can recognize them.

We took a break after a number of hours of filming and went to have lunch at a nearby restaurant. Steve held Chris's arm as they slowly went out of the house and she got into the VW bug with some difficulty. Deb and I got into her car. She turned to me with a shocked if not dazed looked on her face as soon as we were seated in the car with the doors closed.

COLLATERAL DAMAGE

"Oh my God! This is totally Erin Brockovich!" she said.

It was evident that what she had heard so far was beyond what she may have envisioned she might encounter. That is understandable.

Once at the restaurant, as usual, Chris's ability to function began to deteriorate in the chill of the air-conditioning. It was amazing how fast the change in temperature would hit her. Moments after we were seated in our booth, Chris began to look hunched, and she had a pained look in her eyes and she began to jerk a bit.

"Are you getting cold, Chris?" I asked.

"Nyetta," she said, and nodded her head as she wrapped her arms about her.

Deb went out to her car and brought in numerous light coats and sweaters. Soon Chris was wrapped in several layers, and she began to brighten up and join into the general conversation somewhat. Although her now crooked fingers had great difficulty with the utensils and Steve had to cut her calzone into small pieces, she ate more than I had seen her eat in our last meals together over two years or so. To me, she just looked overall healthier than she had a year ago. Her weight wasn't up significantly, but enough that she had cheeks again on her face and even though she still convulsed and lapsed into gibberish at times of exhaustion, I sensed a better communication ability for a longer period of time.

Back at the house we filmed for a number of hours more, and the Mobaldis talked about the proppants that had appeared in quantity in their hot tub. Proppants were the tiny, white, _____ beads that were sent down deep underground with the fracturing mixes and injected with the other substances. They served the purpose of holding the tight sand formations open so as to allow the gas to escape.

Over a span of time, the Mobaldi's hot tub had proppants floating on the surface of the tub and clogging the filter by the hundreds and thousands. They had no idea what they were; they thought it was from the well water.

"They stick to the hair on your arms, they stick to your skin," said Steve.

"They get under your bathing suit, if you know what I mean," said Chris in her strong English accent, with a smile on her face.

"Need I say more?" she continued.

The Mobaldis talked about how the motor burned out numerous times on the hot tub. It was probably because the filter was so clogged with proppants that it was overworked. Eventually they just got rid of the hot tub; it was too much of a problem to keep it operational.

As to why the white beads did not appear in the house water: apparently the filter system inside the house screened them out. That is perhaps partially the reason why the filters had to be changed up to every two weeks, and an extra filter had to be added.

"The water filters got so filled with sand that we could barely keep up." said Steve.

Unfortunately, one thing the filters didn't screen out were any toxic chemicals that also may have leaked into the Mobaldi's water system. And they bathed in that water and drank it. Chris and Steve described the blisters all over Chris's body that developed, the intense redness of her skin and the puffiness that occurred after bathing, and also the sores that appeared in her mouth and down her throat. It was inconceivable to even think about…

"I took pictures of her, I didn't know what was wrong with her," said Steve.

Finally Chris began to get very tired. Steve helped her into the living room where she stretched out on the huge sectional sofa and was propped up with pillows and covered in a white blanket to warm up. Deb sat and talked with Chris, and Steve told me to come with him. We went into a large bedroom, and he pulled some papers from a shelf in a closet.

COLLATERAL DAMAGE

"Here, there is something I want you to see," he said as he handed me a stack of about ten or fifteen neatly typed pages.

I glanced at it and knew in an instant that it was the lawsuit.

"Is it the lawsuit?" I asked.

"Yes," said Steve.

I slowly sat down on the Mobaldi's bed and began to read through the legal document. Steve looked over my shoulder and followed my progress as I raced over certain sections and read more slowly and commented on others.

There were specific dates regarding the experimental fracturing fluids being tested near the Mobaldi's home at the time they lived there.

There was a list of the chemicals involved. It was a long list and it included heavy metals such as _____.

There was a section addressing trespass upon the Mobaldi's property, and there were numerous sections detailing the injuries that Chris and Steve had endured. Three companies involved had merged into one.

I asked Steve about them as I read that section.

"They are very concerned about Chris," said Steve.

I quickly turned my head away from the papers and said,

"What do you mean, concerned?" I was angry, and a bit unsure. Were they really concerned about her health and her as a person, or were they concerned about the fact that her terrible situation had come to the awareness of Astrella and Rice, Attorneys at Law, and there was now a lawsuit and any ramifications thereafter?

Steve looked levelly at me and said in the same steady tone,

"They're concerned—they're very concerned."

I regrouped for a minute; I knew that I was getting angry. I was getting angry about the industry that had done this to Chris, and to countless other people to varying degrees. Exactly what type of concern did they have now? I have met and known too many people who live near gas wells as new and old friends now that are very seriously sick to not be bitter and cynical when I hear that the companies care. I actually hope they do; they need to.

"Steve, you can hear my tone of voice. I don't know what it means. Do they really care about Chris's health?" I asked as I held the lawsuit paperwork in my hands, still reading my way through.

"Yes, they do care about her," said Steve quietly.

I felt a sense of relief when he said that.

That is what the industry needs to do better; really care about the harm they are doing and may do in the future to the people who live around the wells. Not to mention, the land, the water, the wildlife and environment and the air.

Debra and I drove back to Garfield County. We talked about many things; about how Debra had gotten interested in this subject for her film and finally decided to really do it. She had considered several other topics previously, but after a friend of hers nudged her to take the plunge on the gas issues film, she did.

I described to her as best I could how important this past week and upcoming week were to me. Not only had we done the days filming with Lance Astrella and Weston Wilson in Denver, but we had also had a full day at the Mobaldi's and on Tuesday there was the bill signing with the governor. Deb had hired another film professional to assist with that day; it was Dave, whom she had also hired to help film last week in Denver. I knew that Tuesday was going to be a very special day, and that it would take on an energy and momentum of its own.

COLLATERAL DAMAGE

This is a huge time here right now with gas issues, with the passing of the bills, the radio show, the television coverage, and especially the front page article about Dee and Rick and the others in the Denver Post—on the front page! And then today, filming with the Mobaldis, and the governor coming for the bill signings and the event on Tuesday...the press coverage will probably never get this big again, I think.

But one of the biggest things for me was the two calls I got from Lance Astrella this week. After we came down to Denver for the interviews he called me and thanked us for coming, and he said we were doing good work. That meant a lot; more than I can say. Then he called again on Friday, to tell me that he unfortunately wouldn't be able to come to the bill signing in Glenwood. He hoped we would have a good group there for that important bill. I said there would be a lot of us there, as well as in Grand Junction, and that Debra was going to film the signings in Glenwood Springs and in Grand Junction. Also, I said that I had just heard that Kathleen Curry, Harris Sherman, and Governor Ritter had all confirmed to come to the barbeque Tuesday night after the bill signings at the Savage's ranch in Rulison. Lance thought that was great news, and again he said that we were doing good work up here, and that what I was doing was making a difference. To me, hearing that praise from Lance was something I will remember and take pride in for I hope many years. It is hard to explain to someone else, exactly what that means to me. You see, I am a small time landowner who has gotten pretty involved in the issues surrounding impacts from gas development. Lance is one of, if not the world's foremost, experts on the subject. What a week, and there is more to come...

We stopped at the Mobaldi's old house in Rulison on the way home from Grand Junction so Debra could take some pictures of the property and house with the spectacular peeling siding. Huge patches of shaded areas were still brown; the rest was the blue-gray of the metal where the paint had eroded off. Paint hung in feathery blisters from the metal surface. As she filmed quietly, I wandered around the property and tried the different pumps and water valves. It was strange to watch the seemingly normal looking water that flowed from the garden spigot at the front of the house and think that it may contain dangerous toxins that caused Chris's illness. The outdoor pump in the back yard did not turn on; that was the one they had filled their hot tub with.

I went into the huge and now empty garage with the gravel floor. The horse manure from the last tenant over a year ago was still there. I found it strange that someone had let the horse into the graveled area. I guess it made a good weather shelter, in spite of the sharp, rocky footing.

The saddest place I went was up into the barn where the llamas had been. Inside the barn in one corner there was a small rectangular shelter created out of bales of straw. Was this where one of their young llamas had gasped away its last hours before dying of respiratory illness? I remember the heartbreaking time when Chris and I worked together five years ago. She would tell me each day about the baby llama's progressing illness, and it had finally died in spite of the veterinarian's attempts to save it. Then, they lost yet another young one. After that they gave up breeding llamas; anyway, by then Chris was so sick there was no time or energy for that.

Before we drove away from the property Deb stopped and got out for another shot of the house with the for sale sign out by the road in front. As she filmed I looked at the gated-off well site just across the road. I wondered about the emissions I might be breathing as the tanks on the well pad quietly hissed and went about their functions.

We stopped one more time on a very steep mountainside a few miles from the Mobaldi's old house to film two horses grazing right next to a well pad. Across the way, looking north over the Colorado River toward the Roan Plateau, there was a stunning vista of rugged mesas dotted with newly erected rigs. As soon as darkness fell, their brilliant towers of white lights would shine across the wilderness, like so many strange industrial Eiffel towers scattered across rural western Colorado.

Debra and I had dinner at the Elk Creek Mining Company in New Castle. It was late, so late that when I got home it was near ten p.m. I flipped on the outdoor lights and let the house cats in, but the llamas and horse were nowhere in sight. They could do without their ritual small portion of grain tonight, I figured. If they were up by the house and eager by the fence of course I would have gotten a flashlight or lantern out and gone out to feed. But, because the occasion mountain lion frequented the area, I chose to let the big animals fend for themselves tonight. There was enough grazing left in the pasture to occupy them, and I had put out extra hay that morning knowing that it would be a long day.

In spite of my tired state, after unwinding in front of the television mindlessly for a while, I stayed up for long time thinking. Something about the day felt very important, specifically the fact that Debra got everything that she did today on film with the Mobaldis.

Debra and I met for breakfast at the diner in New Castle. I brought in my files and pulled out some articles on the Divide Creek seep and the gas bills for her to read. We went to the City Market next door after breakfast and xeroxed a few pages for her to take. Then she went off to Pepi Langegger's place to film, at the sight of the Divide Creek gas seep. This was the location of the largest finable violation regarding natural gas development in Colorado to this day. Initially, Encana gas company apparently blamed it on "naturally occurring swamp gas." They also claimed that their monitoring equipment would have discovered any problems. I was told that later investigation found there was no monitoring equipment on the well in question. This is just what I have heard, and read in some papers.

I have been on the computer most of the afternoon; this has been a marathon session for me. I guess the timing is right; lots of things are going on, and Al is still out of town on his weeklong raft trip. Today I am happy to be home, do a bit of house cleaning, garden watering, lots of writing, and now spend some time with the animals outside.

May 30, 2007

Yesterday was a big day; a very big day. Governor Bill Ritter and a number of the state representatives and senators that sponsored and worked on the bills being signed across western Colorado came to Glenwood and to Grand Junction for public signings. I had taken the day off in advance; this was a time I wanted to enjoy.

TARA MEIXSELL

After her meetings with the Mobaldis, Debra decided that she wanted to film the entire process of Chris going to the signing of the bills in the park off 24 Road in Grand Junction. She went so far as to hire her friend Dave again, to do additional footage on this hectic day. I was going to drive my truck and be back up in case Chris felt too sick to stay, and also I would drive so Deb could film and interview Chris on the way to the signing of the bills. The bill being signed in Grand Junction relating to gas and oil was Bill 1341, which I refer to as the COGCC reform bill. Since the bulk of the work I did was focused on the health impact bill, which had been incorporated into Bill 1341, the signing of this bill meant the most to me. Now, since Deb wanted me to help with the logistics for the filming of this bill signing, I hoped that Chris would be able to make it through the whole signing. What a shame it would be if we had to leave and take her home if she felt too sick! I put that thought out of my mind, after taking the day off to drive several hours west to attend the signing of the bill, it just had to work out. And typically, Chris was able to summon her strength at least for a short span of time. Especially if no lengthy travel was involved.

The first event of the day was to be the signing in Glenwood Springs at the courthouse, and then we would leave immediately for Grand Junction almost two hours away to get to the next signing with hopefully enough time in between to get Chris to the park. After that, back in Garfield County there was going to be a huge celebration party thrown at the Savage's lovely ranch in Rulison. Ironically, it was just down the road from the Mobaldi's old house.

Although I was planning to greatly enjoy the whole day, the morning was rather strange. I drank coffee rather leisurely and fed the animals, after the luxury of not having to use an alarm clock to wake up for a workday at six a.m. Then I spent a bit of time on the phone solidifying details, and throwing clothes out of the closet trying to decide what looked nice enough for meeting the governor, etc., while still being comfortable enough for the hours of driving that were to come.

Finally I was off, with the animals hayed enough to skip an evening feed.

COLLATERAL DAMAGE

The bill signing in Glenwood went smoothly in a packed courthouse with a lot of familiar press there with their cameras and mikes. Deb and Dave had the big cameras going, and they adeptly moved around getting their shots. I sat with Theo Colborn and her assistant Mary from TDEX, in the front row, and we talked to Kathleen Curry, our state representative, for a bit before the proceedings. She is a person I hugely admire and respect. It was so good to see her looking happy today, at the signing of the bills she has worked so very hard for. Few will really know the efforts she has gone to, and the costs of those efforts.

After the signing ended, I sped off in the truck through a good-sized rain and hail event toward Grand Junction. Just in time, I got to the Mobaldi's and Chris was there looking lovely and dressed in a beautiful black velvet coat with dark red flowers, her face lit with expectation and excitement. We got in Deb's car, and she filmed and interviewed Chris as we drove to the park. Chris gave us perfect directions in her strange accent as we sped through the city. I was determined not to be too late for this, after all the driving we had done to come here. There was no time to spare, and certainly no time to get lost.

As we pulled up to the park, we headed for the big cluster of cars beyond the ball and soccer fields. We saw a large group of people walking down a sidewalk, and we followed them in the car and parked as close as was possible to the gazebo they seemed to be headed toward.

Chris got out of the car with some difficulty, and then steadying herself with her cane she held her chin up and began walking slowly up the cement pathway. We were the very last people in the group, and we had a fair distance to go. In no time at all Chris began to tire and have some trouble, and she had to stop and rest. Deb was filming as we went, and I offered Chris my arm when she began to walk again. The rest of the group was assembling for the bill signing still hundreds of feet away, and I wondered if we would make it in time. The crazy notion passed though my mind that if necessary, we could carry Chris up there the rest of the way. I knew logically that we would not, and that Chris would probably not be able to endure the pain of the physical contact. She couldn't even endure the pain of a typical handshake, so how on earth could we possible carry her?

Nonetheless, I still distinctly recall the seconds of panic when I thought how close we had come to getting her there to the bill signing with Governor Ritter, and that we might never make it up that last stretch of sidewalk in time. God forbid that Chris should collapse, lapse into gibberish, or even worse, get sick to her stomach.

Thankfully after several stops for rest we made it up to the main gazebo area where the signing was going to take place. She walked at a snail's pace and I held tightly to her elbow and forearm, but she made it. We scanned the area near the podium and microphone, looking for a place where Chris could actually view the proceedings. There were no sitting areas that had a view, as a large crowd unfortunately stood between the podium and the numerous picnic tables under the expansive shade roof.

I saw Duke Cox moving through the crowd, and I called over to him and asked him if he could help get a place for Chris up front. He said he would.

Then I saw Dee and her husband Hoff, sitting with several others at a table under the shade roof. Chris finally met Dee Hoffmeister in person for the first time; she had been so looking forward to this. The Hoffmeisters invited us to sit with them, but I was anxious to find Chris a place where she could actually see the bill signing. As the governor's entourage arrived and the representatives and senators took their places near the podium, I escorted Chris over to the edge of the cemented-off area where there was a good view. Several men in dark suits stood there, and they said nothing as we went past them. They seemed to be some type of security detail, probably with the governor.

We got a place in the front area where Chris could lean against a waist high railing and watch the proceedings from a side view. As the governor's speeches went on from time to time I feared Chris would topple over backwards onto the pavement several feet below. Luckily, she did not.

COLLATERAL DAMAGE

Lots of people I knew from Western Colorado Congress and Grand Valley Citizens Alliance were there, and I recognized a fair number of faces from the many hearings in Denver. They were people from environmental groups, and governmental staffers. During the lengthy speeches I mainly focused on Chris, preparing to grab her arm if she began to waver. Finally I assisted her away from the railing when she whispered that her feet hurt her too much to stand any longer.

I took her over to the table close by where the Hoffmeisters had been sitting. Chris basically collapsed at the park table under the shade roof, and moaned and put her head in her hands.

The others at the table exclaimed in alarm that she needed to get in the car and go home, but when I asked her if she wanted to Chris said no.

"Ah wheel beh fahn," she said weakly. I knew she was determined to stay, and that she was not so bad off even though to others it may have seemed so.

"She is OK," I said to the others, who hovered about her in a concerned manner.

A lady who looked very familiar was standing nearby the table, and looking at Chris intently. I wondered who she was, and why she kept looking at us.

In a minute she came over and introduced herself.

"I am Peggy—I am a member of Western Colorado Congress. Are you Chris Mobaldi?" she asked.

"Yas, eh ahm," said Chris.

The two talked for a while, and Peggy was able to understand what Chris was saying in her strange accent. When Chris asked her how she knew who she was, Peggy leaned forward and looked straight into Chris's eyes.

"I heard you speak before, and I have read about you in the papers for several years. I know all about you," she said.

"Ya doe?" asked Chris incredulously.

Then Chris asked her where she lived, and upon hearing that it was Grand Junction she asked if Peggy would come and see her some time.

Chris then told her that she has no friends here, and that nobody comes to see her anymore. Her eyes were filled with tears, and her face contorted with sadness as she said this.

"Yes, I will come and see you," said Peggy, with a smile on her face.

The two then swapped addressed and phone numbers, and I could see that Chris was very pleased with the prospect of a new friend who might visit. It would be a big ray of light in her very limited world. One of the things she missed the most was contact with other people, and it angered and frustrated her that now she had such difficulty communicating with others. She particularly disliked being taken for a foreigner, and being rudely told to go back to where she came from when people couldn't understand her strange accent.

After the speeches wound up, State Representative Kathleen Curry came over and had a wonderful conversation with Chris. Chris thanked her for all the work she had done (in her garbled tongue), and Kathleen said that it was her job to do that as a state representative. Kathleen had heard Chris speak before, and she had met her personally also.

"I'm afraid I haven't helped you much, but I hope that what we have done will keep this from happening to other people," said Kathleen, with a sad look on her face.

Kathleen asked Chris how she was doing, and Chris responded that she was usually in a lot of pain. The two said their goodbyes and Kathleen went on her way back to Garfield County to attend the event at the Savage's ranch. There was something very meaningful and poignant about the interaction between Chris Mobaldi and Kathleen Curry there at the bill signing ceremony. It brought to light that all the work Kathleen has done for years was on behalf of individuals like Chris, and the many more who stood to be impacted from the booming gas development in Colorado in the future. It was obvious that Kathleen truly cared about the issues, and about people like Chris.

And Chris, valiant and brave and determined enough to make the journey to the bill signing in spite of her pain and her symptoms, was an all-too-real example of what horrible things can happen to people's health from exposure to gas exploration near their homes.

It is my belief that very few people in that large crowd around the podium where the governor and the other dignitaries spoke glowingly about the bills that had passed had any idea of who Chris was or what had happened to her.

It is also my belief that some may have wondered why this frail, stooped, and sometimes shaking woman was at such an event out under the broiling sun when she could barely walk, stand, or sit up.

But Chris Mobaldi knew why she was there: to her the passing of this bill represented the price she had paid with her own life as a result of natural gas development. For her, this was a very big deal. Very big.

Within minutes the crowd rapidly dispersed, leaving in groups down the long sidewalk from the shade roof heading back toward the large parking lot that lined the park's baseball fields. The film crew, the Hoffmeisters, Chris, myself, and just a handful of others still lingered. The governor and his people were still there too.

Deciding it was now or never, I gathered myself and walked straight over to Governor Ritter and introduced myself and held out my hand. I thanked him for all the work he had done on the gas impacts bills, and then I said,

"There's a lady over there who would very much like to meet you. She is very sick from contaminated water and air from the gas wells by her house. If you could just say hello to her it would mean a lot to her."

Although his entourage seemed ready to head off and not anxious for another interruption, the governor graciously went over to Chris and talked to her for a minute or two. Deb had her camera going at this point, and she captured the moment on film. I snapped some still shots with my camera too.

Chris was obviously very happy to have spoken with the governor; it must have made the whole event very satisfying and worthwhile for her. On the drive to the park as Deb was interviewing and filming her, she had asked Chris if she would like to meet the governor and what she would like to say to him.

Chris had said that she would like to talk to him if she got the chance to, and she would thank him for passing the bills that would protect people's health from impacts of gas development.

Finally it was just Chris, the Hoffmeisters, myself, and Deb and Dave with the camera gear left under the shade roof. We were all speculating on the best way to get Chris down to the parking lot; there was no way she was going to make the walk there on her own two feet, even with assistance.

I went over to some park employees in their golf-cart-type vehicle and asked if they could help us get Chris to the parking lot. As I asked I eyed the open sides of the cart, and envisioned Chris tumbling out sideways and landing on the pavement. Perhaps I could somehow hold onto her and hold her in...

The friendly park employees said sure they could give her a ride, but they also said that we could just drive our car up the wide sidewalk like the governor's SUV had. So that is what we did, and soon we were heading back through Grand Junction with Chris again directing us on where to turn in her now very garbled and hard-to-understand accent. But, in spite of her difficult-to-understand speech, her directions through the complex maze of city streets were perfect. That is what is so amazing—that even when Chris seems to be totally unintelligible and no one knows what she is saying at all, she is still cognizant inside her own mind. She is just trapped alone in there, with no one to talk to but herself.

COLLATERAL DAMAGE

We went back to the Mobaldi's and Dee and Hoff came. We all had a nice visit, and Steve soon arrived back from work in Montrose. Everyone was in good spirits after the bill signing, and I took more pictures of Dee, Chris, and Deb. Chris and Dee were chattering away like long-lost friends even though they had never met in person before. Somehow the symptoms they both were suffering bound them together in some unique way, like victims from a natural disaster. Finally Deb, the Hoffmeisters and I left late for the barbeque at the Savage's ranch in Rulison. Steve walked out to the road with me where my truck was parked.

"Thank you for taking Chris to the bill signing; it meant a lot to her," he said.

About an hour later, I met Deb and the Hoffmeisters at the Rulison exit off I-70, and they followed me up the mountain toward the party. They waited for me at the Interstate exit, as it was easy to get lost up there on the country back roads. At the last turn before the Savage's ranch we found the cameraman Dave turning around on the gravel road after a lost adventure of his own trying to follow a confusing map of the rural roads.

The barbeque was wonderful, there were so many people there who I knew from all the gas work everyone has done over the last several years.

The sun was setting over the lovely rural landscape, the live band was playing, and the folks were eating, drinking and visiting in a very upbeat mood. I helped Dave and Deb a little bit moving their camera equipment, and then got pulled into conversations with more friends as easily as a fly is pulled into honey.

The governor and Kathleen Curry and Harris Sherman and Dan Gibbs and other dignitaries were there too.

But the overall mood was relaxation and good times with friends—my "gas friends," as I call them. (After all the time we have spent together over the years now, they have simply turned into friends, I suppose.) It was perfect. A huge spread of food had been prepared for the pig roast, and people flowed in and out of the downstairs atrium room that had been laid out as a buffet-style set-up with numerous side dishes, while two women served up the roasted meat. Beverage stations were all around, with wine and beer and soft drinks, and there was a keg of beer in an ice-filled stock tank on the patio. Matt Sura, the Executive Director of Western Colorado Congress, helped me pump up the keg, and then after we each filled our plastics cups, we clinked them together in a toast.

"Congratulations!" I said.

"Congratulations!" returned Matt, with a smile on his face.

How nice it was for once to savor the feeling of having accomplished something positive after so many failures, so many dashed hopes for legislation passing that would support the landowners in the face of gas and oil exploration. Only time will tell how effective the bills signed today would stand to be, but at least for one day and one night we took time out from the work and the uphill struggle to celebrate this rare tangible victory.

A highlight of the evening was the country music featuring Duke Cox, the President of Grand Valley Citizens Alliance on guitar, Harris Sherman, the newly appointed Director of the Department of Natural Resources on fiddle, and a landowner from Dry Hollow and GVCA member also playing the guitar. They played song after song to the crowd. Harris looked like he was having the time of his life, with a big grin on his face as he sawed away on the fiddle up on the tree-shaded second-story deck as the crowd congregated and cheered below.

I got home about ten thirty, and Al was home from his weeklong trip. I was totally wiped out, and he needed to go to bed too. We said our hasty hellos and went to bed. Nothing makes me more tired than driving long distances, and today I had done plenty of driving in both directions across western Colorado. From New Castle to Glenwood, then Glenwood to Grand Junction, then to Rulison, and then finally, back home.

As planned, the next morning Deb came for one follow-up interview that we were going to squeeze in before I had to leave for work no later than 8:15. Unlike the other morning on the weekend when we had time for a leisurely breakfast and I made a nice breakfast of eggs and toast and berries and tea and coffee, this time I had hastily pre-toasted some English muffins and a steeped cup of tea she barely sipped. We were in a rush.

We filmed on the small deck off the side of the house, and Deb began the questions.

As I gave my answers about my thoughts on the bills, and the impacts, I did my very best to speak totally honestly.

When she asked me what I thought about the friends with health impacts, I said that at times it was very sad to hear them tell their stories. And it was, I recalled seeing Karen and Tim Trulove, Dee Hoffmeister, Susan Haire, Chris Mobaldi and others choking up as they described the horrible things that had happened to their health, and how no governmental entity was there to assist them. Then I had some tears as I described that, and a moment later she did too. I paused for a minute, and then looked straight at Deb and the camera to make a statement.

"I know what they are going to say: people are going to call me a whiner and say if you don't like it, then why don't you just go away. They'll say that this industry is bringing lots of money into Garfield County, lots of jobs, and that's true. It's bringing huge amounts of money…

But what about the people who live here, who never knew this was going to happen? What about their lives? It's not easy just to leave your home and job, where do you go? Why should we have to leave?" I said.

(In my mind's eye I thought about the Hoffmeister family, actually three families in two houses, up there on Dry Hollow. There were grandparents, two sets of married adult children, and then two homes of grandchildren. Perhaps if it were just Dee and her husband Hoff they would leave because of her growing health issues. But, because there was such an extended family group there on the large property, it was a huge insurmountable idea to tackle.)

Many of the other neighbors with impacts had left, and more homes were currently up for sale nearby. The Hoffmeisters were some of the last to remain in that area, of those who had come forward with health impact issues from the gas wells on and by their properties.

I told Deb I thought, as others did, that her film was probably the best chance that we had of getting people to pay attention to the illnesses and problems that were occurring. I could see the look of burden cross her face, along with a mixture of gratitude.

"I hope I can do it," she said quietly.

"I think you can," I said.

"I think you are our only hope. I hope that once people become aware that people are suffering from gas development that they will have natural human compassion and something will change," I said.

I talked about community impacts, and then she asked me for a closing statement.

I talked about Weston Wilson's whistleblower action that was ignored regarding hydraulic fracturing chemicals being exempted by the Energy Bill.

Then I had to go to work. She loaded her equipment, and we gave each other a hug, and went our separate ways. She was going back to New Mexico, and I was going back to work after a four-day holiday and week of intense gas activity.

Saturday June 2, 2007

Last night Deb called from New Mexico; she was about forty-five minutes from her place in Santa Fe. I had left her a message on her cell phone the night before: I told her about the little list I had made on the grocery list pad that Al had put up on our refrigerator a year or so before. I rarely wrote groceries or supplies on it, but I had several times used it as a type of to-do or major task list when I was brainstorming ideas.

COLLATERAL DAMAGE

I still had one of the small pages taped to the fridge from a year and a half ago with all of the dog's names on it that had been rescued and transported from New Orleans to Colorado after the Hurricane Katrina disaster. As each dog was either reunited with long-lost owners, or released from impoundment at the shelter in Colorado where they had been initially deemed too dangerous to adopt out, I had crossed their names off the list. Of the original ten dogs, all but three names had been crossed off.

I knew where Buster was, secure in his rescuer's adoptive home, but never had confirmed the whereabouts of Teddy or Bear, two dogs who had remained in the local shelter's hands. After Buster bit the shelter director, to say that communication with the rescuers and the shelter deteriorated is to put it lightly. A full six-month combative period involving legal action, accusations, a lot of press and more ensued. Oh, the memories…

I have a two-inch thick scrap book of all the endless articles, editorials, and letters to the editor that ran in the papers as a reminder of that tumultuous time. I am pleased to this day that to the best of my knowledge, all of the dogs ended up in good situations in spite of having been dealt the fate of being labeled "dangerous and vicious" at the Louisiana Lamar-Dixon facility where they were housed en masse in horrible conditions following the disaster of Katrina.

> Two months ago I began another to-do list. It reads as follows:
> Lance Astrella
> Weston Wilson
> Amy Mall, NRDC Boulder
> Aspen Community Foundation
> John Travolta
> Robert Redford ("Green film channel")
> *Governor Ritter and Kathleen Curry interviews re landowners

The list remained on the small pad stuck to my refrigerator door for some time; writing such a list is my way of keeping track of important thoughts and ideas that spring to mind suddenly.

It is something of a master plan, or long-term goal setting.

TARA MEIXSELL

Sometime last week, in the midst of the flurry of filming with Deb and Dave from Denver to Garfield County to Grand Junction for the gas bill signings, I found myself in front of the refrigerator staring at that small list on the three-by-four-inch pad.

As I read down the list, I realized that most of the items on the list had been accomplished—by one route or another. I also realized there were two more names to add to the list.

As Deb and I talked last night, I told her about my refrigerator list.

"I was standing here looking at the list, and then it hit me that you have gotten interviews with most of the people on it already! Went I wrote it down, I wasn't really thinking about it, but we got interviews for the film with most of them! When I realized that, I got so excited I wanted to call you! We need to add two more names now: Ted Turner and Al Gore," I said.

"There is one more name we need to add to the list too: ITVS. I need them to be interested in the film. Cross all your fingers for me," Deb said.

"I will, what are they?" I asked.

"Independent Television Service, and I just submitted to them," said Deb.

We talked more about the status of the film, the grants and submissions, and about my book manuscript. It is funny that as two people in very different mediums but working on the same subject, we can relate to the progress and challenge of the great task at hand that each of us is doing. One thing we certainly have in common is the dedication to continued forward progress at a steady pace, no matter what. Oddly enough, both of us have been working simultaneously for the last year and a half at least, and both of us have a good amount of work behind us, definitely enough for a first draft of a book for me and an almost completed rough-cut film for her.

Deb then told me that after she left my house on Wednesday morning she had gone to Grand Junction to see Chris Mobaldi one more time.

"I brought her some flowers; she really touched me. I know how much I wore her out with all the filming," said Deb.

I was very surprised that Deb had driven so far to do that kindly gesture for Chris. It must have meant a lot to Chris, I am sure. I could tell that Chris and Deb had developed a very genuine rapport, as she did with all the impacted people she had interviewed around here. But, to have her go one more time over to Grand Junction to deliver flowers and a thank you was very personal and special.

As the sun set over the ranch and the green valley, I went about my chores while on the cordless phone and talked with Deb about plans for funding, and all the great footage she was bringing back with her from this last trip to Colorado.

"It seems like it was meant to be, like there was some force behind this. The fact that the bill signings and everything happened to be going on when I was here was incredible," said Deb.

"Yes, if you had tried to plan or create the events, it couldn't have been better done. The fact that you filmed Chris going to the bill signing in Grand Junction, and how she barely made it through but in the end she got to personally talk with Kathleen Curry and get a minute with the governor was huge. It meant so much to her. It so easily could have happened that she never would have had the chance to talk to them, considering that she about collapsed shortly after she made it up the gazebo where the signing was. That would have been so sad," I said.

We said goodbye, me from Colorado and Deb from New Mexico. Somehow, after the last several weeks, it didn't seem so far away anymore.

I thankfully slept in this morning, until seven thirty. Between gas-related events, driving all over Colorado (again), then back to the never-ending surprises and demands of my working job at Mountain Valley, it was blissful to sleep in. Yesterday we had had our first outing with about thirty clients going up to Carbondale for a community-based day. I was an action-packed day, a very successful day, but it left me very ready for sleep…

This morning after making coffee I went out and fed the animals, looking forward to the unscheduled home days to come. I wanted to do some writing, and Al brought home a lot of flowers and vegetables to plant. The last frost certainly better be over by now.

As he cooked the potatoes and vegetables and eggs, Al mentioned something about an Aspen Public Radio airing yesterday. A lady from Silt had a lawsuit pending against Halliburton for some sacks of chemicals that gas workers had been unloading off flatbed trucks behind her home in a parking lot, and the dust from whatever substance was in the bags had damaged her health. The chemical was barium sulfide—it was extremely toxic and used in the hydraulic fracturing liquid mixes that were injected deep underground with explosives to aid in the gathering of the natural gas from the tight sands formations.

Immediately I knew that it must be the lady who owned the little coffee shop and gift store that I had been told about. Someone had told me that the doctors said that she had permanent lung damage from the material she had inhaled. She was now diagnosed with asthma.

So now there was a lawsuit…

I looked up the number for the store, and when the lady answered on the other end of the line I asked if she was the owner and briefly described why I was calling—I said that I had heard about the Aspen Public Radio airing and I was writing a book about health impacts from gas development. I also said that I was a friend of Nancy's and I believed I had heard about her from Nancy.

Carol, the woman who was ill, told me some details about her situation and seemed quite interested in both Deb's film and my book. She said that Halliburton had been unloading hydraulic fracturing chemicals behind her home and store, and that she had gotten sick and the doctors determined that those chemicals were the culprits.

I mentioned to her that one of my friends was very ill, and described Chris Mobaldi's scenario. Carol asked her name, and said she had heard about her. She also said that she didn't know if she could talk, she would have to check with her lawyer. I said I understood, and that was exactly the way things were with Chris, who also had a lawsuit pending. One of the companies filed against was also Halliburton.

Carol asked me when my book would be finished, and I laughed and told her that I was doing the first edit, but there was a lot of work and inserts left to be done and as I worked full-time I mainly had to squeeze it in on the weekends. I also said that aside from the hard work of editing, I also had to get a publisher.

"I will get it published one way or another," I said.

"Maybe someone with a lot of funding will do it," said Carol.

"Maybe, but getting a publisher is really hard. If I have to I will do it myself. I've done the most of it already. It will get done," I said.

Even as I said that I knew what a task it would be. Lots and lots and lots of time and work...

But, I sit here typing on the laptop out by the back pasture, now plugged into many extension cords that ran my water tank heaters for the llama barn this winter. I am seated on a big plastic rafting cooler beneath a huge elm tree. I love to write outside; it keeps me connected to nature and my animals too. My barn cats come and go from their resting spots in the shade, and the horses and llamas move about in the pasture. The birds are chattering in full force and flying about, and it is a gorgeous day.

Perfect cirrus clouds float in their puffy mounds across the blue sky between the mountain peaks. The oak brush and the elms are almost completely leafed out in lush, brilliant green leaves. Occasional breezes blow through the valley, and the insects are not yet a bother.

This project, this book about gas development impacts has become a familiar part of my weekend routine. I find that if I do not take the time to write on my two days off, it doesn't feel right.

Time to turn off this computer, and spend some time relaxing and gardening and visiting with the animals and with Al.

Saturday June 9, 2007

I went to the store to get coffee filters and cat litter. Al asked me to get a paper too. When I pulled out the Post Independent from the paper dispenser in front of City Market I read the headline and became immediately angry.

"The BLM approved drilling on top of the Roan Plateau in the next upcoming years."

The Interior Department approved that, and the current administration still has the power to get at what they want.

I am furious yet not surprised; the process works so well.

Ask for public comments; take note of the comment, and do whatever the current administration is compelled to do nonetheless.

Why is public comment even asked for? It seems to be a cosmic joke.

I am tired of going to meetings to stand up for things and speak as a citizen when no one listens to the overwhelming consensus of the public. It seems a huge waste of time and energy. When the industries that make such huge amounts of money can benefit for developing an area, no amount of public outcry will stop it if the administration is industry-bent. This is not a democracy at work, it is money-driven interest groups currently in power in the government taking advantage up to the last hour possible to gain the largest profits they can.

COLLATERAL DAMAGE

So: they are going to wreck the Roan Plateau.

The question I asked myself over a year ago still is pending. Do I want to go and see this place that I have heard is a treasure before it is ruined by gas development, or is it better never to know what has been lost?

As I sit here in the barn typing, I am wondering how those people who championed the Roan Plateau issue are feeling today. More specifically, I wonder what they are thinking.

And sadly, I doubt there is anything that can be done to stop the bureaucratic process that will unroll from here. The Roan Plateau is most likely toast, its fate sealed. In my lifetime anyways, and long after that...

Friday June 15, 2007

Finally, Friday night is here. As usual, once the all-consuming workweek begins, a day flows rapidly into the next day then suddenly the week is over. In the back of my mind I have been thinking about this book and the gas issues each night as I came home after work. I didn't however have the motivation to pull out the computer and type; I merely thought about it in the back of my mind. It has become quite hot here, pushing up toward ninety, and when I get home I am spent and want to relax.

But tonight is Friday, and I have the respite of the two sleep-in days ahead of me with no work distractions. So tonight I can spend some time typing without worrying about sleep or work.

Early in the week I learned that Ken Salazar was putting an initiative before the US Congress to block the funding for the money needed for the leasing of the Roan Plateau lands.

I was so pleased to hear this news, and after finding out what we needed to do to support these ideas, GVCA and other community members from Garfield County began contacting members of the House of Representatives and U.S. Congress.

Again, like the 2005 Energy Bill vote, the final decision on these important issues comes down to the nitty-gritty balance of power in United States Congress and in the House of Representatives and Senate. It is frustrating to feel so out of the loop on things that are happening right here in Colorado, things that have enormous and as yet unknown implications for the future.

The Bush government has power over the BLM decisions, and has power over what Congress approves. So it's not over yet perhaps, for the Roan Plateau. It still feels like a long shot though.

Tonight I am out in the big open sided hay barn, watching the last light of the sunset across the narrow valley. When I got home a few hours ago it was beastly hot still, but now it is cooling down nicely. The mosquitoes hatched out yesterday, so today I mixed up a new batch of bug wipe and wiped the horses down. I also stopped at the feed store and got two West Nile Virus shots for the horses.

Amy Mall from Natural Resources Defense Council is coming next week, on Monday. I plan to meet with her Monday afternoon, and Patrick Barker from Western Colorado Congress and I have been talking about interview plans for her visit.

When we talked today, Patrick voiced his frustration over people who will complain about their impacts loudly at one time, then decline to be bothered further for a follow-up interview. I knew exactly what he was talking about, but I am beginning to realize that it is part of the standard operating procedure, for lack of better words.

"People don't want to make trouble," said Patrick.

"I refuse to see it as making trouble—how can talking about people getting sick be making trouble? I do not think about what I am doing as making trouble; I am telling the truth. I refuse to stick my head in the sand," I said.

"Well, that's what Americans do, isn't it? Ignore problems?" said Patrick.

COLLATERAL DAMAGE

Unfortunately, I don't believe there is any need to panic for lack of people willing to be interviewed. Even though people get burnt out with the issue, move away and want to erase it from their daily lives, or perhaps settle legally and sign disclaimers to silence; new people will and do take their places.

That is the frightening part, the unknown people who in the future will have similar if not worse health problems from the impacts and toxins of gas development near their homes.

Saturday, July 8th

Between work and the sweltering temperatures that are hitting over one hundred degrees, I have had a hard time feeling productive lately. Over the July Fourth holiday I never even turned on the computer, which is unusual for me on a day off. Instead, I puttered around the ranch, visited with my friend Kathy who stopped by, and made a huge batch of potato salad with dill pickles, celery, boiled eggs, and garden herbs. I cooked the potatoes in a crock-pot, and then when they were done I chopped them up, sprinkled them with vinegar, and sat in front of the television and painstakingly peeled a pot full of hard-boiled eggs. It was fun to spend a day leisurely, and do home things and not think or write about oil and gas issues. After all, it was a holiday.

Last Friday night I came home from work and there was a message from Deb Anderson on the machine. She had just driven up to Colorado and filmed a press event with Senator Ken Salazar and Governor Ritter regarding the Roan Plateau. She was now on her way to her parent's home for the holiday, and she said she waved in my direction as she drove past on I-70.

That was interesting; earlier that evening I had seen Denver television footage of Salazar and Ritter discussing the Roan Plateau and the impending development of the pristine area. The footage was shot at the Rifle airport, and also during a flyover of the Roan.

Well, regardless of my dilemma of wondering if I should journey up to the Roan and see it before it was ruined; I had now viewed a spectacular portion of it via the television.

Just past the dramatic steep cliff-edge of the plateau where numerous drill rigs already towered, the film clip panned over a spectacular deep and lush canyon where an enormous-looking waterfall fell into a winding streambed far below, and the wild and unspoiled plateau landscape beyond.

As the television coverage continued, it became evident that the United States Congress had failed to support a stay on gas development of the Roan Plateau.

I guess Ritter and Salazar were giving it another plea, of sorts.

I myself was angry, again. And disgusted. The next morning, immediately upon waking, I was still angry about it.

Today, a week later, as I edit and type I wonder about what will come to unfold.

Will the Mobaldis be successful with their lawsuit? I was told that the companies filed against in the suit needed to submit their responses by sometime in July…this suit has been going on for a very long time already. As Lance Astrella had said, it was "moving at a snail's pace."

How extensive will the ruination of the Roan Plateau and the rest of western Colorado be from gas development in the coming years as the huge boom continues? I shudder to think…

Then, on a more personal note I wonder what will become of my writings, and what will be the results of Deb Anderson's film?

I have no idea, but I know that on our very different levels we are both committed if not obsessed with this project.

I myself am currently struggling with how to edit what has been written, and how in the world I will get it published. I wish I could take a few months of from my job to focus on it, but I cannot.

We are going on a weeklong raft trip on July 20th down the Gates of Ladore section of the Green and Yampa Rivers. I will have endless hours of sitting on the blue boat as we float through the red rock canyons. Usually I day dream, stare at the landscape floating by, or read as Al rows the big pontoon raft laden with all the gear.

Some years ago on a raft trip on the Chama River through Georgia O'Keefe country in New Mexico when I was finishing the final edit on my previous book, I scribbled furiously on a spiral notebook as we floated down the shallow river. Back then I had a publisher and an editing deadline.

Since I cannot afford to take a sabbatical from my job, I think I will print off a copy of this 230- page manuscript and my twelve pages of editing notes, and take it on the river. It may become splotted with water splashes, or even end up in a rapid, but I will have ample time to work on it.

Tonight, as I get ready to turn off the computer where I am typing sitting on the floor in the small spare bedroom, the rain is whipping sideways outside the window and the swallows are flying in arcs above the garden through the drops.

The last two afternoons we have had some brief rain showers and cloud cover as a respite from the broiling heat. Several wild fires burned in Rifle since Friday, so thank goodness for a bit of rain.

July 30, 2007

Tonight after 9:30 the full moon rose through the lowest place in the valley toward the east. To my astonishment, the moon was orange, the color of a young pumpkin. I had never seen such a thing.

I called Al, and said,

"Look at the moon, it is orange because of the pollution."

My husband came out and said,

"The cheese is rotten."

Then he went back inside. He is a man of few words, typically.

But he was right; the cheese was rotten. Something was rotten, something was wrong, something was amiss and not as it should be.

Never ever in my life had I seen the moon look the color of a pumpkin.

August 7, 2007

Yesterday while at work I grabbed a copy of the Glenwood Springs Post Independent. Later, when I had time to scan the articles to my huge surprise I saw the headline on page two:

"House approves drilling ban on Roan

By JENNIFER TALHELM
Associated Press Writer

WASHINGTON—Energy companies would be barred from tapping natural gas on federal land atop Colorado's scenic Roan Plateau under an energy bill that passed the House Saturday evening."

After the reading the article I was as first uplifted then confused. What really did this mean for the entire Roan Plateau, and at what stage was the political process in on this issue? Just because a bill passes the House doesn't mean that anything is a done deal by a long shot, and also what about the privately held land and all the rest of the Roan Plateau?

Nonetheless, it was heartening to know that the politicians from Colorado who stood up for the preservation of the Roan Plateau were doing all they could, and more than I knew about.

I suppose the next few days' papers will have more news on it.

If I am lucky, maybe some years from now there will be a part of the Roan Plateau that is worth going to see.

COLLATERAL DAMAGE

On our way back from the raft trip which ended at Dinosaur National Monument in Utah we drove down from Meeker to Rifle a few weeks ago, and a steady stream of pickups and industry trucks poured out from the access road to the Roan.

There is definitely going to be a lot of activity up there, but maybe a portion will be unspoiled.

Bad Times

August 22, 2007

It continues to be beastly hot, and it has been a hectic and at times unpleasant last few weeks.

One night I came home to find damaged fences in the pasture and one of my younger llamas missing. After a panicked search, I located him and spent the next several hours, with Al's help, trying to get Willy back into the pasture.

It turns out that the neighbor's new dogs that had been running at large lately got into the pasture and caused a ruckus.

The following week that unfolded was unpleasant, with calls to the sheriff's office, dispatch, and more. I can live without neighbor problems...

Willy wouldn't get up the day after the chase, and when I forced him up he couldn't walk without a severe limp in his hind end. To say I wasn't happy would be putting it very mildly; I was over-the-top furious and very frightened of a repeat or worse event.

Today I called Chris Mobaldi to confirm her now third address in Grand Junction since their move from the contaminated property in Rulison. It took Chris a while to pick up the phone, and when she finally did I could tell immediately that she was not doing well.

She told me in her strange accent now also even more disguised by crying that she was very sick. She couldn't walk, couldn't get to the bathroom alone, cried all night long because of the pain, and said that the pain was severe and ever present.

COLLATERAL DAMAGE

All I could say to her over and over again was,

"Oh Chris..."

She said that her recent urine test had shown mineral levels off kilter, and that Dr. Gerdes was putting her on more minerals.

"Steve wanted to take me to the hospital last night, it was so bad. But I was afraid; I don't want to go back to the hospital. I hate the hospital, from what happened to me before...I need a miracle to happen....Steve is going to start some far away jobs and he won't be able to help me when I need him anymore. He was working close by and I could call him and say 'Steve, I have to go very, very badly' and he could come and help me," she said, still weeping.

We said a very sad goodbye, and my heart was as heavy as a stone in the pit of my stomach. It was a bad reality check. I told her that many others and I would be keeping her even more specially in our thoughts than we always do.

After we got off the phone I went into the Silt post office and mailed the eight-by-ten photo I just had blown up of Chris talking with Governor Ritter this spring at the Grand Junction COGCC reform bill signing ceremony. She looks so pretty and excited in that photo; she was beaming as she talked with the governor even though she had been about to keel over minutes after we got her there.

A bit later I called Dee Hoffmeister to check and see if she was aware of the annual picnic being held again at the Bell's beautiful ranch just across the road. We talked for a minute, then Dee told me what had just happened with them a day ago.

One of their daughter's dogs, a nine-month-old pit bull named John Wayne, had gone on a ramble and ended up getting stuck in a Barrett Corporation production pit near the house.

Thank goodness that the employees had roped themselves in and gotten down into the pit to drag the dog out. He was covered in black chemical sludge. He was also in very bad shape.

Dee had the wherewithal to call the COGCC and she spoke to several people. The next day Frankie Carver from the COGCC came out and inquired about the location of the dog; she wanted to see it.

Dee said Frankie was very concerned, and reported that a violation had been filed against Barrett Corp. for not having the pit fenced in proper order. Frankie was also concerned about the dog's condition. She said she hoped he made it through the night.

Although the vet had been unable to accept the dog the day of the accident, they took it today.

"The company called and talked to the vet. They told him there was benzene in the water. They didn't tell that to the COGCC or to me; they told me it was water, just water and soap. The vet said that his system was all messed up, and that he had an infection. His eyes are all red…he is such a good dog," Dee said, and then she paused.

"You know, I keep thinking, this has to be ending now, this has to be over. These things keep happening to us, after my first problems then the well blew up and I got so sick and had to go the hospital again. And the bills were huge; we are still paying off those bills, and the company will never pay. Their lawyers won't let them because it would imply they know that they are causing it.

So, if Barrett has to pay a fine, it is going to the county, not to us. We are going to have a big vet bill now too, on top of my bills. This whole thing has made me very angry. I just got so mad," said Dee.

As I drove on Highways 6 and 24 west toward Rifle after we spoke, past the rigs I barely look at any more, (they are just there, more will come wherever they do, and that is what will happen). I couldn't really process the information I had just learned from first Chris then next Dee.

These are people who are my close friends, and these were horrible things happening.

And these things are not stopping; they are going to get worse.

COLLATERAL DAMAGE

Tonight after I got home from work, I let my mind think about things as I tried to relax out in the barns while feeding the horses, cats, and llamas.

I knew, however, that in spite of the effort I was putting into relaxing, the computer was drawing me like a big magnet to turn it on and type something about the sad news I had learned today.

Now it is getting dark out here at the picnic table on the little deck, and I need to go water plants. The crickets are chorusing loudly, and it sounds very peaceful. All the annuals are blooming out, and the tall sunflowers are going to seed.

I can just still make out the colors in the growing dusk.

Wednesday night August 29, 2007

I talked to Dee tonight. She said that she was feeling sick again, and said the well at her daughter's place across the road was being hydraulically fractured and they were flaring again. Then she said that she learned that on a neighboring property owned by a woman who had sold out to Barrett Corp. and moved to England that four more wells were going in. I had heard about that property, it was 250 acres or so and a large number of wells were slated to be drilled there.

It crossed my mind again that Dee and her family should move to preserve what was left of her health. I didn't say anything though; I do not feel it is my place to do so.

"Dennis Webb from the Post Independent called me about the dog, and he asked me some questions. I told him that if I could I would give a medal to the two men who roped themselves secure and went into the pit to save the dog. They literally risked their lives to do that," said Dee.

"Dennis called you?" I asked in surprise.

It had been a number of days since I had called him on the story, so I hadn't heard anything and figured it wasn't going to be covered. Suddenly, I felt a lot happier.

Dee continued to explain her conversations with Dennis, and how someone affiliated with the gas company came by to pick up a copy of the vet bill for the dog's prolonged stay and treatment at the vet after falling into the pit of undisclosed gas industry chemicals.

Dennis asked about the amount of the fine for the violation that had been filed against Barrett Corp. for the faulty fencing of the pit. Dee had no idea, and told him to call Frankie Carver at the COGCC.

I asked Dee if she had called the COGCC about her illness due to the flaring and fracking; she said she and Frankie had discussed it.

We agreed to talk more later about pursuing the installation of an air monitoring unit at their property. They needed one, especially with new wells coming.

"My daughter had to argue with Jim Rada to get the last air grab-sample canister at the county. How in the world can we find out what is in the air after we smell something if we have to go after a canister?" said Dee in a frustrated voice.

We talked a little more, and then said goodbye.

The COGCC Tour: The Fight

October 1, 2007

I haven't written for a while, as I am editing.

I only write if something "big" happens: an explosion or mishap, or a development within the regulatory process or big news in the media.

Today the new and old COGCC commissioners came to Garfield County to do a tour this morning that was run by a handful of GVCA members. The spots visited were heavily impacted areas, with landowners at a few key locations along the way to speak to the commission members.

It was a sizeable group that met at the Silt gas station and assembled for the tour. In addition to the GVCA members, the landowners and the nine-member COGCC commission, a good number of industry people had shown up.

We split up in separate vehicles, and landowners were paired up with COGCC members so as to talk along the route about the impacts and the issues from industry to the area.

We drove across the Colorado River and up the incline into Dry Hollow. I did my best to encapsulate a sense of what was happening to the area there, telling Harris Sherman the Director of the Natural Resource Commission about the people who have been moving away due to the impacts from the gas industry. He asked if the people's symptoms subsided after the wells were completed, and I answered no, they did not. The completed wells and infrastructure emitted fumes long after the rigs were gone, and wells are also refractured and the chemical and toxin emissions associated with that continue also. In fact, one of the processes that Dee Hoffmeister is extremely sickened from is hydraulic fracturing activity in her area. Barrett gas company is aware of that and from time to time they alert the Hoffmeisters in advance of a fracking operation in the area.

As we drove by Dee's house I pointed it out and briefly described her scenario. I said how she had fallen unconscious in her home, and remained so ill for over half a year that she could not live in her home near the industry fumes, and then recently experienced the explosion and fire at the condensate tank 750 feet from her home. I followed that up with the comment that she became so ill from the fumes off the condensate tank fire she couldn't even turn her head. Finally she was carried out and taken to the hospital, where they ran tests on her to no avail.

On Chipperfield Lane we pulled up to the first stop, in front of some nice large rural horse properties. Nice that is, aside from the huge industrial site of the well pad going in just across the street.

Right where all the new activity was going in there was a rustic log ranch entryway to the property, with a sign that had the words "My Heaven" carved into the wood. It seemed odd, with all the newly bulldozed earth for the pad, and all the industry equipment there. It didn't resemble heaven very much, to my eye anyhow.

I was quite sure that this was the property that a lady from England had sold to Barrett Corp., over 200-plus acres. And all this in an area that had already been so heavily impacted with gas development.

The parade of cars parked along the gravel road near the newest well pad and industrial site, and the large group assembled to listen. The landowners spoke about the issues that were bothersome to them. They spoke about the increased industrial activity around their once quiet country homes, about the fumes, and about their concerns for their health.

Industry representatives joined in the conversations and countered things that the landowners brought up.

There was an undeniable tension that pervaded the scene as thirty or more people standing on the side of this country road listened as the huge gray tubular gas trucks rolled by, at times drowning out conversation.

Dee Hoffmeister described some of the mishaps that had occurred by their property, and also talked about her health issues. She told the COGCC members about falling unconscious after arriving home one night when a cloud of visible fumes enveloped their house. Then she discussed her inability to return home for over six months due to the triggering of her illness symptoms whenever she tried to come home. Next, she described the well explosion that happened this past late winter, when she had been so sickened she had to be carried out of the house to the hospital where they ran test after test to no avail.

Dee turned to the Barrett Corp. representative and said

"Did they ever find out what caused the fire?"

The tall thin man from Barrett seemed to looked rather uncomfortable as he answered,

"No, the FBI is still looking into it…"

A few heads in the crowed turned suddenly at the mention of the FBI.

Beth Dardinsky, whose home was just up the road on the crest of the hill, described her fears for her own health as the fumes continued to surround her home and her headaches and symptoms increased.

"I go outside and throw baby powder into the air to find out which way the wind is blowing, so I can report it when I call it in. I am worried about what is going to happen to me, and to my health." Then she paused.

"In another year or two am I going to end up like Dee Hoffmeister?" she asked angrily.

I cringed as I glanced at Dee, who looked valiant but not in good health as she stood at the gravel roadside steadying herself with her cane. Her eyes and face looked fatigued.

COLLATERAL DAMAGE

At one point one of the landowners said angrily that what was going on across the road from their place was going to kill them and their animals. He asked why Barrett Corporation insisted on drilling the well some 300 feet from his front door when they had two hundred plus acres to drill on in that parcel. The tall thin Barrett representative looked uncomfortable and said something to the effect that "the geology" made that site necessary.

Another neighbor complained about pits that leaked undisclosed chemical mixes that were never repaired.

The Barrett official countered saying that people were not calling in to report complaints; they had very few.

One of the solemn looking landowners who had been quite reserved during most of the discussion blurted out heatedly,

"You have beaten us down. No one does anything when we call, so we don't call anymore."

As I stood at the side of the dirt road above the drainage ditch, I picked a spiky invasive weed and looked at it.

It was like being present at a family fight, in a way. This was not what I thought the landowner tour would be like.

It was more like being at a gas hearing at the Capitol, giving testimony while surrounded by industry groups. There was no love lost in this roadside debate; the industry reps looked angry and foreboding, and stood with folded arms along the roadside. The landowners looked angry, frustrated, trapped, and furious. The COGCC members took it all in, and I myself was unhappy that the landowners were being picked at by the industry instead of being able to tell the COGCC what their issues were. In retrospect, however, I do believe there was something valuable in the COGCC members experiencing the hostile dynamics that prevailed so often between the gas company representatives and the landowners who suffer the impacts of the drilling on and around their properties.

The tour continued, and we went up over Jenkins Cut-Off down into the Mamm Creek area to one of the well sites on the Pitman's place.

It seemed that the discourse at that sight was equally back and forth, and became just as heated at times.

The Pitman sisters, long time cattle ranchers on whose property that well pad was situated, talked about their concerns over the negative impacts to their property, the environment, and people's health. Although years ago they, as mineral owners of the large ranch holding, had agreed to allow gas drilling, that was when the spacing unit were at ___ per ____-acres. Since that time the spacing has been changed to ____ per acres. The changes have been gradual, and the impacts that occur now with the dense spacing are totally different. The Pitmans have publicly stated that they never would have agreed to lease their mineral holdings if they had known how the spacing units were going to change.

The land on the supposedly reclaimed pad site looked like an abandoned gravel parking lot. It was a barren dirt patch festooned with invasive weeds around the industry equipment.

Rick Roles, a close neighbor, was also there to speak to the COGCC members. He talked about the venting of methane from the tanks at his well pad, where the methane blew off straight out into the air.

He also talked about the misters that had been used at the evaporation pits on his property that blew the liquids from the pit into the air and onto the ground. Then he brought up the subject of hydrogen sulfide exposure, and several of the industry employees heatedly denied the existence of hydrogen sulfide in this area.

Rick stated that the labels on the tanks right in front of us listed hydrogen sulfide, and then he described the drills that he had been trained in as a gas field worker to deal with hydrogen sulfide exposure.

Beth Dardinsky walked over toward the equipment on the well sight, and peered at the labels on the tanks. One of the industry reps angrily said that people had no business being near the equipment, indicating that Beth shouldn't be walking around the pad on the Pitman sisters' property. One of the Pitman sisters asked in a deadpan tone exactly what were they supposed to do when they had to retrieve their cattle off the well pad, as it was not fenced.

Tensions were extremely high as we left the well pad on the Pitman's property. The level of anger has escalated far beyond what it had at the first stop—it was downright ugly at this point.

When we parted from the commissioners as they continued on to an Encana property for an industry tour, we headed back to our vehicles at the gas station in Silt.

What happened next was surprising.

A man from Encana who had been on the tour came over and talked to Patrick, Liz and myself. He had a friendly smile, and displayed none of the crossed arm hostile stance that we had seen in front of the COGCC members. We talked for a long, long while. He talked about coming together with solutions, and continuing to have dialogues. We discussed closed loop systems, and best management practices.

He mentioned that he lived here with his family, and planned to be here for a while.

"I have to live in this community; I don't want to not be able to talk to people," he said.

After close to what seemed to be an hour, the representative from Barrett arrived at the gas station in his car and he came over to us.

"You're not ganging up on this guy, are you?" he asked in a joking tone.

"No, actually it's quite the opposite. He came over to have a positive discussion with us," I said. Again we discussed the issue of hydrogen sulfide emissions from the wells. The industry reps again said, no, around here there were none. I asked why then the placards on the tanks said there were? I also asked why then were some people being overcome by noxious fumes to the point they fell unconscious? They also agreed it would be nice to find out why...

Our conversation continued for a good time longer.

Now, it is almost five p.m.

It is raining, I am going to go out and feed the horses and llamas.

For some reason I cannot quite catalogue what I thought of today. I suppose I do not have a deep enough knowledgeable sense of how a committee such as the COGCC can understand a tour and the public comment that they saw and heard today.

Saturday October 20, 2007

The last month or so, a number of things have happened regarding gas industry mishaps, accidents, and even a death. As I read the headlines and articles, I somehow didn't feel the same urgency as I would have a year ago.

An Encana employee at a well pad site was crushed to death by heavy equipment when they were moving the rig. He was from Texas and in his forties.

An industry truck carrying fracking liquids rolled of the road south of Silt and spilled fracking chemicals. The quotes in the articles said that the liquids were not harmful. They also said they were undisclosed—meaning not identified. (Who, then, I wondered, could accurately say they were "safe" if no one had to know what was actually in them?)

A co-worker of mine called me and told me that the tests from their newly drilled water well (just four miles or so from the Debeque Black Mountain waste pits facility) had come up with contaminations, and that the lady from the testing company advised her not to drink the water. The contaminants included benzene, toluene, and a number of heavy metals.

COLLATERAL DAMAGE

She faxed me the report and I then faxed it to a specialist for analysis.

Her husband was now hauling water for house use in their brand new Debeque home, where they had a young son and a newborn daughter. They had just moved in recently and had only been drinking water brought by the Culligan man, as the water from the tap stank, looked milky, and was very high in salts.

Yesterday was the end of a long week at work. I had been filling in for my boss who was on a well-deserved two-week vacation. I had had a bit of a cold and an earache and plenty of work-related challenges to deal with daily, and I was very ready for a weekend.

On the way home from work I stopped at a store in Glenwood and glanced at the cover page of the Glenwood Springs Post Independent that was laying on the counter by the cash register.

One of the headlines read that the head of the COGCC was leaving office.

Unfortunately, I failed to take the paper with me; I will get an old copy on Monday.

As I drove home I was curious as to why Brian Macke from the COGCC had really chosen to leave. After the passage of bill that changed the make up of the COGCC from five industry people to only ___ out of nine, did he feel pressured out?

Had any of the landowner issues, problems, illnesses, press coverage and complaint played a part in all that?

The answer had to be yes.

It all continues to feel rather distant, even though it is certainly a sign of progress for the landowners to come perhaps.

I suppose the sheer vastness and power of the gas industry's impacts on this once rural area seem too overwhelming for anything to change.

Nothing will change what has happened to Chris Mobaldi.

And likely, little will be done to help Rick Roles, Susan Haire, Dee Hoffmeister, and the many others out there living in the gas fields who are seriously sick and debilitated.

No wonder I am becoming numb to it.

Nonetheless, I have hope that the films and the press and this book will help in a new way to bring light to the seriousness of these issues.

Last night after getting home from work I made dinner then took a shower. As I came into the living room I saw that Al was watching the Erin Brockovich movie on TV.

I became riveted on the remaining hour of the movie, and at times felt with certainty the exact same issues were at play here in Garfield County—only with a different energy company in a different context.

The next morning I was surprised by a question that Al asked me.

"So is there going to be a case here like the one in California?"

I turned to look at him, as I stopped folding the laundry on the bed. Without hesitating I said,

"Yes, but probably not for about ten years or so. It will take that long for enough people to get sick for a class action lawsuit."

Then I continued folding the laundry and he became busy in the living room.

COLLATERAL DAMAGE

Later, it struck me how matter-of-fact I had felt as I answered his question. How impersonal and distant it felt to think about people who in the future may be developing cancer and other debilitating illness; illnesses that will take years to show their symptoms.

People who might end up like Chris Mobaldi—or worse...

This was not my own theory; I know I have heard it from numerous experts already. The long-term effects of the toxins from the gas industry on the local population remain to be seen, and the prognosis is not good.

Congressional Oversight And Governmental Reforms Hearings: Washington D.C.

Sunday October 28, 2007

Today I feel like I need to pinch myself, to wake up from the total surprise at the information I heard yesterday.

After an early start to the winter with mid-October snow and cold and damp weather, this past week had been quite nice. Sunny crisp days and pleasant temperatures prevailed, and some of the yellow and mustard-colored leaves still hung on the trees.

Al was gone; back up at the property in Fairplay doing work on the house he now had rented out. He was spending the night at my dad's ranch house where I used to have the livestock.

As usual, I had grand plans of all I could accomplish in the two-day weekend before returning back to work on Monday. There was the ever-present laundry, winter and summer clothes to sort through, housecleaning, barn chores including moving some hay into the barn, writing, and relaxing.

One priority I had my mind set on was riding the horse Doc Holliday. The ground was dry, and his leg injury that had been wrapped in bandages for over a month was finally healed up. The fall was my favorite time to ride, and this summer I had done little riding.

Before going out into the pasture I put a call into Deb Anderson in New Mexico to touch base on the progress of the film. I wondered if she had any new news…

COLLATERAL DAMAGE

We talked for almost an hour, and Deb said she was close to finishing her one-hour-long rough cut of the film. Now she was working hard to meet deadlines for submission for next year's television line-ups. She was sending it to public television and also venues such as Frontline. We talked about the Mobaldis, the court case now set for next June, and much, much more.

Deb said she would send me a copy of the film so I could comment on it, but I couldn't show anyone else. I couldn't wait to see it…

After the phone call I grabbed my riding helmet and the bridle and went out to catch the horse. After luring Doc and Buddy with a bag of carrots I caught Doc with a piece of twine and then lifted the bridle over his head and slipped the headstall behind his ears.

In spite of his long vacation from being ridden, he was great. There is nothing more relaxing than riding; my mind goes into a place where all my focus is upon the moment at hand. Although trotting in circles in the dry back pasture above the steep creek bank is a far cry from the level of riding I did in years past, it was still pure enjoyment.

I came back into the house about an hour later and the red light on the message machine was blinking. I reran the message and it was Steve Mobaldi saying simply,

"Call me back."

I called promptly and Steve answered.

"I have something to tell you," He said.

"What?" I asked, bursting with curiosity. Had they settled out-of-court?

"We're going to Washington!" Steve said in a jubilant voice.

I was stunned.

"What? Washington? What for?" I asked.

I could not get my questions out fast enough.

Steve filled me in on the details.

He and Chris had been asked to testify before members of the Governmental Oversight and Reform Committee.

Their expenses and airfare were being paid for, and Gilad Wilkenfeld from Henry Waxman's office, Amy Mall and Craig Dillon White from NRDC were involved in arranging it.

After flying from Colorado they were going to dinner with a group including Gwyn Latchet from OGAP.

The next day they were going to speak before the committee at the Rayburn House Office Building. Steve said that he had five minutes to deliver his speech. Other landowners with negative health impacts from gas and oil development were also going to speak. Susan Haire, who now lived in Texas, was going to speak also.

I was almost rendered speechless by the news.

"Oh my God, this is big! This is big, isn't it?" I asked.

"Yes, this is big!" said Steve in reply.

"Amy must have presented her report then, and this must be a reaction to it," I said.

"Yes, she did," said Steve.

"The day after one of the hearings in Denver I came back home and was all fired up, and I called Weston Wilson from the EPA to ask how we could try to undo the exemption of hydraulic fracking chemicals from the Safe Drinking Water Act.

He told me to contact Henry Waxman's office, and that's how I got in touch with Gilad Wilkenfeld. Patrick Barker and I had a long conference call with him, but I never knew it might turn into something.

He said that there were so many projects being presented at that time," I said.

How vividly I still recall that morning, sitting in Patrick's apartment overlooking the Colorado River and out toward the peaks of the mountains near Aspen. I remember how hard I had tried to convey the seriousness of the health issues surrounding hydraulic fracking chemicals: describing Laura Amos's case and Chris Mobaldi's among others.

"So is this about hydraulic fracking chemicals, about undoing the exemption?" I asked.

"Yes, it is," said Steve.

"I am so happy, this is such good news! This makes all the work seem worthwhile, after all this time! I am going to stay happy all day! Oh my God!" I was still stunned.

Steve said that the meeting was confidential, that I was to tell no one.

I found that curious, as I would have assumed that all governmental meetings were public record.

"Will the press be there? Can't Deb be there?" I asked. I knew how incredible it would be to have footage of this for the film…

"I think press will be there, but maybe afterwards. I'm not sure, but you can't tell anyone about it," said Steve.

"What does Lance think?" I asked.

"We had to ask the lawyers if we could go of course, and they said yes," said Steve.

I began to laugh.

"Of course they said yes! How could they not want you to go to Washington?" I asked.

Then I immediately realized that if the lawyer had been in the final stages of a delicate negotiation for a settlement, perhaps the Mobaldis would have been told not to speak.

I thanked Steve for calling to tell me about this hugely exciting news, and I swore I would not tell anyone.

Then I talked to Chris.

Her speech was quite hard to understand, but I managed to get everything pretty well although I had to ask her to repeat some things.

She was very excited about the upcoming opportunity to tell their story, to tell of what happened to her.

"Chris, are you going to speak?" I asked.

"Ah duhnt know; may spetch es so bahd this dies…" she said slowly.

"But Chris, you always do such a good job of rallying your strength when you have the chance to speak; you should say something if you can—the people need to hear you," I said.

We talked a bit more, and one thing that Chris said sticks with me.

"Et neds toe bey told; mo pay-pul weel day from these," she said.

The Mobaldis will call me when they get back from Washington.

It is rather difficult yet again not to let my hopes get too high, but at this point I want to believe that something positive and tangible will come out of this.

Saturday November 3, 2007

Finally, after the busy week there is a bit of time to write about what went on this week. As I went about my regular hectic workdays on Monday and Tuesday, it was in the back of my mind that the

Mobaldis and others were in Washington D.C. to testify about gas-related health concerns before the House Committee on Oversight and Governmental Reform. I knew that Amy Mall's report was also being presented, and that the entire event was combined.

Tuesday afternoon I got a message from Matt Sura asking if it was OK to give my name as a local contact to a Denver Post reporter who might want to talk to someone from this issue about health impacts. He anticipated articles would be released after the D.C. hearings.

I left a message at the WCC office that I would be willing to talk to the reporter.

The next day, the same day as the hearings, I got a call from a reporter from the Denver Post.

He said he understood I was cataloging information re health issues near gas and oil development. I said that I was keeping track of what I heard about, but I cautioned that I was not a scientist, just a landowner.

He said that was fine.

I gave Steve a brief description of the symptoms that many people seem to have in common: the bad headaches, the nausea, the exhaustion, dizziness, pains in the joints, hands, and feet, and more. I also told him about Dee and Susan's episodes of being knocked down by fumes, and the goings-on at Dee's with the well blowing up and her collapse after those fumes overcame her.

On Halloween just as I left work I got a call from Amy Mall, who was still in Washington.

"Tara, you're going to kill me! I was so busy I totally forgot to call you about the hearings!" she said.

"Don't worry about it, it doesn't matter. How did they go?" I asked.

Amy filled me in, and gave me the sense that the committee members had taken the testimonies from the landowners, the scientist, the environmentalist, and the physician seriously.

"How did Steve's presentation go?" I asked.

"He did great, at one point he got choked up and almost started crying…the room went silent when he spoke," she said.

"It always does, when the Mobaldis speak," I answered, remembering the numerous times in large hearing rooms that I have heard the sad and shocking story of Chris's health issues as those in the audience listened in stunned silence.

Amy continued to tell me what she thought the results of the hearing might lead to—not immediately she said, but perhaps next year.

"Does it include changes in exemptions for hydraulic fracturing chemicals?" I asked.

"Yes, it does," she answered.

"That's how I got involved with Waxman's office in the first place: to ask about the possibility of reintroducing Weston Wilson's whistleblower action on the EPA report that deemed the hydraulic fracturing chemicals safe," I said.

Amy said she would send me all the links to the meeting, plus a copy of her report.

After I got home I called Debra Anderson in New Mexico and filled her in on the hearings.

She was eager to see if she could get footage of the hearings, as Chris Mobaldi was the focus character of her film.

The next day I grabbed a local paper, the GWS Post Independent and to my surprise I saw the headliner article on the front page.

COLLATERAL DAMAGE

"DRILLING HEALTH CONCERNS DEBATED

Testimony in Washington suggests harmful effects

By Dennis Webb"

As I was busy with no time to spare, I scanned the first section of the article briefly, then tossed the paper into the truck. I would read it later.

There was no free time at all that day until I was driving home. I called Patrick Barker of WCC and touched base with him. We had been out of touch for a while.

We discussed the hearings, and the press response.

"There are lots of articles, the Glenwood Springs Post Independent, The Denver Post, Rocky Mountain News, the Montana? Gazette, U.S.A. Today, and probably more," said Patrick.

"What? U.S.A. Today?" I exclaimed, as I was feeding quarters into the newspaper boxes in front of the New Castle City Market. I had stacks of the three Denver papers piling up on the front seat of the truck.

When I got home I called Steve and talked about the trip to Washington, and also told him about all the newspaper articles so far.

"It was incredible," said Steve when I asked him what it was like.

How I wished I could have been there...

"Well, all the publicity certainly shouldn't hurt the lawsuit," I said.

"No, it shouldn't," agreed Steve.

Later that night I finally read the Glenwood Springs Post article. It was the longest of all of the articles, and for some reason I was holding off on reading it.

After reading and rereading the article several times, I folded the paper and set it on the carpet with the others by the sofa.

I was pleased. Susan Haire had said exactly what I felt to be the issue: the EPA has been stripped of its power to do its job of protecting environmental and public health.

And now, no one was protecting environmental and public health.

Also, no one seemed to care.

November 10, 2007

It is another in a series of unbelievably warm weekends, despite the fact that Thanksgiving is just around the corner. I cannot recall it ever being so warm this far into late fall; usually we were contemplating ski trips, shoveling snow, and dressing in thick winter layers to fend off the cold at this time of year.

Instead today I am outside in the barn in shirtsleeves contemplating the need to run the hose out to water some plants and shrubs as the ground is dust dry. Perhaps I will do that after I finish typing. I need to take advantage of the balmy weather and fill the llama barn stock-tank before the hoses all get rolled up for the winter and hung in the basement, and the outside spigots get turned off. After the big freeze comes, if we're lucky the back pump will still bring water up, and I can haul the warm hoses out each week and unroll them and fill the tanks. Otherwise it will be the cumbersome and time-consuming ordeal of hooking hoses up from inside the house and running them out through cracked windows to fill the tanks.

Although I gripe, I realize that I am very fortunate to have water at all, unlike the Amoses' neighbor's that went dry after the Amoses' water well exploded when the nearby gas well was being worked on.

And unlike some of my friend's situations, our water doesn't fizz, stink, show discoloration, or have an oily film on top of it. Our water just ruins coffee makers (I need a new one every year, despite frequent cleanings with white vinegar) and leaves a white residue on everything it touches.

We are still waiting to hear back on any conclusions from my co-worker's water test on their well water that showed levels of a variety of toxic chemicals and heavy metals.

Although the lab that analyzed the water sample sent the results, surprisingly enough they didn't send any comment with it, or any data to compare it to.

So the results have been sent to a research scientist, and we are waiting to hear.

This week out of the blue I got an e-mail from Joe Brown, the film producer of *National Sacrifice Zone*. He filled me in on the latest goings-on; his film was being shown this coming _____ in Denver at the Denver Film Festival. Also, the film would be shown this weekend in Crestone Colorado at that town's film fest. Joe was proud to say that the film won the third place at a Colorado Environmental Film festival.

He also said that some people from the Rifle area who believed they had symptoms similar to those of the people in his film had contacted him, and they were looking for someone to film interviews so they could send it to 20-20. Ironically enough, it was twenty people that were reported to be interested in talking about their health concerns.

Joe asked if Deb Anderson was still working on her film, and might be interested in following up on this.

I e-mailed him back, and said I would pass it along.

The other interesting news that Joe mentioned was the upcoming showing of Zach Fink's film at the Denver Film Festival. Zach had filmed with a number of landowners here regarding gas drilling impacts, including Dee Hoffmeister and the Amoses. I heard that he filmed at the Amoses' on the day they moved out of their home, and that was before they had reached a settlement with Encana.

Earlier this morning I called Deb Anderson and got caught up to speed on the film and other events. We discussed the recent Washington DC hearings; Deb said that she watched the two and a half hour footage of it and it was very good. She said that one of the senators brought up Laura Amos, and asked about her. He wanted to know if it was true that she could not testify due to disclosures she had signed with the gas company. I recalled that I had heard through the grapevine that Laura couldn't testify without being subpoenaed.

Deb seemed somewhat surprised.

Sunday November 11, 2007

Today it is becoming overcast; it is not quite as warm as yesterday. Ed the farrier came this morning and the horses got their hooves trimmed. Now it is chilly as I sit on the hay bales in the shed and type, with the big striped cat Mister snoozing beside me, curled atop a pair of old black and red plaid woolen pants that have become cat blankets.

There most likely won't be many more chances to type outside this winter—this is a fluke. Before I go inside tonight I better finish constructing some hay bale cubicles in the stacks in the barn for the cats; they are going to be looking for warmer places to sleep very soon.

Yesterday after talking with Deb I got motivated by some of her good news. It was obvious that things were moving right along for her, and she had gotten some very good funding news recently and was awaiting more news this week and in December. Although Deb was working out of pocket, the final costs for the film were going to be substantial.

"I don't think you're going to have trouble getting the rest of the funding." I said to Deb.

"No, I don't either, once funders see that you have a few names already behind you, then they're more likely to jump on board too," she said.

COLLATERAL DAMAGE

Deb was planning another trip up here to do more filming, this time with a woman whose family had received payment for damages to their water well from the Divide-Creek-seep-related problems. The woman's father had passed away not too long ago, from pancreatic cancer.

He was the man who discovered the original gas seep location when he was down checking on the watering access for his cattle on Divide Creek.

The woman's name was Lisa; I had met her about a year ago at an election celebration party in Rulison. She had told me at that time how her father had drunk water from Divide Creek, and that they had filmed him standing in the creek when the gas bubbling up from the seep was lit on fire. He was a Native American, and I got the impression from talking to Lisa that her father was a deeply spiritual man and that their land was very connected to their spirituality.

"We will never leave our land here," she had said.

As Deb and I talked, we both agreed that it was a good thing that four films were now being done simultaneously on the same issues regarding natural gas development. It spoke volumes that more than one person came up with the idea that this was an issue whose time had come, and the time was now.

"It wasn't just someone's Sunday night whim," I said.

"You know Deb, what is funny is that I started working with Joe Brown and his film and I wrote all about that in the book. I was so excited! I had them stay at my place, I bought food for the weekend and we went out and filmed and got sunstroke for several weekends. I would have done anything....Then when he was rethinking his plans and was going to put the film on hold, I was devastated. I remember telling him that if nothing else he could consider passing his footage off to you; I had just met you. But Joe e-mailed me back later and vowed that even if it was going to be a 'basement film', he was going to complete it soon. And he did! With very little money or experience; I can't believe that he did it!" I said.

"He did a great job, and he was so sweet about it. When he said that he wished he could do more, but now at least one thousand more people know about this issue from watching his film, that was great. He has done an incredible job," said Deb.

"Well all three of you are in my book, first Joe then Zack Fink, then you. It was all a progression," I said.

Then Deb told me that I should plan to release my book at the same time as the film, as they were so closely related.

"It all ties in together," she said.

As I heard her say those words I knew that it was true; yet I know I also was envisioning the huge amounts of work that would be involved. This was not the first time I had ventured into publishing, and unless you are very well-heeled it is a daunting process to say the least.

Nonetheless, I knew what a huge boost it would be to be able have a connection between Deb's film and my book as I went searching for a publisher. There was already one house in particular that I had found through conversations with other publishers.

Although the many, many steps that would be involved loomed in the way, from what I had learned about this publishing company and what Deb was saying it all made very good sense.

"It always helps to be connected to other things," said Deb.

I knew she was right.

"I am not going to worry about it; I will just trust that it will happen. I am writing about what is happening here, what is going on on the ground, with the people. And all the politicians and the scientists need to know what is really going on on the ground," I said.

We agreed to be in touch, and to meet up when she came here in the near future. I told her that Dee would like to see her too. Unfortunately, Dee was very ill with a sinus infection now. When I spoke with her on the phone several evenings ago she sounded terrible, and was having coughing fits.

People from Channel 5 in Grand Junction had arranged to film her last Friday, but Dee was too sick to do it.

"They should have filmed her anyways, and shown how sick she was! She says that the baby was really sick, and that she can't be anywhere near anyone who is ill because her immune system is shot," I said angrily to Deb.

Deb agreed.

Yesterday after our phone call ended, I went out and typed in the shed. As I typed I knew that Deb was in her film room, editing away at her final product. We both concluded that we inspired one another, but I have a sneaking suspicion that she works longer hours on it than me.

Nonetheless, I knew that she was out there many miles away, working on the same thing that I was. I felt in good company.

Now there are just a few hours of Sunday afternoon left. I need to get out there and accomplish some things before the sun goes down. How these weekends that seem so long on Friday night fly away so quickly, I have yet to understand.

Nonetheless, I got some good typing in both days, and that is the way a book is written. Several pages at a time; at a steady and consistent pace…

But, now it is time to work in the barns and move the hay and spend time with animals and relax.

November 25, 2007

Yesterday I went to Silt to the feed store to get grain; I was going to get enough to cover the time we would be away for a good part of December in Mexico.

When I was at the counter ready to put in my order for eight bags of four-way grain mix, I noticed that the woman in front of me was Karen Trulove. I had not seen her for a very long time, over a year, since we were up at her place filming with Joe Brown for his film *National Sacrifice Zone*.

After I put in my order, I went out to the lot where Karen was getting their flatbed pickup loaded with long posts, I was sure for their new corrals where they had moved some time ago to Carbondale, away from the drilling in Silt.

We talked briefly, as she was busy loading.

Karen said how she had still not been feeling too well, and that she felt she had aged about ten years from being near the drilling. She asked about Dee Hoffmeister and her face fell when she heard that Dee was still there and didn't seem to be doing well.

"She needs to get out of there," said Karen.

I agreed.

"This whole thing is going to boil up, you know," said Karen, looking at me straight.

"I know," I said, with equal certainty.

Karen told me she was living in a trailer at their new property, and she sounded very happy to be away from Dry Hollow and the fumes and the drilling.

As I drove away, I had no doubt that Karen's prophecy was true, but would anybody hear?

January 1, 2008

It is New Year's Day, and we have been home exactly one week now since our vacation to the Mexican Yucatan. Perhaps I am still basking in the warm after-glow of memories of the sixteen days spent in the far-off tropical country where the clear turquoise waves, "olas," rhythmically crashed against the white coral beaches. Somehow in the countless days there of swimming in the wonderful warm ocean waters, and sitting for hours-on-end just watching the waves rolling in, I achieved a quiet peace that has been missing from my life for some time.

We stayed at a huge and luxurious hotel smack dab in the middle of Cancun's hotel zone for the majority of the trip, and then stayed the last two nights in a small and modest hotel in the old Centro part of the city where the locals live.

Although high-end hotels are not exactly our first pick of vacation accommodations, Al's time share often lands us in some pretty splashy places where our Teva sandals and wet swimming attire are not the norm in the spacious glittering lobbies. Nonetheless, we always manage to have a fine time and mix up resort days with trips off the premises to more remote locations.

Using some good guide information we hit the dusty highway down south and traveled day after day into the quieter stretches of the Riviera Maya. We visit our favorite snorkel spot north of Akumal, and spent the day floating in the lagoon in the company of absolutely breathtaking schools of fish of all colors. And once again, just as I did last year, I was fortunate enough to spend time in the presence of a large school of angelfish, the size of dinner platters, and all swimming parallel to one another in the deep underwater canyon.

Another day we ventured down a small dirt road off the highway below Chemuyal and came to the isolated tortuga-nesting-sanctuary beach. At times we were the only people in the whole bay, aside from the local family who lived there at the headquarters caretaking the beach. Although they spoke absolutely no English and my Spanish (although inspired) is fairly sparse, we worked together cleaning beach trash for part of an afternoon. The Hispanic man remembered me when we returned another day; his face split in a wide smile and he spoke in rapid Spanish of "basura" (garbage) and his "ninas" (children).

TARA MEIXSELL

How I love being in that part of Mexico, near the beautiful ocean, even when it is stormy. One night it rained hard and the wind blew noisily in the night against the window of the high-rise hotel. The next morning some large windows in the hotel lobby facing the beach were smashed, and debris from a hotel far down the beach had washed up; entire decks littered the cement wall against the beach, and a good deal of the coral sand had been washed away by the currents, revealing dangerous limestone boulders where we had been swimming and boogie boarding just the day before.

Shortly after arriving at the hotel I struck up a conversation with the beach lifeguard, asking about the rip tides off the beach. He was friendly and spoke good self-taught English (he kept a Spanish-English dictionary in his pocket, and frequently looked up the English translations of words), and we all became friends and enjoyed many hours of lively conversation with him, both at the hotel after the beach was closed and then we also spent a few nights in town together after his work shift ended.

His name is Morgan, and he was a rare person to meet. In asking about his interests he told us that he enjoyed reading philosophy including Hegel and Kant, and that he liked Chopin, Mozart, Bach, modern music, rock-and-roll, and currently was very interested in Gregorian music.

One day on the beach I was describing to Morgan the work I was doing regarding impacts of natural gas development. I told him about some of the people who had health impacts, and described in awkward but understandable Spanish what happened when wells were drilled right by people's homes. I told him of the press, radio, newspapers, television and of the documentary film producers who were working on the subject.

At one point he asked me what I was doing about it, or was I just complaining against it?

"Son tines peoples jus complains, complains, complains about son-thing...but they dons do any-things about its," he said.

COLLATERAL DAMAGE

I thought for a minute, and then I answered him.

"Most people are blind to it, they don't see it; they are blind…" I put my hands over my eyes and looked toward the sky as I spoke.

"They don't have time to see it, they are too busy with their lives, their families and jobs, and watching TV. I am trying to make people see, and think about it," I continued.

Now, days later I continue to think about Morgan's question. What was I doing about it? And the larger questions loom: is it doing any good, and has the struggle been worth it?

I cannot ignore the fact that again, just as last year while on vacation in Mexico, I have returned to Colorado with a conscious change of attitude regarding my involvement with gas issues, and this book.

Last year I realized that I needed to stop writing in continuing directions as the saga will never end in my lifetime, and that I needed to begin editing what I already had.

Now, a year later I know that I need to find a publisher or get some assistance of a professional sort to take the ramblings that I have worked on so diligently to the next phase. I knew that even before I left. I have worked hard and very diligently for a year editing, but I need help from outside.

Last night I called the Mobaldis to wish them a Happy New Year, and to be sure that they knew they were in the national news in the Parade magazine insert to the Sunday paper just one day earlier. National news!

We had a good phone call, and they relayed to me details of Chris's most recent illness and hospitalization, when her feet and hands turned black and she became severely ill after taking an antibiotic following some dental work. She was better now, thank goodness.

"What accent has she been speaking in lately? I wonder if I can speak Spanish to her." I asked.

"She's mostly had a real strong English accent, then of course there's the French and Russian," said Steve matter-of-factly.

I threw my head back and laughed; I couldn't help it, it just sounded so absurd.

"Of course, the French and Russian....I didn't know about the Russian," I said.

When I talked to Chris, she indeed was speaking in a very pronounced English clipped accent.

We wished one another a happy New Year, and discussed our hopes for the future. They included continued and improving health for Chris, success for Deb Anderson's film (and eventually, with luck, my writings) and lastly a successful settlement of their court case. The date of June 2008 seems just around the corner, and one ever knows when an out-of-court settlement will occur, either.

I confessed to Chris that while in Mexico this year I had come to the unconscious but strong conclusion that I needed to step back somehow from the gas work. This didn't mean that I intended to stop; I just needed to get the focus back on my own life more. I needed to get rid of the anger born of helplessness, the bitterness and frustration that become so all-consuming. I had to trust that things were going to work out, and the ball would continue to roll forward.

A few months ago a woman I know from Grand Valley Citizens Alliance who had just sold her home south of Silt where they had lived for years with some pretty nasty impacts from the wells both on and surrounding their large acreage had said something similar.

"I have become a very angry person, and that is not who I am. I just need to get away," she said.

"You know, even if my book never really gets published, it doesn't really matter. I have been so lucky to come into contact with the people I have, like Lance and Theo and Deb, and all the work has helped. And there are more good things to come..." I said.

COLLATERAL DAMAGE

"Yes, there are!" said Chris, with joy and certainty in her strong English accent.

Only time will tell...

January 12, 2007

Although it is my plan to stop adding to this text, there are times like today, when something so important happens that I cannot help but write about it. I know that as time passes, and years on down the road I will want to remember things like the joy I felt as I talked to Deb Anderson this afternoon.

Except for one quick e-mail, we have not been in touch in a long while due to travel and holiday breaks for both of us. So today, after a string of messages left on answering machines, we finally caught up with each other.

To spare Al having to listen to yet another lengthy gas-related phone call in our tiny farmhouse, I took the phone as I often do out to the now frigid greenhouse. As Deb and I chatted and discussed the latest happenings, water dripped slowly from the Plexiglas roof onto the cement pavers of the floor as the sun began to briefly melt the deep covering of snow on the greenhouse roof. Outside, the fresh banks of newly fallen snow glittered like diamonds in the sun. Inside, despite the freezing air, pots full of cilantro, chards, parsley, and beds of straggly mint grew from the dirt raised-bed.

While we discussed many late-breaking events both in Colorado and New Mexico gas-related politics, and the goings-on at the latest meeting we had attended in our respective states, the biggest piece of news Deb told me caused me to break into a huge smile.

In Santa Fe, where gas development was encroaching into some very high-end real estate, the fight was growing rapidly between the landowners and the gas companies. Also, apparently the state's recent efforts to ban pits didn't sit well with the industry—they were threatening to sue the state.

But, here is the news that Deb told me that is so exciting—in just a few weeks a private invitation-only screening was being held at a theater in Santa Fe of her film *Split Estate*.

"Oh my God Deb! This is so exciting, I am going to think about this all afternoon, this makes me so happy!" I exclaimed.

Immediately my fears of Deb's film not reaching the large audience that I so hoped for vanished.

"I'm a little nervous because there is still so much work to do…" she said.

"I wish so much I could be there! I want to see it, not just the film but also the whole thing, the event. You must be so excited!" I said, imagining even as I spoke of a softly lit lush Santa Fe theatre with deep red velvet curtains full of affluent and important people, and Deb being there introducing her film.

Deb (who I know as a casual woman wearing jeans with hair arranged in a hasty ponytail as she squinted behind the camera), would be dressed for the occasion (for some reason I envisioned black something), and I knew how her eyes and her whole being would be lit from inside with the importance of the event. Oh, how I wanted to be there, and to see it all!

Then a crazy notion hit my brain, even though I had used up all my vacation time this year (over half of it going to gas and oil events—none of which I regret for one second): all I needed to do is get to an airport and fly to Santa Fe for the day. It was conceivably possible although not necessary realistic. After all, there were two airports just a few hours from here, and anything is possible.

We talked about the possibilities, although it felt like a long shot, of my going there for this private showing. If not this one, then there would be another, Deb was sure.

A memory suddenly came to my mind.

COLLATERAL DAMAGE

"Deb, I remember talking to you along time ago, about six months ago, about imagining the premiere of the film. I just knew it was going to happen, that there was going to be an event, and how it was going to be. When I asked you if you ever imagined it, you said you didn't dare to. Now I know it is going to happen, I am just certain. Do you ever think about it now?" I said.

"Well, maybe in the very back of my mind," said Deb.

Today I know that regardless of the very slimmest of possibilities of me actually going to this private showing of *Split Estate* that a great thing was going to happen. And it didn't matter if I were there or not, not one bit. The showing was going to take place, and the film was beginning to do its work with the one-hundred-and-fifty privately invited citizens.

Then Deb and I talked of deeper and more personal subjects, of our thoughts on how we had come to work on this issue and the connected energy and efforts of so many with whom we had come into contact.

"You know Deb, we work so hard and it seems so few tangible good things happen that it is important to be happy when they do. Like when you came up here and filmed the bill signings—that was huge. Sometimes we have to celebrate the good things and believe they are going to lead to something. I am going to be happy all afternoon about your film," I said.

Deb laughed.

"Tara, I love to talk to you, you make me feel so inspired," she said.

I knew exactly what Deb meant, for she does the same for me.

As I contemplate the writing that I have done on the same issue, this book in its first stage and the stack of audiotape interviews I have yet to incorporate, it is so uplifting to know that the film is on its way.

It is so helpful to have a friend out there in the world working on the same thing, in a different medium, and watching their project, like a jet on the runway, warming up to take off and fly high.

And also to know, that this friend has just as much confidence in my project.

It is hope, and confidence, even though I don't know where the path will go.

The important thing is that I do not have a doubt.

Tucked into my mirror's edge in the bedroom is a fortune from a Chinese cookie I got some time ago.

"Nothing in the world is accomplished without passion," it reads.

Another fortune I saved reads,

"Deep faith eliminates fear."

As I venture outside soon now after writing, into the first sunny day after several weeks of snowfall, my joy will follow me closely like a happy shadow.

When I talked to Deb today, before we said goodbye, I said,

"Deb, now you know you are not just sitting alone in your basement with a reel of film, wondering…does this one mean more to you than the other work you do, because you produced it?"

"Yes, it does," said Deb.

Now it is time to bundle up in warm layers and enjoy the ranch before the sun disappears behind the mountains in about one hour.

There are feet of snow to dig away from the tarp-covered haystacks; horses, llamas, and barn cats to feed and visit with; and barns to clean.

Feb 8, 2008

I got a call from Lance Astrella this morning; he had a question for me about the Mobaldi case. There were some contacts that he wanted to make, and he wondered if I could help. It was my greatest pleasure to do so, and I knew the people he wished to talk with.

The situation involved an incident that occurred some years back near the Mobaldi's house when there were some pretty big problems going on with the gas wells in the area.

The lawyers were interested in connecting the timelines between the appearance of Chris's symptoms and the blow outs at the nearby gas wells, and the subsequent contamination of some of the adjacent water wells.

As Lance talked, I hastily scribbled some notes on a piece of scratch paper.

During the call Lance said that the Mobaldi's case was looking pretty good, as long as it didn't get thrown out of court by the judge. He said it was likely going to jury trial as slated in June.

I was stunned by the news.

Never, never did I think it would ever go to a jury trial!

When I left work a few hours later, I drove around the block to a park near a school and parked the truck. I dialed the number of the landowner the lawyers needed information from about the well blow-out on their land in 1996. I had met her several times, but doubted that she would know me by name.

To my great relief she was extremely receptive and said that Lance should call back the next day when her assistant (who I also knew) would be there and she could get her hands more quickly on the needed information.

I thanked her for her willingness to assist, and she said,

"You can call me anytime."

When I had described to her the reason for my call, and that it involved a lawsuit regarding her former neighbor Chris Mobaldi, she had quickly said,

"I know about her."

I called back the law office and left a message that the calls would be welcome.

Now, what I am going to say next is very personal.

I sat in my truck on the side of the road, in the snowy suburban neighborhood near my office.

I held the steering wheel in my hands and tried to process my emotions; it was a combination of high hopes that the Mobaldi's case would be successful, happiness that I had been asked to assist in assembling proof that could be helpful, and the dreaded fear that the case might get thrown out of court.

The pendulum swings…

That night I had a long phone call with Sara Wylie from MIT: she had completed reviewing my manuscript.

We talked about options from here, publishers, putting it online, and the work that still needed to be done.

What a day…

Mobaldi Legal Case Ends

Saturday March 29, 2008

Today I had the phone call I have been anticipating for two years. I called the Mobaldis and talked to Chris, and after a few minutes of chatting in her strange accent, she told me the lawsuit was over.

My heart sank with disbelief; this was the worst we ever could have imagined. Chris was in tears as she relayed to me what had happened, and she said

"Our hopes were so high for this, things seemed so good…"

I was just stunned; apparently the case had been thrown out of court. The worst-case scenario.

So many things went through my mind as I sat numbly, holding the phone and listening to Chris crying as I sat outside the farmhouse on the picnic bench.

How could two years of legal negotiations end like this? While one steeled oneself for this, it was unimaginable when it happened.

Chris and I talked a bit more, and I said that regardless of the lawsuit, Deb's film and my book were still going to tell her story.

"Chris, when we were down at your house last summer filming, after we left the house Deb said to me 'this is so Erin Brokovich!' and she had a totally shocked look on her face.

"Then she said 'When I got my grant from the State of New Mexico to do the film, I knew that I was committed to doing it…but now I really know that I have to do it.'

COLLATERAL DAMAGE

"Meeting you made her want to tell the story of what is really happening around the gas wells, and that is what is important, that people hear what is going on," I said.

While I believe that, nonetheless the words sounded hollow in my ears as I said them. I was still in shock over the news of the failed lawsuit.

Chris said that Steve was at work, but he was coming home soon and then going grocery shopping.

I asked her to have him call me when he could.

Then we said goodbye, and I hung up the phone with a heavy heart. A very heavy heart.

Later that day…

A few hours later, I was just turning on the computer to insert the sobering end to the "Chris Mobaldi Story."

It had to be done—after all, so much of this book has focused on the long and slow progress of the Mobaldi case.

As the computer was powering up, before I could type one word, the phone rang. It was Steve.

"Did Chris tell you I called?" I asked, wondering if she also told him what she had let me know about the lawsuit.

"Yes, she did," said Steve.

We then began discussing details of what happened, and it began to dawn on me that I hadn't gotten things straight when I talked with Chris.

The case had not been thrown out of court.

Instead, the decision had been made to accept a non-negotiable settlement (amount yet to be disclosed) and not proceed to trial.

The Mobaldis did not even know the amount of the settlement at this point in time; the attorneys informed them that the check was in the mail.

Again, for a second time today, I was completely stunned—a total turn of events!

After hearing the devastating news just a few hours earlier, it was as though a magic wand had been waved and all had been fixed. Of course, I did not know the details of the financial implications (and risks) of accepting a settlement as opposed to going to court, but the bottom line is that a settlement was reached. And most importantly, the case was not thrown out of court!

The lawyers informed the Mobaldis that if they went to trial that the law firm would potentially be liable for hundreds of thousands of dollars of legal fees incurred by the opposition over the past several years.

The decision was made to accept the settlement, and end the case.

So, this afternoon, as I sit at the computer looking out over the mountainside while white clouds blow across the blue Colorado sky, I am content.

I knew today would come: the day I would get a phone call saying that the suit was over.

I never expected that it would come in this way, and I have no idea of the overall scope of how successful or unsuccessful the outcome actually was monetarily.

But that is inconsequential; after thinking for several painful hours that the case had been dismissed from court, now I know that an out-of-court settlement has been reached.

That is validation, not only for the Mobaldis, but for all the others suffering with health impacts from natural gas drilling.

MIT And The BBC

April 12, 2008 7:20 a.m.   Logan Airport Boston

Last Friday I got an e-mail from Laura Amos. She sent me an attachment; it was an itinerary for the Disruptive Environments Symposium at MIT. As I scanned the impressive schedule of renowned speakers, including scientist Theo Colborn and my friend Laura Amos, I felt a deep pang of regret that I would not be going. When I first met Sara Wylie, one of the MIT graduate students who helped create the symposium, she had invited me to participate. After I had contacted MIT in search of options for graduate student assistance with completing my own book, she had found me through a contact from a professor I had spoken with. That was over a year ago, and I had so looked forward to the possibilities of attending. But now it was so late; I had let it go too long.

But, when I got home from work that night I went into autopilot and I spontaneously decided to just go ahead and do it, however irrational it might seem. I just really wanted to go, and for many reasons. One important reason in addition to attending the symposium itself was that it was a chance to see my mother and father. They lived in Sudbury, Massachusetts where I had grown up, about half an hour from Boston and MIT. And MIT was very much part of my family's life story. My brother Tim graduated from there with a degree in genetic studies, and my father, who was a ballistic missile scientist and physicist, had taken classes there and also worked at MIT's Lincoln Laboratory research facility when I was a small child.

So I picked up the phone and dialed Sara Wylie's number in Boston. To my surprise she answered immediately, and she very happy to hear that I wanted to come. She said I didn't need to worry about enrolling for a spot online—she would take care of it. I then called my mother and father, and Laura Amos to let them know that I was after all going to go!

COLLATERAL DAMAGE

Now I just needed to book a flight, and pack.

The next morning Laura called and said that she and Sara had talked, and they wanted to see if I would do an introduction for Laura. Of course I would! This meant that I would actually be part of the symposium panel, too. I was thrilled!

I took a notebook and pen out to the big barn and sat on a bale of hay and cranked out a double-paged speech for what seemed like two hours. I'm sure it wasn't near that long, but I filled an impressive number of pages....I went in the house, called Laura, and read her my speech. She liked it, and gave me some feedback also.

The next task was to go online and book a flight and arrange for travel and a hotel in Denver. I was pleased that even at the last minute I got a great price on a round trip ticket to Boston, about $450. Things were going well.

Sunday I packed my suitcase, got as organized as I could, and prepared to leave for Denver straight from work.

SIX DAYS LATER

I am waiting for my flight back to Denver after attending the MIT Disruptive Environments Symposium. After the hasty decision I made just one week ago to come, I am so very glad that I did...

I know that I will never have this kind of opportunity again; things went very well. It will take time to absorb all that happened, but even after only three hours of sleep last night I still recall the incredible events of the previous four days. Especially yesterday, when Laura Amos and I spoke during our panel, which was first of the day. And it was a very long day...

We spent many hours in the hotel room the first two days we were there, rewriting and critiquing each other's speeches, editing, crossing out, and doing timed run-throughs.

I sat on my bed, and Laura sat in the chair across the room surrounded by piles of notes, scribbling furiously. In between we would talk about gas issues, the Mobaldis, other friends, politics and more.

On Wednesday we went with Theo Colborn for her endocrine disruption speech at Tufts University, then the next day it was off to Harvard Medical School in Boston. What an impressive campus....I chose not to even attempt to get a chair at that event, the security was pretty tight, like you were going into the Pentagon or something. Metal detectors, guards with wands, the whole nine yards. And the audience was packed with esteemed physicians, scientists, and Harvard medical students.

I cut loose for the afternoon and jumped onto the subway for a day of revisiting Boston. When my brother Tim was an undergraduate at MIT, I spent many days running around Boston, the Back Bay, Boylston Street by the Prudential Center, Beacon Hill, Hay Market and Faniel Hall. It was a fabulously sunny day; it couldn't have been more beautiful for walking through the city. It was pushing seventy degrees and I was warm in my long-sleeved dark sweater. After leaving Harvard Medical School campus in Boston, I got on the "T" subway and headed for Copley Square near the Prudential Center. My first stop was at a copy store where I Xeroxed my speech for the next day. I was worried about losing the one copy that I had, which would be disastrous after the hours I had put into it. After that, I then visited landmarks around Boston, poked in shops, admired the architecture, and had a nice lunch at a Quincy Market restaurant of fresh cod and coleslaw. The subway makes it so easy to get around town, and I was able see most of the places I had hoped to with no difficulty. Although it's been a good number of years since Tim's days at MIT when I spent the most time in Boston, the city still seemed familiar and home-like.

Later I went back to the hotel and regrouped and met Theo and Laura. Then we were off to the first evening of the symposium at the MIT museum.

Scientists, grad students, professors, activists, and journalists all mingled over a dinner buffet and then the first panel began. It focused on global warming, and after four twenty-minute presentations there was a question-and-answer panel. I found the speakers to be well polished; they had their talks down and no one ever referred to any notes. New York Times reporter Andrew Revkin spoke about the changing role of media in this technologically fast changing time, describing the trend toward Internet and away from newspapers. A world renowned MIT scientist (who I recognized from a television report) spoke about his timely prediction of the flawed New Orleans levy system just weeks prior to the Hurricane Katrina disaster, and a top governmental dignitary of Papua New Guinea spoke about the complexities and politics of deforestation issues in his country. Hmm, Laura's talk and my eight-minute introduction were going to be different. Well, we were who we were, after all...

That night, back at the hotel, we were at it again: reviewing our speeches and tweaking them. We were talking out things we had inserted; we debated over word choices to the point that it almost became meaningless. After all, did it matter if Laura said the word motherlode two times in three sentences? Would it really be better to use a different word? How about epicenter? Finally we agreed that we were done, we were just undoing and redoing things that were really fine the way they were. Late, after 1 a.m., we put in a wake-up call for 6:30, set the alarm as a back-up, and finally went to bed.

Five hours later the phone rang loudly with the wake-up call, and we were up fast. I made coffee in the room with the tiny coffee machine that only took special coffee pods. Even though I had planned ahead and brought good coffee and filters to supplement or replace the typically poor hotel coffee, there was no way I could use them.

I ran to the deli in the conveniently located Star Market next door and brought back two more cups of coffee. We dressed, got organized, and met Theo to leave for the first panel of the day. It was starting with us...

As the room in the museum began to fill, I spied my mother and father in the back. I ran over and gave them both a kiss and a hug, and told them that I was speaking first. Then the symposium began.

After I was mistakenly introduced as Laura Amos, I went up the podium and leaned into the mike:

"I'm not Laura Amos," I confessed.

"I'm here to introduce her…" and then I launched into my speech.

As usual, when speaking in public about gas issues, my adrenaline began to rise, but not to the point that I began to shake or choke up too much. I did my best to look up at the audience intermittently, and followed the text on my papers with my finger running down the side margin to help me not lose my place.

I did choke up a little when I came to the end, when I was introducing Laura.

When I was done, I walked back to my seat in the front row. The panel moderator, Professor Kim Felton from PT Rensselaer leaned over and whispered, "You did a great job!"

I leaned back in my seat with great relief, and then focused on Laura who was making her way up to the podium.

After Laura came Theo's scientific presentation and PowerPoint with the data on the toxins and their known effects.

Discussing how she couldn't get the list of chemicals in the fracking mix from the local fire chief, she reported their conversation:

"'I'll end up down a bore hole if I give them to you', he said."

Theo: "'My life is almost over; it's time someone shows them for what they really are.'"

(I think this rather chilling piece of information stunned others, not just me.)

Although the day was almost gruelingly long, all the panels continued to be very engaging. About 5:30 the symposium ended and the panelists and MIT students and professors were going to reconvene at an Italian restaurant for an ending dinner.

The restaurant was very small and L-shaped, and we packed the place. I grabbed a seat near the door, and sat at an empty section of the table. I wanted to just end up with random people as dinner companions, and see what happened.

I think in retrospect now that fortune was smiling on me…

As a group of us walked out into the rainy dark Boston night, we continued to talk. I will never forget when Kim Felton, the Rensselaer professor said to me,

"You are my hero."

We were standing on a curb along the sidewalk, headed to the Miracle of Science bar where some of us were going for a drink before leaving for our different distant locations. As we stood waiting to cross the street on the dark downtown back road surrounded by puddles, I was honored and humbled at the same time to hear such words from Kim.

Some of us said our final goodbyes, and Laura and I, along with the handsome and hilarious African American Professor from Dartmouth who had been part of out panel, joined some of the MIT graduate students who had put together the incredible Disruptive Environments Symposium.

Although both Laura and I were flying out early the next morning—and I literally had just a handful of hours before I would need to get in a taxi in the dark and go to Logan airport to fly home to Colorado—I am so happy we went! We had a hilariously fun time, swapping stories, joking, and the Dartmouth professor insisted that someone dance with him. The MIT students were out in full flare, and the place boomed with music. Laura was stunned over the students. "They're so geeky!" she whispered to me, her eyes wide with disbelief. The graduate students said that some people intentionally dressed geekily when going to the renowned MIT bar. Many awards for clashing plaids and fashion failures could have been handed out in that crowd.

We regretfully left the group who were headed off to a club for dancing, returned to our hotel the Meridian (formerly The MIT Hotel) and got to bed under the blankets with calculus formulas printed into the material. Oh so MIT...

April 16, 2008

Now, after being back in Colorado for four days I am beginning to come back to everyday reality. But yesterday I got a phone call in the morning that surprised me. A woman with a very strong English accent identified herself as being with the BBC. They were doing a film piece on natural gas development and she wondered if I had a few minutes. I said I certainly did, and after leaving my office and getting to a private location we talked.

They already had interviews set up around Garfield County, but they wanted a few more, specifically with people with health issues. I gave her some stories, some names and numbers (a few which I know by heart, and are people who are now among my close friends).

Zoe, the producer from the BBC was very interested in Chris Mobaldi's situation, but I told her that they were in the final phases of a two-year legal battle, now in negotiation phase. It would be doubtful if they could be interviewed.

Yesterday afternoon I called Steve, and he was in the hospital with Chris waiting for her to be disconnected from her IV so she could go home after her nine-day stay. Her blood levels were wildly off; the physicians suspected toxins in her blood, and they conducted internal examinations and finally did repair work for numerous hemorrhages. She was being referred to a specialist in Denver after this. I did not even want to guess at the medical bill, and she was uninsured after all the problems that came before.

Today I had numerous phone calls, from Rick trying to find Dee's exact address to meet the BBC for an interview, then the BBC wanting Lance Astrella's number, then calls to and from Lance's office confirming his interest in talking to the BBC.

COLLATERAL DAMAGE

Ironically enough, I was in my boss's office, with all my co-workers who had worked with Chris immediately prior to her pituitary tumor surgery over seven years ago. We were having a staff meeting, and I apologized ahead of time about phone calls that might come during the meeting, but I said that they were about the gas issues, Chris, and the BBC who was currently here filming. Their first question was about how Chris was doing.

I gave them an update on her recent hospitalization, and her release from there yesterday. They were saddened, concerned and shocked.

Then I told them that the Mobaldi's two-year legal battle was coming to a close, and they were receiving a settlement—although no one except the Mobaldis, their attorneys, and the companies involved would know the amounts of the settlements.

One of the staff members said,

"Well, money will never buy back her health."

Then conversation continued about how staggering all of Chris's medical bills must be, and how fortunate she was to have had such a dedicated husband by her side through all of this.

At five p.m. I talked to Zoe from the BBC again. They had a hectic filming schedule now, and she didn't stay on the phone for too long. She did say that they were shocked that no one was paying attention to things like the Hoffmeister's condensate tank explosion effect.

"There were toxic carcinogens burning, and all that black smoke! I read those hazardous materials placards right there on the tanks!" she exclaimed

I did my best to explain how well the industry deflected these issues: deny, deny, deny.

And if no responsible entity collected proof, then it was easy to deny. At least it seemed to be working quite well now, and has been for many years previously.

The last three nights I have been unable to let go of the events of the MIT symposium, and I have reviewed some of things that occurred there. Over, and over, and over…

Now tonight, I am tired and need to get some real rest.

So after I got home, I watched some national evening news as I like to do, and as it was cold outside tonight I delayed going out to the barns to feed the animals. Yesterday had been sunny, in the seventies. Tonight it snowed lightly on the way home. That's the Colorado mountains for you…

I scanned through the channels, and by chance came to a channel showing the beginning of a broadcast from the BBC World News.

I didn't watch the show; I had to go feed. I turned off the television, put on my barn coat and shoes and a ski hat, and filled the bucket with feed and a smaller container with dry cat food.

Climbing over the fence into the pasture, a thought struck me. I realized that after all this time of hoping and believing that this story would go national, it had skipped a step and gone international. The BBC World News!

I poured the grain into the bins for the horses and the llamas, then put out the cat food in the enclosed barn. In the big barn where I often write at the haystack, I threw flakes of green and crumbly alfalfa hay over the corral panels to the horses and llamas. The barn cats gathered for pats and attention and hugs as they do every night.

It was one thing to process the events in Boston and MIT last week, and now the BBC on top of it.

Who knows, maybe their piece would be small or big, but I realized now that that didn't matter.

The world news had come to cover this issue, of the little people here in Garfield County dealing with unrecognized health impacts from natural gas development. They were the forgotten victims of today's exploding natural gas boom.

As I completed my rounds in the barns, I also knew that I was still the same person, living the same simple life and that would not change. And for some strange feeling, that felt OK. This is where I want to be, and this is the life I want to have. If the BBC World News, the universities, the press, the documentaries can help us keep it, then we will be immensely thankful.

But, if our way of life is destroyed by the energy industry, then the story needs to be told.

Stake Holder hearings

Formal hearings regarding the statewide COGCC rulemaking were slated soon at the Denver Petroleum Club. Western Colorado Congress and the Environmental Coalition asked Dee Hoffmeister to be one of the few landowners who would testify before the COGCC that day. I was asked by both Dee and Frank Smith of Western Colorado Congress to assist her in writing a speech, and Dee also asked me to travel to Denver with her for the hearings. I agreed to both, and one evening after work I drove over to her place on Dry Hollow South of Silt and the Colorado River to begin the draft for the speech.

As soon as I stepped out of the truck in that hot and sultry evening, I detected a whiff of something. Dee came to the door and let me in, looking a bit wan and worn out as she smiled her ever-present smile. She explained that they had been fracking around the neighborhood and that the odors had been bad, and she had not been feeling well. We chatted for a brief bit, and I toured the now totally redone house that the Hoffmeisters had up for sale. Things had been repainted, tidied up, and the house had a generic new look instead of a homey lived-in feel. Dee explain that the realtor had advised them that this was the best way to sell a house—remove evidence of yourself and your family's existence so the person viewing the house could imagine themselves in the space.

We sat in the living room and I pulled out a notebook and began tossing out ideas for the content of the speech. Dee nodded in agreement and threw in comments, and I scribbled an outline down. Then we moved to the kitchen, and really began to grind the thing out. At one point I began to feel lightheaded and rather disconnected. I stopped for a moment and commented that I felt a bit strange. Then I began to nervously wonder if it was the Dry Hollow air getting to me—Dee had said the fracking odors had been bad that day—and I remembered the time we were filming up at the Smith's house with the French reporter and we had both gotten sick from the odors wafting off the frack pit they were pumping. That night I had been ill well into the evening; the worst I have ever felt.

In spite of my lightheaded feeling, I powered on, and after less than two hours I left with a decent draft created to send off to the Environmental Coalition attorney who was going to review it for Dee. We agreed to call prior to the date for the hearings at the Petroleum Club and arrange for carpooling together in Dee's vehicle.

A day later I raced home from work, finished throwing some respectable hearing clothes into a suitcase: a summer suit coat and nice dress pants, and a button down shirt. After a quick run-through of the barns, I jumped in the truck and we were off a short bit later in Dee's big SUV heading East on I-70 to Denver. We finally found the hotel where we grabbed a late dinner in an extremely noisy restaurant bar at the hotel. It was nice enough, and food was decent, but it was overbearingly loud with music blaring out of the sports-bar area adjacent to the dining area. Mike Freeman, a lawyer from the Environmental Coalition eventually found us after making cell phone contact. We began reviewing the draft of the speech over our unfinished plates of food, almost shouting at one another over the din to be heard. I finally suggested that we reconvene to the room to continue working, and thankfully we did…

It was a grueling evening; Mike kept asking Dee to reread her prepared draft from memory, over and over and over. I finally thought she was going to smack him if he said,

"Let's do it one more time…"

I suggested that we draft an outline of bullet points that Dee could reference during her testimony—they would serve as beginning points for her to jump off from though her testimony. It was a good idea, and we soon had the bullet points and we called it a night after running through that a few times. Mike asked us to come to the Coalition office at 7 a.m. the next morning in downtown Denver where we would again practice run-throughs of the speech prior to the hearings. After Mike left, Dee looked squarely at me and said,

"I was going to strangle him if he asked me to read through it one more time—I was getting close!"

We wound down, chatted a bit, and set the alarm early then fell into our hotel beds.

The morning came fast, and we were up and running and quickly off for downtown. We found the office, did our last four or five run-throughs of Dee speaking off-the-cuff, then together we all walked to the nearby Petroleum Club. We entered the posh building, and made our way to the huge auditorium, replete with huge chandeliers dripping yards of crystals. The Petroleum Club apparently had cash to spare...

I recognized different faces, both on the dais where the COGCC members sat behind their microphones, and in the audience of the huge auditorium. There was no love lost between some who sat in the audience, either.

After lengthy testimonies prior to hers, Dee's turn finally came before noon. She began her speech with an easy confident tone; she was doing great. Then she got to the hard part, where she talked about her health deterioration after being overcome by fumes and falling unconscious. Struggling not to weep, she continued to describe her long-term health problems, and how her quality of life and health had been permanently altered. The speech ended after a rather lengthy Q-and-A from the COGCC members.

The hearings were breaking for lunch, and I had a strong urge to get out of that huge hearing room and not return. Hearings with the industry are not always a pleasant experience. I did not need to hear anymore.

We walked down the sidewalk toward a restaurant destination where we were going to have lunch with a number of the lawyers and others before we returned to Garfield County. What a relief to have that over and done; those hearings are grueling.

A tall, handsome, African American gentleman in a nice suit came up behind Dee and said

"Excuse me—"

Dee and stopped and turned around to face him.

He looked straight at Dee and he said rather softly,

"I am _____ from B.P. I want to apologize to you on behalf of the industry—what happened to you should never have happened to anyone. I am sorry you had to go through that."

Dee thanked the man, they exchanged a few more words, and then we continued on our way.

That was a very memorable moment; things like that never happen. It was a rare event.

The industry apologizing to the injured landowner who had lost her health and wellbeing after being exposed to fumes near the gas wells and pits...

It was an acknowledgement of the reality of the human cost of natural gas development. And an acknowledgement by a member of the industry reaping the profits for what they were doing so close to people's homes.

And it was a moment of rare compassion—reaching across a great divide.

Rulemaking—The Legal Action Committee

March 6, 2009

What a week. After some confusion over the date of the hearing on the rulemaking that was going to be held at the Capitol, finally plans began to fall in place. The House Bill 1341 that was passed in 2007 and signed by the governor in 2008 was finally going before legal review. The hearing would be in the supreme court chambers at the Capitol in Denver. The commission reviewing the proposed revised oil and gas regulation promulgated by HB 1341 was the ten member Legal Action Committee comprised of state legislators.

Plans were made for landowners with health impacts to be at the hearing to give brief statements. But overall, it was expected to mainly be a hearing involving legal experts representing either the industry or the environmental and landowner groups.

In typical fashion, exact times were not pinned down literally until the day before for the beginning of the hearings. In a hasty scurry, the group driving down from the Western Slope of Colorado finally left at 6:00 a.m. I rode down with them, happy to finally be on the way. The evening before I had drafted the three-minute speech I intended to deliver if needed. I focused on describing the seriously impacted individuals living near gas development, giving about one paragraph each to the health problems endured by Dee Hoffmeister, Susan Haire, Karen Trulove, and Rick Roles. I asked for the passage of the revised gas and oil rules by the Legal Action Committee. Time would tell if I actually would read the speech before a body, more often than not my testimony was not called on. It was the impacted landowners who needed to speak. But, if they were unable, I was always willing.

COLLATERAL DAMAGE

Susan Haire Babb had intended to fly to Denver from Texas to testify, but due to a sudden illness the day before the hearings she could not come. Dee was out-of-state at a family wedding in Louisiana, and both Rick Roles and Karen Trulove could not get away from their ranches. Nonetheless, Rick signed a letter in support of the passage of the revised rules, and supplied a copy of his blood test that showed findings of benzene, toluene, xylene, and more in his system. Karen spent hours at her computer writing a letter in support of the need for and the passage of the rules.

Several experts intended to come, including one of the doctors and one of the lawyers who had initially testified for House Bill 1341 in 2007. Many calls had gone out to others, but I really had no idea who we would actually see once we got there. Nor did I know if the Capitol would be flooded with industry workers, as the hearings in Grand Junction this summer had been.

After arriving at the Capitol we found our assembled group in a large hearing room in the basement. The room was full of members of environmental groups, citizen's groups, and outdoorsmen's groups. There were well-dressed political types who seemed to be directing things, and we all sat in the rows of chairs as the day's itinerary was described. Industry had not brought in their workers at this point in the day, and we were told that the hearings were basically going to be legal in nature. If testimony was given that did not fit the nature of the proceedings then that testimony could end up damaging our chances for a passing vote for the rules. The many landowners with health and other impacts seemed a bit stunned and disappointed that they might have driven such a long distance for nothing.

The agenda for the hearings was lengthy, and it was sure to be a long day likely to last well into the evening hours. We split into informal groups and had directions given to us on how to go about most effectively testifying if needed. Although we did not expect to be called, we worked hard on the brief speeches we were prepared to deliver, crafting them to address the legality of the passage of the gas and oil rules. We were instructed that this was not a hearing about the need for the rules themselves—which is what we usually testify on.

TARA MEIXSELL

What ensued after that was for some a typical day at the Capitol, perhaps. But I do believe that even for those who are on the inside track there was something a bit special about it. What began over two years ago was now coming down to the last days, and regardless of your part in it—be it impacted landowner, environmental professional, stakeholder, lawyer—it was coming close to the finish and today was going to be a crucial day in determining the success and passage of the hard-fought-for rules that were designed and developed to protect public and environmental health.

For the next hours the group either remained in the basement area, listening to the hearings three floors up in the supreme court chambers by computer Internet links, or actually going into the chambers themselves and listening. Members of the press came and did their work. As the day promised to be long, I was in no rush to leave our basement headquarters as we were assured that staffers and others would come and inform us as to the goings-on upstairs. Also, we had the list of presenters up on a chalkboard, and it was a very long one. As the hearings began somewhat before noon, we began tracking the progress of who was speaking on the list. That way we knew when it would be time to go up and hear the people we were interested in hearing speak, or even more importantly—be aware of when we would speak.

That day I saw friends from all over Colorado who have been so instrumental in these rules, and many people who used to live in Garfield County but now live elsewhere. They are people I met at numerous gas-impact-related meetings, and people whom I have spent many hours with as we traveled together from the Western Slope of Colorado (usually at some dark, pre-dawn hour) to a hearing at the state Capitol. Somehow, these people who were once complete strangers become allies and an important part of a support system as we navigate our way through the at-times formidable legislative hearings. I don't quite know how to describe accurately the relationships I have with these people: it is not as personal as a regular friendship; it is something more that unites people together who believe in fighting for justice and a cause that benefits the future. There is something very beautiful in that, and something that is inherently good.

COLLATERAL DAMAGE

In the afternoon I went into the supreme court chambers and took a seat. People sit at the aisle seats of empty rows so they can come and go, and many people and legislators were going in and out of the chambers continually. Also, many were texting, on laptops or other devices. There was a lot going on behind the scenes…

The chambers were not as large as those where Dee and I had gone in Denver at the posh Petroleum Club for her stakeholder testimony this summer, but they were more beautiful in an historical way. There were panels of incredible stained-glass windows with Colorado scenarios depicted. The sun was streaming through and they were illuminated brilliantly. It was a high-end room, but I was focused on the hearings and my potential three-minute participation in it.

I went up to the third floor to hear Lance Astrella's testimony and questioning, and also to be ready for our Grand Valley Citizen's testimony. Three of us were going up to speak together.

When our time came, mine was the first name to be called. On autopilot, but calm, knowing that I had a good speech in my hand, I took a seat and spoke into the small microphone. I asked for the passage of the revised rules, and addressed specific rules that would reduce VOC emissions near schools and homes that have caused so many to become so ill. I referenced the rules that asked for no drilling within 500 feet of water supplies and baseline monitoring of the water supply. I asked for passage of the rule that would require disclosure of the hydraulic fracturing chemicals used in the event of an accident or a mishap so that physicians could know how to treat the patient (see speech of March 6-09/Capitol).

I asked the committee if I could give them further comment in written form, and I was allowed to do so. Additionally, Rick Roles's letter, his blood test containing benzene, toluene, xylene, and other chemicals—and also the NRDC's "Drilling Down" report regarding Rick—were provided to be entered into the public record.

After the three of us from Grand Valley Citizens Alliance spoke, we left our seats and walked down the aisle toward the exit of the State Supreme Court. Lance Astrella was sitting by the door we were headed to on a seat next to the aisle. He caught my eye and mouthed the words "good job!" and smiled. I cast my eyes up toward the ceiling of the chambers, in as much to say let's hope we did what we intended to…and didn't stray too far from the legal nature of the hearings today. I was extremely happy that Lance indicated we did well—that made me feel good about things.

On the way back to the Western Slope we talked about the day, the events that had transpired, and what would be the next plan of action as the House and the Senate did a similar legal review of the revised rules. Snow was falling lightly on the Continental Divide and the roads were slick, but we avoided getting caught in blizzard conditions, thankfully. We passed on having dinner in Denver, and ended up stopping at a gas station where a few of us got food items. What had initially been planned as a two-day trip with a night of hotel and relaxation between the lengthy commute from Denver to western Colorado again turned into a marathon day of driving, waiting, stress, and testifying—and gas station dining. Been there, done that…many times now.

When we were about an hour from my drop-off point we got a phone call from Denver. Frank, the Western Colorado Congress organizer who received the call and was driving the van, shouted "Yes, yes, yes!" I was in the third row back in the van, and I leaned my head over the seat in huge relief. We hadn't been set back today—the Legal Action Committee had voted down party lines six to four in our favor. They also had accepted none of the twenty potential objections to the rules proposed by industry. Apparently industry did not put forward all twenty before the committee.

The time flew and I was back in New Castle, in my truck, and soon home at the little farmhouse. Al greeted me and we hugged, and I said, "You would have been proud of me today." I know he would have. What I do and where I go are places different from what my husband's interests are, but I know if he had been there when I spoke my brief piece at the Capitol today, supporting passage of the gas and oil reformed rules he would have been proud. I am also so grateful that I have a husband who supports me in what I do for this cause.

COLLATERAL DAMAGE

Yesterday, in the basement of the Capitol, a box of T-shirts were being passed out. The box emptied quickly. The T-shirts said "I survived the COCGG Rule Making." There was a picture of a rig on the back. Regardless of how the final votes come out after the House and Senate legal review—I too have survived the COGCC rule revision. So many of us have given all we can give to this effort. It has to pass…

March 29, 2009

Over the next several weeks those of us from across Colorado who wanted the gas and oil rules to pass monitored the progress of HB 1292 as it made its way through the political machine. Next stop was in the House of Representatives where the bill passed in a great majority. That was such a huge relief, but we knew we had to keep the pressure on, just as industry was doing. Television broadcasts ran frequently, discussing the two sides of the issue. The landowners and environmentalists wrote scores of letters to legislators, first the House representatives, then next the Senate members. Also, e-mails went out to various forms of the media, from newspapers to television reporters and more.

I contacted Judith Kohler of the Associated Press who covers gas and oil news stories, and let her know that there were landowners from other areas in Colorado who had bad things happening from the wells around their homes. Judith had been ill and missed covering the March 6th hearings at the Capitol, so she missed the chance to interview them in person. I sent her the contact phone numbers for her to get hold of those landowners; to the best of my recollection some people actually had structures imploding from the gas.

Last Saturday Al came home from a day skiing and tossed a paper on the table. He showed me an article describing a Colorado landowner who was igniting the water coming from his tap like a blowtorch—amazing.

The next morning I called Judith's number at the AP and left a message that she would get the following day. I left the contact numbers again for the Fort Lupton landowners, and referenced the article printed in yesterday's Glenwood Springs Post Independent.

The woman from the Associated Press told me that actually Judith had written that article.

All day long on the weekend I wrote e-mails to each Democratic senator asking them to please pass HB 1292 unamended and stated my reasons for supporting their passage. I also sent five letters to the editor to numerous Colorado papers asking for the same. Monday the bill was slated to hit the Senate floor, and a big fight was predicted. The strain of working every weekend day, and every workday evening on supporting the bill, plus the continually prolonged timetable for a final decision was beginning to take its toll on me. Again the bill was delayed—on Monday I found out it was postponed one more day. Great, one more day of hurry up and wait, and one more day of coming home from work, and then working on supporting the rules. More hours at the computer, more e-mails, more contact information to the press. As I worked that night, powering through sending and receiving gas e-mails, I let out a cry of delight when I opened and read an e-mail from one of the key environmental supporters.

It turns out that Judith Kohler had gotten into contact with the Fort Lupton landowners, and yes, they were the same people who were lighting their tap water on fire. Additionally, several television stations had picked the story up and filmed footage was being shown around Colorado of gas-laden water igniting like a blow torch as it was lit by a cigarette lighter. Channel 4 had a story done by Shawn Boyd, the very same woman who interviewed Dee Hoffmeister this summer when we went down to Denver for the stakeholder hearings at the Petroleum Club before the COGCC. Dee had been the one landowner representing the Environmental Coalition. The e-mail I had sent several days earlier had been to Raj Chohan, Shawn's husband who was also a reporter at CBS 4 news. He had come to Rifle to interview Rick Roles and Dee Hoffmeister shortly after the huge fire at Dee's home from the condensate tank explosion.

Suddenly, I realized that I may have played a part in getting this story out at such a crucial time. The next morning Frank called from Western Colorado Congress in Grand Junction and told me a landowner had just called the office to let them know that Fox News had just broadcast the story of the flaming tap water. That was national news!

A bit later I called Lance Astrella's office and we spoke at length. I knew his firm was representing some of the landowners with the gas contaminations affiliated with the flaming water problems. As Tuesday progressed, I knew that the rules were being scrutinized simultaneously at the Capitol by the Senate, and I was a nervous and exhausted wreck. It was hurry up and wait drawn out to an almost tortuous extent.

Midafternoon as I was leaving the office to go to a work site, the phone rang when I was getting in my truck. It was Frank Smith from Western Colorado Congress, and he was screaming with joy.

"They passed!" he exclaimed.

"What!" I said to him, intensely relieved but also confused, as I knew they had to go through three stages with the senate, and we had anticipated that it would take several days at least.

"I just got a call from Chuck Mallock (the Environmental Coalition's lobbyist at the Capitol), and he told us that they went through the second hearing in the senate, and they passed with no amendments. The second hearing is pretty much where it all takes place."

"So the third hearing, is that just a rubber stamp?" I asked.

"Pretty much, then it's off to the governor's desk to get signed."

I was so, so, incredibly relieved—and a bit stunned. This is what I had been hoping so much for, and I could not believe it went this well. I also could not believe it powered through the senate in less than one day.

I went back into the agency and told our executive director Bruce Christensen the news—Bruce was also Glenwood Springs Mayor and he wholehearted supported the passage of the rules and better health protections for all who lived here. Bruce was equally as excited as I was, and then I went to one of my work sites and literally floated through the rest of the afternoon.

Later in the day I got a return call from Tresi Houpt, one of the Garfield County commissioners and also a recently appointed member of the revised COGCC (the revision a result of the passage of the initial HB 1341, which all of the impacted landowners had been so instrumental in pushing through).

I asked Tresi if she had heard the news, and to my surprise she hadn't. I told her about the call regarding the rules passing the second senate hearing, and she was extremely pleased. We talked at length, and I filled her in on some of the events that transpired at the hearings at the Capitol on March 6th. What an honor it was to be the one to tell Tresi that the rules had passed their critical mark, and victory was imminent.

After getting home, my previous exhaustion turned to exhilaration as I called some of the people I have worked with regarding better protections from the damages associated with gas production. How wonderful it was to tell those people such as Dee Hoffmeister, Rick Roles and others who have suffered so much—and who worked so hard time and time again with the press and the government—that the rules passed.

E-mails and phone calls continued for several days conveying the news, and I am so thankful that we succeeded in making some legislative improvements that will better protect future air and water quality for not only Colorado, but hopefully our nation.

These rules by no means are going to right all of the wrongs that come about from gas and oil development impacts, but it is a step in the right direction—and their passage shows that at least this time, in this multi-year effort through the legislature in Colorado—our government does care that people and the land are being harmed by poorly regulated energy development.

It is the ant against the elephant, and the David and Goliath story.

COLLATERAL DAMAGE

Energy development is important, and as this fossil fuel is being harvested it is also important that it is done in a responsible manner. My hope is that lessons will be learned from so many grievous mistakes made in the past, and that those whose health and lives have been compromised will not have suffered in vain.

Hollywood Here We Come

July 18, 2009

Ten days ago I opened e-mail from Deb Anderson. After a period of not hearing much about the progress of her film, suddenly here was a huge update. As I read the e-mail I began to cry. Al, my husband was sitting nearby in the living room and I gathered my emotions and explained to him the reason for my tears.

"Deb's film can be submitted for an Oscar; she is in a film festival in New York and Los Angeles!" I said.

We continued to talk about the details, and I was overwhelmed by a huge sense of hope, mixed with relief and validation of a sorts. After years and years of work on the book, work with the media, work with numerous documentary film makers (Deb being the one I had worked with most closely), now something was breaking through…

I knew from the content of the e-mail that Deb was going wild with doing everything she had to do to be prepared for the New York and Los Angeles screenings, and all the many details that were unknown to me. There were going to be two screenings a day in both cities, from August 7th to August 13th. That would be a total of twenty-eight screenings in all.

I was just so happy that she had at least gotten this far—what a huge achievement that was in itself—let alone what the future might bring. This was not just about a movie perhaps getting an Oscar; it was about telling what happened to people living close to natural gas wells. It was about Chris Mobaldi, Dee Hoffmeister, Karen Trulove, Rick Roles, and so many others. The most important thing is that is also about the future: will this snapshot of those individuals in

COLLATERAL DAMAGE

Garfield County help prevent similar tragedies elsewhere? All we can do is hope…

A day or so later I knew that I had to go to the film festival; it was a once-in-a-lifetime opportunity. I sent Deb an e-mail, and began doing computer searches for flights and hotels.

I have to go.

As all this is going on, I am patiently waiting to hear back from the first publishing company I sent a query letter to for my book. Somehow the movement of Deb's film *Split Estate* finally coming to fruition is very satisfying, and I am feeling confident in the face of huge obstacles. Publishing is no picnic, and I am well aware of that. Nonetheless, I am greatly heartened by this wonderful turn of events.

Today I called my parents in Massachusetts and told them the news, and they were very happy to hear it. I will confess that I feel a certain guilt for not putting my money and time into going to the East Coast to visit family—they understood that this would be a once-in-a-lifetime chance that would never come again. Just like the trip to the MIT symposium was, and that time I got to see Mom and Dad.

Well, now I am waiting to hear details. The end of the DocuWeeks Film Festival will be on August 13th, and that is the date I plan to target for my trip. How strange it is to suddenly be planning a trip to Los Angeles—destination Arc Light Studio in Hollywood on Sunset Boulevard. I guess I will dust the hay off and go see what it's all about. Even though I'll have to miss a raft trip this summer to pull this sudden trip off, I know it is going to be worth it. Maybe I can take a taxi to the ocean during the trip—I would like to. I love the ocean, and it would help make up for missing a long raft trip—what I have done every year for so long; my time to unplug from civilization. Dee Hoffmeister might come, too—after all we have done together, I hope she does.

July 26, 2009

Oh, what a week! Deb Anderson is on a massive—but short—effort to raise the $36,000 the film still needs to make it to DocuWeeks and deliver it to Discovery's Planet Green Channel. The technical work required to get the completed film screened in both New York and Los Angeles and deliver it for broadcast is expensive. It costs more than my five years of college did, I am realizing.

Well, every night after work this week in the scorching Colorado heat of July, I powered up the computer and did what I could to help send the call out to the network of gas-minded people and groups from Colorado and beyond. Simultaneously I was working on this book, and I called the first press I had a query letter with. I was excited to give them an update about the success of the film at this point, and told them that it could now be submitted for an Academy Award. I also let them know that I had been honored in the credits as both a consultant and an associate producer for Garfield County, and that I was making plans to attend the last several screenings of the film in Los Angeles on Sunset Boulevard at Arc Light Studio. "This is going to help my book!" I told Brianna, my contact at the press in Vermont. She advised me to send her the links to *Split Estate's* website, which contained all the information about what awards the film has already won, and also about the upcoming submission for Oscar contention.

I typically scan my numerous e-mails each time I go online, and since I sent the query letter out several months ago I keep an eye out for the publishers e-mail in my inbox. Right now things are hot and heavy with the local and national gas impacts issues; I get upwards of twenty new e-mails daily. From hydraulic fracturing regulations to Antero's plans to begin work on two-hundred gas wells inside a retirement community's boundaries—the e-mails are flying. As I worked my way down the list, opening and reading important e-mails, and deleting the junk ones, suddenly I noted the subject of one e-mail. It had an attachment and said "query letter." Then I realized it was an e-mail from the head publisher at the press I was querying.

Knowing immediately that it would either contain good or bad news, I continued checking my other e-mails. Either way, the news would stop me in my tracks. If it were bad, I would have to absorb that fact. If it were good, I would have to take a deep breath and get ready for daunting phase two of the submission process.

COLLATERAL DAMAGE

Once done with e-mails I wanted to open, I got to the publisher's message. With huge relief I read that they wanted a second submission; two sample chapters or the whole thing, plus a synopsis.

I turned the computer off. I needed to regroup, and absorb the news. I had planned for this, but I needed time now—at least a day or two. Then I would get to work again.

Several nights ago a reporter called from the Glenwood Springs Post Independent—he had heard that I was writing a book about gas and oil impacts. We talked at length, and I immediately jumped to the subject of Deb's film. I later sent him links to the website, complete with film trailers, pictures, and loads of information.

Yesterday, I was awakened from a dream by a phone call (a strange and vivid dream of gas- related political hearings, hotels by rivers, and traveling across Colorado for gas and oil issues and hearings—I recall there was a governmental headquarters for some politician, in some remote Western town near Utah. It was in an old-time gas station, sparse and dusty, and I was taking a nap in a very plain bed under a brown blanket. I was very tired and I just wanted to stay asleep.) My friend told me that an article about the film was in the paper. Apparently I was in the article, too. I was surprised; the reporter got the piece out in two days. That was fast...

Groggy from my Saturday morning dream, I got up and began the day. I had no intention of driving even six miles to get a paper. I knew I would see a copy. It was my day to avoid the truck, stay home, regroup, do ranch chores, and relax.

A co-worker called and asked excitedly, "How many copies do you want?" referring to the paper. She was a good friend who I often confided in regarding efforts—successful or not—relating to gas and oil work.

Next I made coffee, and called Dee Hoffmeister to discuss plans for flying to Los Angeles for the film. She had just returned from a long out-of-state trip: a family reunion. I had spent hours the previous weekend with Expedia getting itineraries and hotel and flight packages.

We planned to fly out of Western Colorado as opposed to Denver, it was cheaper and even with a stop in Phoenix it was faster. That four hour drive to DIA in Denver was not much fun.

Today Al left in the morning for Fairplay some four hours away. Our old house was rented out, and the dishwasher was broken. He had to go try to fix it. Also, he was going over to my dad's ranch house to move some furniture and belongings we had left there, where I used to keep my horse and llamas; the beautiful eight and a half acres that were abutted by a 2,000-acre corridor of Nature Conservancy property. There were endangered species and rare peat bogs in that parcel—Dad bought it for those reasons. In addition to seeing all the way to Pike's Peak, the ranch also had fabulous water rights and there was no gas and oil in the area. I always keep that in the back of my mind. Currently we had it as a rental property.

Today I went to town briefly, then came home and began to tackle the second phase of submissions for the publisher. I made good progress, but really think I need to talk to the publisher. I feel that I have done draft after draft after draft of the layout, potential chapter breakdowns, synopsis, etc.

It is a complex process.

Nonetheless, I got good work done, and after a few hours of working on submission material, I switched to this writing—of what is going on now, and the unbelievable idea that we are going to Los Angeles (and Deb too both New York and Los Angeles) for the film festival.

Although I feel a tinge of guilt for not going with Al to Fairplay for a grueling day of work on the rental and the ranch and hours and hours of driving, I am grateful to have the whole house to myself to write and listen to the music which has accompanied me so many hours and years through this book. Joe Cocker and Chopin, a strange mix perhaps, but for me very effective at enabling me to write.

This morning I rode the horse Doc, trotting in circles in the pasture. As usual, after I jumped off and petted him at length, he followed me up to the house where he received a few more carrots. Now, hours later, it is raining and nicely cooled down from the lengthy—and for me unbearable—heat wave that goes well into the nineties.

I am glad that I have gotten this writing time in, and now I suppose I should think about laundry and ranch chores before Monday morning rolls around tomorrow; oh so soon.

August 1, 2009

Yesterday while at the office I got call confirming that the film had safely passed the important hurdle of getting enough money to ensure it could be screened at DocuWeeks, both in New York and Los Angeles and that it could be delivered to Discovery. I was overjoyed by the news—and it meant that I was safe to book a flight to Los Angeles. My co-workers rejoiced with me once I showed them the film's impressive website, and all its accomplishments to date. They knew I went to Denver and elsewhere over the years to work on gas impact issues—but this was different.

A day earlier I had a call with the publishing company, discussing details for the second submission. We had a good call, and as I outlined the realities of where the project was—plus my timeline with the film activities—I felt a sense of relief. They were not expecting my material immediately, so I had more time to polish my outlines and do a decent second submission. They had no problem with getting it after the film festival.

The week's toll of working every night on fundraising for the film, and planning for my second submission to the publisher for the book had left me exhausted. I left work a few hours early, a combination of exhilarated and totally spent of energy. After making a few stops at stores, my last being in a grocery store where I shopped on slow autopilot (and I usually enjoy grocery shopping), I was finally and thankfully at home in the little farmhouse under the huge elm trees. Not much else got done that night, and I slept in this morning.

Today I again went online and checked into hotel and flight packages, wondering who would actually come. Dee couldn't, Rick might, and there was a chance of a few others coming. I had to make some calls and confirm things both with Deb and the landowners from here. If others were able to come, I needed to book a room with enough beds. My goal was to get a reasonably safe and clean hotel, fairly close to Arc Light Studios on Sunset Boulevard. That way the transportation to and from the showings twice a day would be a doable thing; not multiple long and expensive taxi rides.

This afternoon after a trip to town for an appointment, I told some friends about the film—they were eager to help donate money to assist with airfare to fly "the cowboy" (Rick Roles) to Los Angeles for the film festival. I thanked them, and directed them to the website. The donation information was all there. After returning home, another long call with a landowner who might go to the screening, but it sounds doubtful. I told her I didn't care if she called even after I was in Los Angeles; I would be happy to share the room and it would help on the cost too. Now I knew I had to pick a room with a few beds. Potentially someone might end up on the sofa, or even on the floor....We didn't have money to waste, and after all the years of traveling across Colorado to hearings, dealing with all types of media, hours and hours in vehicles on highways, hours of waiting for television crews, hours of being challenged by industry lobbyists and the like, we pretty much knew each other quite well.

Even now, after years of this, it is still interesting to ponder: These people are my friends; they know me and I know them deeply. But we are not regular friends, we are people brought together by an issue we have in common. There are no dinners with families, no cookies and coffee, no holiday card exchanges. Instead there is a shared deep devotion to a cause, and an intimate knowledge of each other's realities. You learn a lot about a person after hundreds of hours spent in meetings, discussing issues, dealing with every known form of media, and burning phone time.

Just before turning on this laptop I went into the barn and sat on a pallet with the barn cats, forked hay to the horses, and wrote drafts for chapter breakdowns for the book. It was hot but not too hot, and I swatted at the flies that wanted to feast on my ankles. Nonetheless, I love to write in the barn. It is where I wrote a good deal of the book.

The flies annoyed the horses, too, and I applied fly wipe before I came into the house.

Right now things are moving fast. I have to make financial decisions, and also weigh the need to go to Los Angeles for the film. And yes, there is the book. In more ways than one this reminds me exactly of how I reacted to the MIT symposium. At the drop of a hat, after opening the itinerary for the symposium, I had known I had to go.

This is exactly the same. Now I have to turn off this laptop and look into booking a flight and hotel to Los Angeles for Deb's film.

As I told my boss yesterday, when I come back I know I will come back to face reality—there will be barns to clean, the winter hay to buy and put in the barns, and the gas disposal well going in across the street. This is not going to solve all the problems, but it is so important that this film is telling the true story of what happens to people living near gas and oil wells. So important!

August 9th 2009

Tomorrow, after going to work in the morning, I will leave early enough to come home, do a last run through the barns and leave everything set up as well as possible for my four days away. Then, I'll go meet Rick at Dee's house in Dry Hollow and head to the Grand Junction Walker Field Airport about an hour west from here on Interstate I-70.

To the best of my knowledge all the necessary arrangements have been made; I ended up booking a nice hotel in Santa Monica right on the beach so we can relax and enjoy the ocean, in between the screenings of the film in Hollywood. It will be so much more pleasant to walk out of our hotel anytime at all and be right there at the beach, instead of somewhere in downtown Hollywood. I have my lists written of what I need to pack, and I am making little piles of things so I will be ready.

This morning my friend Dana came over and helped me do some work on the cats and horses, plus she is going to stop in midweek and help out when I am gone so Al doesn't have to do it all. We spent some time talking in the yard about the film, and the upcoming trip to Hollywood. When we finally came into the house, right by the sliding glass door of the living room a small and brilliant green hummingbird lay beside the glass. The bird must have flown into the glass, but luckily he was still breathing. I ran in the house and got a shot glass and mixed humming bird food into a small bit of water, and found an eyedropper. Dana gently lifted the tiny glittering bird into her hand and began offering it the red liquid. Amazingly, the tiny bird began to slowly drink, sticking his thread thin tongue out to lick the liquid at lightening speed. But, he still remained very still and she carefully fed him for some time from the eyedropper. We decided to let him rest in a well-fenced and cat-proof vegetable garden while we continued on with our business. Dana laid him gently onto the ground beneath some large plants where he was shaded. He squawked a bit, and flapped his tiny wings, but then became still. We would come back and check him after the horses got their vaccinations done.

Some time later, out in the barn, I heard a loud bird commotion coming from the fenced garden area where we had left the hummingbird. I took off running, and was surprised to see several brown humming birds on the garden fence, making quite a racket. The little green humming bird was calling back, and as I stepped within a foot or two, to my immense delight (and relief) he spread his wings and flew into the air, sailing past the blooming six-foot-high sunflowers and off into the great blue Colorado sky. I had dreaded the possibility of the opposite outcome. It is very hard to nurse injured birds back to health, but in this case luckily he was just stunned, not severely injured.

Our yard is heavily visited by humming birds of all types: aggressive gold-backed rufuses that dive bomb the other species and fiercely guard the ruby-colored feeders of sweet liquid that hang from the elm trees, ruby-throated humming birds, and emerald-green glittering hummers. From now on, anytime I see a green one whizz by, I'll wonder if it is the little one we helped out today.

After Dana left I went down into the basement and located my suitcase and what I hoped would be a small enough carry-on bag. I dragged them up the steps and brought them into the house. Soon my

COLLATERAL DAMAGE

slowly growing piles of things will be put into the suitcases—everything except the nice skirt I bought yesterday and another dress I have yet to hand wash. As usual, I know I will bring too much; every time I travel I always do. Nonetheless, so what if I have the room in my $30 extra suitcase? I'll stuff it full. This time I will not forget my sandals (like I did last winter on our vacation to Mexico), and as usual I will bring along coffee, filters (yes, I am a coffee snob), some snacks, writing material and lastly this laptop—along with all I will need. Maybe I won't write at all, but in the back of my mind I hope I do take a quiet break in between all the excitement to get my thoughts down.

Just like today, as I write here with unpacked bags behind me, I may want to capture this particular moment in time and in my life to remember it when it is still fresh. The last several weeks have been very exciting and at times tense. Now as the day of our departure is just a day away, I am feeling relief. The journey is beginning—the last minute plans and packing become the first part of the journey. What you take along (or forget) will affect what you do on your trip in some small or big way.

I am bringing a little surprise for Deb; I just tucked it into my hand bag and wrote a card to go with it. I can't wait to give it to her, at the right time in Hollywood. I know she will either laugh, or cry, or both…

Now it is time to turn off this computer, go outside, clean the llama barn, visit with the horses and cats, and then keep packing. Maybe not exactly in that order.…Also, we have to roll a big round bale out into the pasture for the animals to feed off; they about finished the last one off. Al made a loaf of rhubarb bread with the last of the crop, and right now the house is full of a smell of coconuts from the oil he used. It is a perfect August afternoon in Colorado: sunny blue skies, not too hot to be out—plus, the sparkling green hummingbird flew away.

August 11th, 2009

Chez Jay's—Ocean Boulevard, Santa Monica, CA

Well, we are finally here. Yesterday I went to work in the morning, and finally the much-anticipated trip really began. After checking the lists for the essentials, (bathing suit, laptop, camera, itineraries, and more, we were off to fly out of western Colorado on a nice cheap non-stop flight. It was Rick's first plane trip, and we spent most of the two-hour night flight swapping stories and laughing as the sparkling lights of the western cities glowed in the blackness below the plane. At one point I saw the colored lights of Vegas, I was pretty sure. We landed at LAX and fairly easily got a cab with a friendly North African driver who got us to Ocean Lodge Hotel on the waterfront at Santa Monica Beach. Our key was taped inside an envelope for the after-midnight check-in, and also the key was to room 217, not 210 (which I had booked). Soon we were in a room—a bit old and kind of small—but clean. At about midnight, the three of us went to the renowned but secret and self-proclaimed-dive bar/restaurant Chez Jay next door for a drink. We walked in just minutes before last call, and I had a Blue Moon and Rick and Dee had 7-Ups. The floor was littered with peanut shells, and the dark red walls were covered with old-time pictures of movie legends and celebrities. Soon we went to bed, or in my case, tried to. Dee and I shared the king bed, and Rick (who usually slept in a recliner due to the pains he endured) did his best in the double bed in the other room. I had called for extra pillows to be put in the room, and a coffeemaker, but they were missing. Rick can only sleep in a propped-up position due to his physical condition—but he thought with enough pillows he could get by. Although it had been quite a day of planning, anticipation, and travel, I was unable to fall fully asleep for what seemed like hours. In the night a car alarm went off near by, and I heard the AC going on and off. But in the am I had been in a deep-enough sleep to actually wake up; and knowing that we were finally here in California I was up for the day.

We set off in search of coffee, and as all savvy travelers do, asked people on the street where to go. We walked the promenade, replete with hedges shaped like dinosaurs spouting water amidst a pedestrian-only mall lined with trendy shops like Diesel, and ethnic restaurants of all types. Even at that early hour it was a people-watching mecca. A local recommended a good breakfast spot, "The Deli," where we had a high-end but excellent breakfast and many, many cups of coffee and tea. The prices were high, but the food and atmosphere and service were great, and they had a fabulous bakery. I knew I had to go there the last day and get a bag full of goodies to bring home to Al…

COLLATERAL DAMAGE

Santa Monica was fun to explore, and we admired the towering palm trees and other trees that were strange to us but beautiful. Hedges of pink and purple bougainvillea were brilliant along the roadways busy with fancy cars: Mercedeses, Lexuses, and others. The locals moved about in their fashionable clothes, tanned and stylish and trendy—and the street bums too were tanned but scroungy and carried backpacks taped with silver duck tape as they scoped the tourists for a hand-out possibility.

We found a big grocery not too far away and got waters, snacks, fruit, cheese and provisions to stuff in the small hotel fridge. Then, after learning about the local buses that were cheap and easy, we came back to the hotel and officially checked in. Due to booking us in the wrong room, the friendly front desk clerk gave us a free upgrade to a room with three beds and away from noisy Ocean Boulevard.

After moving all our things and newly acquired groceries, we scoped out the beach just a block away and checked the water temps then went up onto the oh so touristy Santa Monica Pier. We actually had a good time, bought some drastically reduced straw shade hats and a rolled bath mat to take to the beach. For the third time that day I told curious people why we were in LA, and all the people were quite interested in *Split Estate*, DocuWeeks, and learning more about the film. I gave them the website, and jokingly introduced Rick and Dee as the movie stars I was traveling with. Although we were a pretty humble-looking threesome, everyone seemed interested in the issue of natural gas development and the impacts that could happen as close as 150 feet from a person's home—on your own privately owned property. When they heard that the film could be submitted for an Academy Award in the seven-day DocuWeeks Film Festival currently running, they treated us with great respect and wished us luck.

We are taking a break for a bit, came back to the hotel, and planned to regroup, then go into town for dinner. I am writing for a bit before memories of the past two days slip away in the greater excitement to come (as memories always do if not captured at the moment). Plus, I am strategically hiding from the most intense sun of the afternoon before I go back down to the beach with my new hat and beach mat to enjoy the ocean before we head to find dinner somewhere. I want to enjoy being at the beach amidst the excitement of the upcoming film screenings and events. After all, there is no beach to gaze at and walk in back in western Colorado. And it will be four months before I can enjoy the tropical waves of the Yucatan again on our yearly trip.

The beach here is clean and wide open around the carnival atmosphere pier, and the views of the mountains to the north by Malibu rising from the sparkling water and crashing rollers are somehow strangely beautiful and different from any ocean I have spent time near. The closest view I can recall would be of Acapulco, where the mountains also rise from the sea. But this was not the same, as I knew I was in Los Angeles, California. That thought then led me into realizing that our unlikely trio was heading into the heart of Hollywood the next two nights for the film screenings—and the unknown.

I came over here to Chez Jay's to write on the laptop in the cool, dark red ocean-themed bar. I can see the intensity of the California sun burning the air outside through the one front half-door and small rectangular window above the bar. Dee is resting her feet (and mind I sure am, too!) in the hotel room, and Rick the cowboy is out taking in the sights and people of Santa Monica Beach. I almost want to laugh at the sheer thought of it: Rick from Rifle wandering around in the fast-paced California beach scene. I am so glad that I am not here alone, luxuriating in a solitary room. It is perfectly fitting that the three of us are here together for this strange and once-in-a-lifetime adventure.

After all the countless days spent in interviews with the press (newspaper, radio, television reporters, international press, and documentary film producers), testifying at hearings in governmental buildings, and enduring heckling and hateful stares from the oil and gas industry groups as we spoke of the horrific impacts that gas wells near homes had likely caused, this was a hard-fought-for few days to enjoy. And they will be days to remember, and treasure.

I know they will not happen again. I also know what we are all going home to Garfield County; this movie—no matter how big it goes—will not fix the problem. But, what puts a good feeling into my heart both about the book and the movie is that they will certainly help.

COLLATERAL DAMAGE

I cannot wait to see Deb, and if I think that being at the beach in a retro Santa Monica Hotel, writing in an old movie star hangout with red-flannel-and-rope-lined walls dimly lit with ship lanterns is fun—I can only imagine what the next two nights on Sunset Boulevard in Hollywood will be like. Talk about being a fish out of water…

But the very best part will be both seeing the film for the first time, and seeing Debra Anderson enjoying the event as the producer. I need to get the list from the fridge into the frame I bought just yesterday on my way home from work as I prepared to leave for the trip (that now seems like weeks ago).

Neither Rick, Dee, nor I have seen the completed film.

Just seeing the short trailer on the website made me go cold—the end moments were of Dee and her husband helping a weeping Chris Mobaldi down the walkway from the bill signing with Governor Ritter. Even though I had been there myself, seeing it again made my heart go sad and cold.

The quiet oceanside Chez Jay restaurant is now gearing up for the dinner crowd, and the waiter is laying silverware on the red-and-white-checked, cloth-covered, and soon to be candlelit tables.

Now it is time (no, not to be with the horses and barn cats) but to see Dee in the hotel room, and change into beach clothes, grab my new straw mat and hat, and head to the beach to the north of the pier and enjoy the California mountain views behind the waves crashing gently onto the beach.

August 12, 2009

Chez Jay Restaurant—Santa Monica, California

Last night we all had a good time downtown again; Dee needed to buy shoes that she could bear to walk in as she was in pain from too much walking in her sandals to keep up with us as we raced around the promenade and beach area. Finally we ended up at a nice Italian restaurant and ate great food, in spite of a huge family of kids screaming nearby the entire evening.

In our new free-upgrade room (with one more bed so we each had our own bed in which to toss, turn and kick all night), we settled in for a much needed sleep; late, likely after midnight.

The next a.m. I managed to make coffee (that I brought from home, with a filter), but encountered a slight problem: no cups in the room. Rick and I ended up drinking milk and coffee from washed out plastic water bottles, while Dee stirred her tea with her glasses (no spoons in the room either).

This a.m. Rick and I went to an incredibly huge and beautiful farmers market—it ran for blocks and the flowers and produce were absolutely incredible. It made me homesick for the Northwest where I had lived in those lush valleys for years. In dry western Colorado, there is no way we can grow things like that. We bought fruit and flowers for Deb (that currently are in a coffee pot in our hotel bathroom sink, waiting for the trip to Hollywood tonight.)

I went onto the Internet this a.m. to get the exact time for the screening at Arc Light Studio tonight, and I also checked my hotmail and got an e-mail from Deb saying there had been a problem: she was not flying in until the afternoon of the 12th (today) due to work that needed to be completed in NYC for Discovery's Planet Green Channel.

I e-mailed Deb all of our cell phone numbers, plus our hotel number and room number in the hope that she would call prior to this evening. Then we called the marketing director, Diane, in Santa Fe to see if she knew more about Deb's plans. Diane didn't know much more than we did, except that DocuWeeks would have one week of daily showings in Manhattan. That was such good news for the film!

When Rick and I got back to the hotel after having a fabulous lunch at a small Greek restaurant on the promenade, Dee was still in pain and had not left the hotel. We sat on the beds and talked and laughed for hours and watched Tim's Tools on TV—waiting for a call from Deb.

COLLATERAL DAMAGE

Rick was concerned about his new cowboy hat that he bought yesterday on Santa Monica Pier—while it looked whitish when he picked it out, somehow he failed to note the faint pink tinge it actually had. This seemed to be bothering Rick, as he usually wore a much beat-up and dirt-encrusted cowboy hat back in Rifle. Currently, Dee and I are sitting in the Chez Jay restaurant next to the Ocean Lodge Hotel; Dee finally managed to get down the steps and go in search of food. We are waiting for a cell phone call from Deb, and planning in a few hours to take a cab into Hollywood to finally, finally, see the movie. None of us has seen anything more than a short version of the first draft or the trailer currently running on the *Split Estate* website.

Rick is apparently out somewhere in Santa Monica—likely on the beach somewhere, trying to take the pink tinge out of his new cowboy hat. For some reason, he stubbornly refuses to buy a new five-dollar suitable hat; he is intent upon changing the color himself. When he said he was going to get chewing tobacco and spit on the hat, I told him that I wouldn't sit with him tonight if he had splotches of tobacco on his hat. Then he said he was in search of tea to dye it. Rick is a stubborn man. I am sure he is at this moment having a wonderful time here in Santa Monica somewhere on the beach doing his old-time cowboy best to authentically turn his pink hat brown—or at least light brown.

You have to trust me here, I am not lying: just a moment ago, Rick himself strolled into Chez Jay. It turns out he ended up going all the way back to the Vons store on foot about twelve blocks away, bought black tea and a spray bottle and dyed the light pink hat in the hotel shower.

So, aside from not having a clue where our film producer Deb might be at this moment in time—or what state she is in—we will be fashionably presentable for tonight's screening, if we actually make it there.

No, we will make it there, after coming many hundreds of miles to be here. In spite of the great fun we are having together, the real reason for this adventure is to be here to support the film. And we intend to do our very best to do so.

TARA MEIXSELL

Chez Jay Restaurant, Santa Monica, CA

August 13, 2009

Last night we took a taxi into Hollywood, through rush hour LA traffic. How strange it was to see the city landmarks passing by that I recognized vaguely from TV. When we turned onto palm-lined Sunset Boulevard I immediately thought of the scene from Annie Hall when Woody Allen and his LA-smooth friend in the white pantsuit were cruising in the huge convertible on Christmas day. As we drove we talked with our friendly taxi driver, Eddie, who was from Rio De Janeiro. He asked us what we were doing in LA, so we told him why and gave him a brief introduction to the realities of the impacts from gas and oil development (done badly) as we knew from western Colorado. He, like all the other locals we met and talked with, appreciated the serious nature of the subject and he said he looked forward to checking out the movie web address, which I handed to him on a scrap of paper. Finally the taxi climbed the incredibly steep boulevard up to the Hollywood Hills, and we neared the Arc Light Studios and the impressive cinema district. Within moments to my delight I saw that *Split Estate* was in white lights on the huge electric marquees outside the theater, along with the names of the other films being shown that night. It was almost time for the screening, so I ran inside and soon had our three reserved tickets in hand and we headed toward cinema thirteen on an upper floor. Rick helped Dee walk; she was having a terrible time with back pain. We went into the semi-darkened theater, with a few people sprinkled about the massive cinema that must have had a seating capacity of well over 100.

Quickly we greeted and hugged Deb, and Rick handed her the bouquet of colorful fragrant flowers we had bought her earlier in that day at the farmers market on Arizona Avenue in Santa Monica. Then the large theater darkened and the film began to roll on the huge screen.

As the story unfolded, tears occasionally rolled down my cheeks and I knew that Dee was wiping tears away. In the scene where Dee broke down crying as she described the horror of what she and her family were enduring surrounded by wells and their toxic fumes, I reached over and put my hand on Dee's for a moment. I just had to…I could tell that she was shaking with emotion as she watched.

I will not attempt to summarize the content of the film, but I can say that while being so happy to hear the hard truth being told about the gas industry's impacts, simultaneously it was difficult to realize that it was true that the industry was ultimately in control and governmental or political efforts to turn the tide were near to impossible against this extremely wealthy and powerful industry. Impossible, that is—I realize yet again as I type here for the last afternoon in Chez Jay on the red-checkered-clothed tables in the intimate bar café beside our hotel—except for the act of getting the truth of the situation out.

And oh, Deb had done an incredible job of doing so! The beautiful cinematography, the interviews with landowners and experts and industry players told the story so well, and I recognized many of the locations and film clips from the innumerable hearings I had been in attendance at down in Denver. In fact, I had provided the film tape for two of those hearings, having convinced my friend Nick Isenberg, a professional filmmaker to come along to the Capitol hearings and to the air quality hearings held in Denver.

We had an intimate Q-and-A session with the few who were there for the showing, and Deb sat in a black director's chair answering questions into a handheld mike. It was tomorrow night, the wrap night for DocuWeeks, when the larger group was coming to the screening and to the Q-and-A and after-party at Magnolia restaurant on Sunset Boulevard around the corner.

Deb invited us to sit in the theater restaurant and have a drink and some appetizers in the almost comically huge round dark-blue booths. We sat and visited and laughed for some time, swapping stories, answering questions, and just having a plain old great time. It was so good to see Deb, and I was very interested to hear the details of the inner makings of getting the film to the festival, and what would transpire from here. Deb seemed animated and beautiful as always in her quiet way; I had a hard time fathoming that she had spent the day in an airplane flying from the film festival in NYC where she had been for the past few days while simultaneously completing the necessary technical and business work to prep the film for submission to the Planet Green Channel (today!), and participating in DocuWeeks in Los Angeles, too.

Deb very graciously insisted on driving us all the way back to our Santa Monica hotel even though she was staying just a block or two from the Arc Light Theater. We drove on the now almost-deserted wide boulevards down out of Hollywood, and I looked with great curiosity through the tall hedges into the lit mansions (or what you could barely see of them) in the ultra-posh and lush Beverly Hills neighborhoods. The great trees were thick with foliage of some type against the dark sky, and they hung over the roads as though protecting them. The private drives that turned off into the dimly sparkling darkened estates epitomized massive wealth and luxury and privacy. What a distant reality from the lives we lived on our ranch properties in Colorado…

With relative ease, we got back to our Ocean Lodge Hotel just by the Santa Monica Pier. We bid each other a good night, and as Rick helped Dee climb the steps up to our second story room 106, I stood in the parking lot and watched Deb's small black rental car drive toward Malibu on Ocean Boulevard. We would meet again for the final showing.

This morning Rick and I walked south on the Malaccan and discovered some new areas, and watched the surfers riding the low breakers that crashed onto the shore off the bay. It was an interesting and more low-key area than that around the pier, and I planned to revisit the beach later in the day.

We bought some water and sundries at the first convenience store we had found near our hotel, and went back to see Dee who was struggling with immense back pain unfortunately. After perusing the yellow pages finally we made an appointment with a close-by chiropractor for her, and we called a taxi and went to town. Rick was already strolling the streets, and he came and met us at the chiropractor's office. I went on a tourist adventure then and shopped and took in the sights while Rick helped Dee get a taxi back to the hotel.

COLLATERAL DAMAGE

Soon I met them at Chez Jay for lunch, and we made plans for the rest of the day before we would head back to tonight's final showing of the film in Hollywood. After returning to our room, Dee prepared to relax her back for a few hours, and I got the laptop (that I stupidly forgot to recharge yesterday after draining the battery!) and headed to Chez Jay to write. Rick went on a ramble, and I was happy that the friendly waiter helped me plug the laptop into a convenient outlet in a booth. I told him briefly about the film, and what I was writing and why we were here. He was interested, and I passed him a napkin with the website written on it.

"The two people who I just had lunch with here are in the film, you'll see them if you watch the trailer," I said.

So, now I am winding down what is likely my last installment from Santa Monica at Chez Jay. The lunch crowed has thinned out and the place is almost empty. It is the lull between lunch and dinner—the time I have enjoyed each day since Monday hiding from the beach sun and writing quietly about our Hollywood adventure. The room is too claustrophobic to spend the entire day in, but Rick and I are being thoughtful of poor Dee who is laid up with her back pain after the first wild day of pounding the pavement a bit too hard in the wrong shoes. Although she bought some gym shoes on the 3rd Street promenade eventually, the damage had been already done.

Today is the last afternoon here, so I will go and drop the laptop off in the room shortly, check in with Dee, then stroll about and visit the beach before we have to get dressed and ready to go back into Hollywood. We arrived here about midnight on Monday, and now it is Thursday night. But all three of us agreed even after our first day here that it seems as if we have been here for a much longer time. In spite of our less-than-posh hotel (we are calling it retro), and Dee's back mishap, we are enjoying being in this oh-so-different environment. And what is pulling our unlikely adventure together is the backdrop of the film. How important it is, and even more so how wonderful it is to be here while Deb enjoys her well-earned moment in the sun…

Last night after we left Arc Light Studio on Santa Monica Boulevard, and Deb was headed down the sidewalk to find her rental car in the underground parking garage, Deb—dressed in a pinstriped dark suit coat, pointed heels and designer jeans—spun in a celebratory pirouette as she held her bouquet in her left arm. She gave a shout of joy; this was a moment to remember!

Well, it is about three and I need to get going if I am going to have a bit of a stroll, enjoy the oceanside ambience and beautiful town before we go to the film.

It is time to say a final goodbye to this fabulous Chez Jay, my unlikely writing studio this week on our Hollywood adventure. I hope these words I have been writing will help me remember: the red interior, the soft ship-lantern light, the wood chips and peanut shells on the floor, the huge stuffed fish behind the bar and the three model schooners in the private dining booth where the big stars hang out in the back, the photos and sea-memorabilia-strewn wall, and the friendly wait-staff discreetly serving the well-heeled Hollywood celebs looking for an few hours of anonymity. It is certainly different than the llama shed back at the ranch in New Castle!

Now I will go wander for my last afternoon hours in Santa Monica, where the porches and convertible Mercedes zoom down Ocean Boulevard while tourist and locals alike parade down the sidewalks to see and be seen—holding the occasion tiny pooch on the forearm like a designer handbag to show off. Somehow, Rick, Dee and I don't have the LA look—I am sure we stick out like sore thumbs—but at least we are real!

And back pain aside, we are having blast and tonight is yet to come!

August 19, 2009

It is Sunday morning, and I am having my first semi-normal day in week. Even though we arrived back in Colorado on Friday evening, things have been more than hectic. Instead of unwinding yesterday, I spontaneously decided to go with friends to the Obama rally in Grand Junction. So yesterday was another day of travel and strangeness...but before I get to that, I need to relay what transpired in our last day of the film festival in Hollywood.

Thursday afternoon I went down to the beach for a while to unwind and have some time alone. The three of us were having a great time together, but it was good to have a bit of space. Also, I wanted to spend a bit of time taking in the sights of the beach. I crossed the street in front of Ocean Lodge and went down a side street that went directly downhill to the promenade along the beach. Every form of recreation was being enjoyed on the cement pathway: bikes of every shape and size, skateboards, roller blades, joggers and walkers. The really athletic types wore tight spandex suits and their skin glistened with sweat as they performed their ritual workouts. I stopped at a very expensive hotel and had a cold drink in their patio restaurant, surrounded by wealthy guests who sipped drinks and ate salads and tapas in the lovely semi-outdoors patio. Lush trees hung over the room and I noticed a tree with green lemons growing right beside my table. I could have reached out and picked five or ten from my seat without even getting up, but my better manners prevailed and I restrained myself. I figured other guests would enjoy looking at them as much as I did—and besides, I had a bag of lemons in the room that I bought at the farmers market on Wednesday morning.

I left the restaurant and walked out to the water's edge to watch the waves and the surfers. Then I changed directions and headed north, past the pier and into the heart of town. After checking my cell phone for the time, I finally headed back to the hotel to get ready for the evening. It wouldn't take me long to get dressed, but I still had to put together my gift for Deb. I stopped in the lobby and got two pieces of scotch tape that I would need.

When I got into room 106, Dee and Rick were dressed and ready. I carefully assembled the gift and wrapped the frame in tissue paper. It went into the gift bag that had a few other mementos for Deb that I had bought over the last several days in Santa Monica, plus a thin blue scarf from Ecuador that I had gotten in Glenwood at the Splendor Mountain Spa when I had my massage several weeks ago. That had been a fortuitous decision, because if I had not gone in there I never would have discovered Allegiant Airlines—and I never could have afforded the plane tickets to fly all three of us out to the screening.

The clock was ticking and Eddie our cab driver from the previous night would be picking us up at 6 o'clock. I tied up my hair and washed my face, threw my sleeveless floor length summer dress over my head, and made myself as presentable as I ever get. I even applied the oh-so-rare brushes of mascara without making a huge mess of it or poking myself in the eye.

Rick popped back in the room from the balcony and announced that Eddie was there. I was surprised—it was just twenty minutes till six—he was early! We went and greeted him, and the first words out of his mouth were,

"I saw the film, and you were in the movie!" he exclaimed, gesturing towards Rick and Dee.

"Yes, I told you they are movie stars now!" I said jokingly, with a huge smile on my face.

Then I ran back to the room to get the last-minute things done, and soon we were off driving into Hollywood. This time Eddie gave us a special tour of Beverly Hills in the light of day, so we could see the neighborhoods we had driven through last night around midnight.

We drove past the infamous Hotel California where John Belushi had died, and Eddie pointed out the drive that led to the estate where Michael Jackson had recently died. A woman in her twenties pulled out of one of the posh neighborhoods in a sports convertible. She looked bored and anxious to go elsewhere, and I wondered what the details of her life were....Even the super-wealthy are still people, and they must share the same emotions as the rest of us do: frustrations, stress, happiness and misery, the complexities of families and relationships and ambitions and expectations. No amount of money can take those things away.

On the long cab-drive Eddie told us more details of his family's life in Rio De Janeiro, and about his wife who was a pharmacist studying to pass her American tests. Language was her main barrier; she had to learn to translate the prescriptions and medications in English. Her training had been in Brazil, so she needed additional courses to be certified in the United States. She was currently working at one of UCLA's medical schools.

Eventually we turned onto Sunset Boulevard and crawled along in LA rush-hour traffic. Along the way we took in the sights and people as they transitioned from work toward the night scene. Beautiful people, men in expensive suits and women in ruffled mini-skirts and stilettos met and exchanged hugs and cheek kisses in front of the clubs. We were up on the Hollywood hill, and LA sprawled below us to the west like a mirage.

Five minutes before screening time Eddie pulled up to the Arc Light Theater with its signature white dome. It had taken us and hour and half to get there—and we were less than ten miles from the hotel! What everyone had said about LA traffic was true after all, at least at rush hour.

We hurried into the theater to pick up our tickets and get up to cinema 13 on the second floor. Rick stayed behind to help Dee who still had trouble walking and I literally ran upstairs to the DocuWeeks information center. I got the tickets, and picked up some cards, fliers, and hand-outs about the movie.

Suddenly I saw Deb—she seemed somewhat distracted, and I told her that I had the tickets and Dee and Rick were coming. She looked at me and held out a DVD of the film and said,

"You need to give this to Duke. He's going to give it to Obama."

I just looked back at her and attempted to process what she had just said. Vaguely I recalled that Obama was coming to Grand Junction, Colorado soon to speak. I also realized that Duke and I no longer live in the same town or even county, so this might be a bit more difficult to do.

"When?" I asked.

"Saturday," she said.

I took the DVD of the movie and put it into my purse. I then knew that I had to begin making a plan. Tomorrow morning I would make some calls and figure out how to make it happen.

"I'll get it to him; we're flying into Grand Junction tomorrow night," I said.

But now it was time for the final screening, so off we went down the dark entrance to the huge theater. Deb held a large white cup of tea and confessed that she was very tired. No wonder: she had been in New York all week dealing both with the film fest and the final details for the Planet Green Channel. Even this morning she had been working on that.

To my indescribable immense joy, the theater held quite a crowd! I was so happy for Deb, and for us! Last night the almost empty theater had been like a slap in the face—even though we were riveted on the incredible film. In a strange way it was perfect—we had a private screening first (and a bit of a disappointment) —then the final big night was a hit!

Tonight we sat about seven rows back with Deb. This night I saw the film in a deeper way: I had imagined that it would seem repetitive after having seen it just twenty-four hours earlier, but it was like seeing it through new eyes. I was able to appreciate the expertise with which Deb and her technicians had edited and formatted the soundtrack and the visuals—plus the impact that the movie delivered—right down to the last seconds. It was a beautifully done piece of work. I was so proud and happy.

After the film there was a Q-and-A session, and Dee and Rick were invited to come down to the front of the theater. Dee and Deb sat in the high director's chairs and Rick stood beside them, along with an activist from California. The audience was very engaged, and the session was active and lengthy, but finally the DocuWeeks staffer said that we had to wrap it up—the next film was set to begin shortly.

COLLATERAL DAMAGE

In a celebratory mood a large group of us left Arc Light Studios and went down Sunset Boulevard to a nice restaurant bar, Magnolia, for the after-party. What a great time we had; people were so busy having impassioned discussions, exchanging contact information, ideas, strategies, stories and more, we never even ordered food until well after midnight.

I took photo after photo with the digital cameras, trying to get as many pictures of the people there. Deb would like that later, I knew. Although I had barely mastered the techniques of Al's new camera, I was having a great time moving about the lounge and framing, zooming and snapping frame after frame.

At one point Deb said,

"I'm so tired, but I can't leave until I talk to Keith."

She gestured toward a man with silver hair and glasses who was sitting on a stool at the bar with others.

"He's the one who helped me start my career, he was so kind to me," she said.

Then I knew we needed to order food. Deb could not run on exhaustion and exhilaration and pressure—she needed food.

I went up to the bar and ordered three appetizers: calamari, a cheese plate with fruit, and meat skewers. Others were also finally ordering food. Soon the group was dining on the large sofa area around the big square table in the lounge. It was just what the doctor ordered, and soon we were all fired up again and riveted in intense conversations.

I met Keith, Deb's mentor, and we must have talked for close to an hour about a myriad of subjects. He was a wonderful, imaginative and creative man, and from what Deb had indicated a hugely talented film editor.

Finally the party ended, we all hugged and said our goodbyes. Rick and Dee and I walked up the semi-deserted Sunset Boulevard resplendent with neon and towering buildings. At one point I look up at a street sign and said,

"Sunset and Vine—that's famous for something, isn't it?"

I began to laugh, and I said,

"Well, it's not quite Rifle, is it?"

Rick and Dee and I laughed as we walked toward Arc Light Studios to hail a taxi. But first I had to get to an ATM machine. There were some noisy partiers doing drunken antics on the sidewalk. One young man did a handstand on top of a cement garbage can next to the first ATM. I chose to walk a bit farther and withdraw my cash in a more private area.

Within seconds a taxi turned the corner as I waved it down. We got into the yellow cab driven by a small dark-skinned African American who was talking on his cell phone. I heard a glass bottle breaking behind us and laughter from the curb. Apparently the partiers were getting a bit wild. The driver almost pulled away from the curb before Rick was all the way in the cab, we had to shout to him to wait—as he still had the cell phone pressed against his ear.

Soon we were zooming through LA on La Brea, and the driver never put down his cell phone that he was chattering into in a foreign language. We went through low-rent neighborhoods, with sad looking people standing on the curbs in front of run-down establishments and chain-linked vacant lots. It was like the epitome of the opposite of what we had seen last night on the road home through the wealthy part of Beverly Hills.

To my right I saw a sign that said "Pinky's" and a large crowd waiting outside the small building. That must be the famous Pinky's hot dog stand. I gestured at it and told the others to look.

The driver finally disengaged from his cell phone to look my way and say,

COLLATERAL DAMAGE

"Yes, Pinky's"

Then I asked him where he was from. Somalia, he answered. I asked him how things were in Somalia.

He looked straight at me for a second and said,

"It is over. Terrible things are happening."

He then continued his quiet and mysterious African cell phone conversation as he tore through intersections at a high speed in the seedy LA neighborhoods off La Brea. He definitely abided by the rule that yellow meant to go fast, and he even kicked it up a notch by driving even faster when the lights were red...

We eventually got onto the five-lane freeway and sped toward Ocean Boulevard and our hotel. As I got out of the cab, I knew that the driver was living in a world we could not even comprehend: the atrocities of Somalia, his family, his village, the killings and more....To him, we were probably just a bunch of Americans dressed up for Hollywood and oblivious to the ugly realities of the world.

Back in room 106 we all attempted to wind down, but for me it is always difficult. Every night I had been up late, unable to fall asleep.

Finally I took a long hot shower, and Dee and Rick were already in bed. Rick was in his room with the door pulled close, I could see the flickering blue TV light through the thin crack around the door.

The shower had wound me down, and I dried off and put on my nightgown and robe. I went into the room and thought to check the door. To my surprise, we had failed to lock the deadbolt, and door was even a bit ajar! I pulled the door shut and flipped the deadlock, but suddenly the handle was shaking and someone was pushing the door from the outside.

I was terrified—and I began to pound on the door and scream,

"No, no, no!"

Then I screamed repeatedly for Rick. I thought that someone had gotten the door open, and they were coming in to either rob or kill us, or both.

To be perfectly honest, I don't know what exactly transpired next, but it turned out that it was Rick who was trying to come in—he had left his room while I was in the shower to go out for a last smoke on the balcony.

Well, all the relaxation I had acquired from the long hot shower was gone; now I had adrenaline coursing through my body on top of my excitement of the entire evening's events.

We laughed and laughed and laughed—then Rick went to his room and poor Dee had to listen to me talk as I attempted to wind down. Finally after three a.m. I attempted to go to sleep, with my earphones pouring Mozart into my over-stimulated brain.

The next morning about seven a.m. I awoke, and pulled back the curtain and looked out onto the foggy Ocean Boulevard where the endless cars drove by. Our last day...

I got up and made coffee, and went down for a last walk on the beach. Down on the bike path the regular sportsters were out doing their thing, and numerous dogs were confidently walking their owners—the leashes gripped firmly in their teeth. They knew who was really in charge.

I snapped some photos of landmarks, palm trees, and dogs.

Then back to the hotel to get ready to leave.

Eddie came at eleven; our flight left LAX at 1:30. We drove by the famous Venice Beach and the colorful neighborhood, and soon we were hugging Eddie goodbye. He had been so generous and genuine to us. In fact, so many people we had met had been the same. When they asked what brought us to LA, we told them. Then I handed them a scrap of paper with the website. They all took us quite seriously—strangely and ironically, because the people from Garfield County seem to be blind, deaf, and dumb to the realities of what is going on in their own back yards. Go figure...

After getting through the lengthy security, and waiting at the gate for an hour or so, we were soon on the plane flying back from LA. I was tired but happy, and I dozed and listened to the Eagles' "Long Road Out of Eden" CD through headphones. It seemed very appropriate.

Duke was waiting near the gate, and we had a quick hug and hello, and I gave him the DVD that I had kept in my handbag since Deb gave it to me the night before in LA. There had been a surreal sense of importance since that DVD was put in my hand...just like the surreal sense of hope the landowners out here have—if they have any left at all, and frankly, most don't.

Nonetheless, I will not forget the fact that I never let my black handbag with the DVD out of my sight for that night and day once it left Deb's hands, and as we flew from LA to Colorado, it was under my feet and under the seat in front of me. I did have hope in the power of the message of the reality of the film, and going to the very top of our government—the president—it was a big deal to consider.

Hoff, Dee's husband, met us at the airport. We grabbed dinner at the Olive Garden, then did the final drive back home. By the time I got into my white pickup truck I was very, very, tired.

Soon I was home, checked on the horses, llama, and cats, and briefly told Al some of the highlights and gave him the gifts I had bought for him.

There was a t-shirt from Santa Monica Pier, the end of route 66. Also, the bag of lemons from the farmer's market, plus a bag of Cajun-spiced sliced almonds, and a bag of cookies from the fabulous Broadway Deli on the Promenade.

In spite of my exhaustion, I still had trouble going to sleep. I listened late into the night to music through the headphones while lying in bed.

The next morning, I woke up about nine or so. Suddenly I knew I wanted to go to the Obama rally. The night before, I had talked to Leslie Robinson, my friend who had tickets to the town hall. She was going in the morning to be part of the rally. I called another friend and arranged to be picked up in New Castle.

Soon we were off, strangely headed (again) to Grand Junction. We stopped at the peach festival and I bought a t-shirt that said "Homeland Security" above pictures of well-known Native American chiefs. Below the photos it said "fighting terrorism since 149'". I needed a t-shirt; I knew the rally would be very hot, and I had failed to grab something cool to wear.

The rally was pretty crazy, and the Republicans were on one side of the street and the Democrats on the other. Police walked up and down the middle of the street to keep things calm, and for the next three hours we rallied. There were shouts of hate, accusations, chants, singing, and passion. When Airforce One flew over we all went crazy, screaming and jumping and waving our posters toward the huge white and blue plane in the sky that carried Barack Obama. We didn't know if his entourage would drive by the rally, but we saw his plane. I hope he saw us, and knew we were here for him.

I was so glad that the organizers on our side had music. At one point when things were getting ugly, I went to an organizer and said,

"Put on some good music right now! Otherwise, we are going to keep fighting with them."

Immediately Beatles music permeated the scene, and the angry screams were overpowered. The opposition retaliated with air horns, but eventually they seemed to unravel, and trail off to their vehicles.

Again, I arrived home tired and very sunburned this time. My throat was sore from the chanting—Oh-Bah-Mah! —over and over, and more. Before I took a bath I called Leslie to find out how the inside had been, and shared how the rally had ended. She had been there at the beginning. I had no idea where the DVD of the film was that I had kept under my feet on the plane from LA, safe inside my handbag. I had no intention of losing it.

Strangely enough, even if Duke didn't succeed in getting it to Obama's people yesterday in Grand Junction, I am confident that the president will see it soon. I know the route to take to see that that happens.

Well, I haven't done laundry yet, and I have been typing here for hours. The cats are on the porch waiting for me to come out, and I need to clean the llama barn at the very least. It is a beautiful day, autumn is in the air and it is pleasant—not too hot but very gusty winds.

It is time to go out and see the animals and do chores. I am home—tired, sunburned from the rally, and processing this whole incredible week. Who knows what the future brings…

The Big Time

One should never make assumptions in life, as things can change.

I thought that my active writing in the present was finished in this book—even as days and hours of grueling final edits continue, sometimes seemingly endlessly, and with a fair amount of pain and struggle.

Strangely, the closer I got to the finish wire of this book, the more the paralysis took over—a strange and unusual phenomenon for me, as I have never suffered from writer's block.

Suddenly, this was different, and I was paralyzed with the pressure of the need to finish. I had imagined it might even be nostalgically pleasant—but it was not. It was pure torture for a few awful days.

The last nine days have been a roller coaster of emotions—and before that memory slips away, I will jot down a few last pages in this book that has consumed my life for over four years.

I cannot omit the sad fact that nine days ago we learned that an old close friend and ski buddy from the years we lived in Breckenridge had very sadly left this earth in a very tragic way. As I told a close friend shortly after learning the news,

"I feel like a page from the photo album of my life has been torn away."

To finish the final edit I took three days off work while Al was away on the first chilly raft trip of the season on the Salt River in Utah. I had strategically planned this, as I knew that solitude in our small house was a rare and necessary commodity for completion of the final edit.

COLLATERAL DAMAGE

But, once alone in the quiet house, I choked on the pressure. Luckily after a day or two of hard hours of struggling, I got back up on that horse for two marathon writing sessions to complete the final edit and compression of files. Fourteen- and fifteen-hour marathons.

The last day I was spontaneously asked to participate in an NPR radio show focusing on natural gas issues. Hmmm, that was a break of sorts but not really a change in subject.

Finally, at about 5 a.m. on April 1, 2010, I was finished. Or so I thought...

It literally looked like a strange bomb had gone off in the house: there were manila files, white papers, pages of edit notes (many versions, all strangely similar) everywhere. The kitchen table was completely covered, as was the living room sofa and even the master bedroom and bed. Both computers were running, and multiple flash drives lay on the tables and desks.

Finally, I was done for all practical purposes—and the sense of relief and completion was huge.

I now realize that until this manuscript is sent for the final printing, I will never know if it is totally finished.

Yesterday morning a group of handpicked landowners and activists met at a house in Silt in the heart of the gas fields to talk with Denver Mayor John Hickenlooper, who is running for governor in the upcoming election. Rick Roles came over and picked me up, and after throwing everything I needed for the multiple events into the crowded cab of my small Ford Ranger we headed out for the day, which would end with meeting up with Deb Anderson for the awards ceremony in Beaver Creek for the Vail Film Festival. She was flying in from New Mexico that day, and we couldn't wait to see her again...she is a very special and sweet lady; all the landowners truly love her.

Unfortunately, Deb told us that we had to drive to a theater in east Vail to pick up the tickets we needed to get into the awards ceremony that was being held miles away in Beaver Creek. We parked the truck in the underground deck, then proceeded to get lost multiple times trying to find the ticket pick up location.

At one point we went into an information center, and to my huge surprise I saw Deb being interviewed live on a screen televised behind the desk.

I grabbed Rick's arm and said,

"Look, it's Deb!"

We got directions, then went out into the maze of shops and restaurants and finally found the theater. By this time, the clock was ticking closer and closer to 5 pm, the beginning time for the award ceremony. We tore over to Beaver Creek on I-70, finally found the Vail Plaza, hastily parked in yet another massive underground parking lot, then ran off to try and find the venue. After a seemingly endless series of bad directions from multiple people, after going in and out of the same buildings only to be directed back to the one we just left—in massive frustration and fear over missing the awards ceremony, I got a concierge to take us to the venue.

Finally we were there, and soon we were hugging Deb. We brought a beautiful yellow bouquet for her that Rick picked out, with yellow and brown orchids and mums and greens. She left it in a coat check office, and off we went into the huge theater for the ceremony.

How relieved I was to finally be there! I was exhausted after a very grueling and sleep-deprived week and I leaned my head against the back of the theater chair and kicked my shoes off. My feet were hot from having literally sprinted though the resort during the last ten minutes.

The presentations for each category were short, only the first-place winners received awards and came to the podium. Second and third were recognized verbally.

The presenters were witty and engaging, and the audience was very involved. It was a surprisingly massive theater, and it was packed.

COLLATERAL DAMAGE

Suddenly, the Activism category was announced. I braced myself and wondered: was this our category? I had never heard of that category before—usually it was simply "Documentary" category.

Before I had time to think, I heard Debra Anderson's name and *Split Estate* being named as the first-place winner for activism films.... Oh my God!

I watched Deb get up, and then she gestured quickly for Rick and me to follow her down to the podium. We followed her down the steep carpeted theater steps, then up onto the stage by the podium. I gave Deb's I-phone to the presenter and asked him in a hushed whisper to please take a photo.

Deb spoke a few simple sentences, introduced Rick and myself, and gave her thanks for the award. Simultaneously I pulled Rick beside me and put my arm around his shoulders as we stood by Deb for a quick photo. I realized that I was just wearing black socks; my shoes were discarded up at my seat.

I stepped up to the mike and simply said,

"Thank you."

We floated back up the red carpeted stairs to our seats, totally overcome with joy. I could have wept.

The filmmakers beside us gave their whispered congratulations.

Within a handful of minutes, they too won first place in their category. We decided that we were in a very lucky row of that theater.

The awards continued and were very engaging, but I was itching to call a few very special people and share the news right away; I couldn't wait. I borrowed Rick's cell phone and ducked out into the lobby. The cell phone had no signal there in the basement area of the huge hotel, so I went upstairs in my black socks and out the revolving glass door into the valet drive area. I sat on some cement steps where the road came in, and it was snowing on me.

Finally, I got reception and dialed Lance's number at the Denver law office.

I left a very brief message for Elisa and Lance telling them that Deb's film had won first place in the Activism category. I could hear my voice choking up with emotion as I finished the message.

Next I called Al's cell phone and left a message—he must still have been on the way back from Arizona.

Then I went back inside the hotel and downstairs through the opulent lobby and into the darkened huge theater. I didn't go back to my seat, but instead sat on a step in the aisle next to the row. I felt antsy, and soon went back outside and left Al a message on our home machine at the farm.

After the ceremony ended we went to dinner in a noisy but nice and rustic restaurant, along with an interesting and intelligent filmmaker named Rick who struck up a conversation with me in the lobby of the theater as I waited for Rick Roles and Deb to come out. They eventually came out, after all the ceremony ended. They had my clogs and my coat which I had abandoned on the floor under the seat.

Dinner and extended conversation ended, then we walked across the cement sidewalks of Beaver Creek—a seeming Disneyland of artistic opulence and wealth.

I have never really seen anything like it...

I recalled having had a vivid dream of running through posh hotel lobbies and high-end stores in a very similar setting, while at a gas and oil event...

As Rick and I frantically tried to find the theater yesterday, I recalled that dream and I was stunned. I had turned to Rick and said,

"I dreamt about this!"

COLLATERAL DAMAGE

And that dream was quite a while ago—maybe six months to a year—but it was still very vivid, especially in this setting that it so resembled.

As we walked across the dark and beautiful cement areas between the hotels and stores of the posh resort, a group of festival goers asked us if we were going to the party. I had been told that the big wrap party for the festival was going to be the next night, Sunday. So this was an unexpected surprise...

A few in the group seemed hesitant—I was not. I followed them up the escalator quickly; I was going to take this opportunity and see where it went. I had a feeling that one of the group who invited us to come with them was Jane Seymour, who had been a major panelist that night at the awards ceremony, and had won a lifetime achievement award. I thought it was, anyways...

But that was not the real reason I went. I went because we were here at a rare event, and I intended to make the most of it. I tend to be stubbornly independent, and my husband can confirm that for certain. This was a once-in-a-lifetime opportunity—I intended to take full advantage of it...

Rick—my "handler and chauffeur," (and many other things at that point in time), who had my truck keys in his pocket—could do whatever he wanted.

I can get anywhere I need to be, and get home from there too.

We had just won a first-place award; I was going to go all out...

A woman from the group right in front of me stopped and waited for me to catch up with her. She fell into step with me there in the dark plaza and asked about the film. She said she wanted to see it. It was Jane Seymour, the renowned and beautiful movie actress. What I need to say is that what overrode her fame in my mind was how genuine and sweet she was—how real.

I find that heartening in a very special way, even a day later.

I will not lose that realization, it is important.

We made our way into an absolutely enormous ballroom-type place, and again—posh is the theme word here at Beaver Creek.

The evening transpired, and we had a wonderful time. People came and talked to us, they congratulated us and wanted to discuss the issue, and we fell into lengthy and interesting conversation after conversation.

Many e-mails and phone numbers were exchanged on business cards, coasters, and scraps of paper.

Finally, Rick and I got into the white Ford Ranger heaped with coats and more, and we left for home. It was late. We talked, and listened to Pink Floyd on the CD player, processing what had taken place that night.

Sometime after 4 a.m. we got to the ranch and we hugged goodnight and said goodbye. This night felt very different; and we have been on many gas and oil trips: to the Capitol, to Hollywood, etc.

This was different.

This was validation; this was getting the message out.

Al was up; we visited for a while. I slept five hours, and then Al went skiing at Sunlight. I called my family and talked to my brother and sister-in-law and niece. It was Easter, and Charlotte told me about her Easter basket.

Earlier I called Rick and we talked about choices for the back cover of the book. I thought I had a final pick, but I wanted him to decide.

I knew I wanted him to be on the back cover, I just knew it.

Eventually as I did a few ranch chores and pondered further choices on the final questions still left before sending the book off for printing, I realized that I needed to write about these last several days—more specifically, yesterday.

COLLATERAL DAMAGE

So I went into the shed with the laptop and typed for a while until a neck ache prompted me to go park at the kitchen table—the most comfortable place to write.

I have been here for hours; I don't count them. About an hour ago the phone rang, and I answered. It was the Associated Press, and a friendly female reporter asked me about the film's success at the Vail festival. We had a lengthy conversation, and I let her know that I was doing one last entry in the book, relaying part of what transpired last night.

Now I believe I am finished—with this writing.

I am not, however, finished with my mission…

Sixty Minutes

April 6, 2010

I knew from Deb Anderson that the news show Sixty Minutes was interested in her film—and in the issue.

Last night I got a call:it was with Sixty Minutes from Queensland Australia. We had a long and good conversation.

The energy companies were there, preparing to drill in the area near the Great Barrier Reef, one of the world's largest geological treasures.

I hope they can help with making improvements for the future. Too many have suffered huge life threatening and ruining transgressions as a result of having the bad luck to be living near gas and oil development.

It is beyond sad.

My hope and the reason for my five years of writing is that they will learn from the mistakes, and develop energy in a more responsible way in the future—everywhere on this planet.

What has happened to the landowners here is truly unconscionable.

They have been the forgotten collateral damage.

COLLATERAL DAMAGE

Arbany gas kick 03-09-2004

aerial view - Garfield County Colorado

TARA MEIXSELL

drilling rig in action

Divide Creek Seep ignited -Silt, Colorado

COLLATERAL DAMAGE

Magnall Blowout

TARA MEIXSELL

rigs in Garfield County

COLLATERAL DAMAGE

Trip to Hollywood for Split Estate in DocuWeeks Film Fest in 2009,
Deb Anderson - Producer, Rick Roles and Dee Hoffmeister, landowners

Vail Film Festival, First Place Activist Category - Producer
Deb Anderson, Rick Roles, Tara Meixsell

TARA MEIXSELL

Vail Film Festival 2010 - Rick Roles, Deb Anderson, Producer of Split Estate, Tara Meixsell - consultant and Garfield County Associate Producer of Split Estate

Lance Astrella of Astrella Law Office - Super Attorney

TARA MEIXSELL

Debra Anderson, Dee Hoffmeister and Chris Mobaldi - day of COGCC Reform Bill signing

Chris Mobaldi and Governor Bill Ritter in Grand Junction at signing for COGCC Reform Bill

COLLATERAL DAMAGE

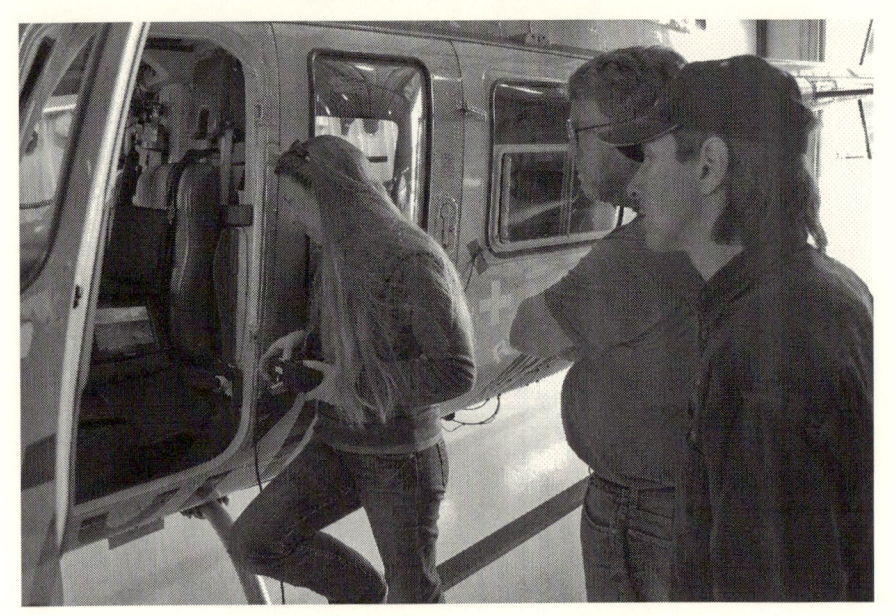

Film Producer Deb Anderson - Split Estate

Attorney Lance Astrella

TARA MEIXSELL

author Tara Meixsell - Collateral Damage

Made in the USA
Lexington, KY
12 December 2011